An Introduction to
COHERENT OPTICS
and HOLOGRAPHY
SECOND EDITION

An Introduction to
COHERENT OPTICS
and HOLOGRAPHY

SECOND EDITION

George W. Stroke

Department of Electrical Sciences
Electro-Optical Sciences Center
State University of New York
Stony Brook, New York

 1969

ACADEMIC PRESS   New York   San Francisco   London
A Subsidiary of Harcourt Brace Jovanovich, Publishers

ACADEMIC PRESS INC.
111 Fifth Avenue, New York, New York 10003

United Kingdom Edition published by
ACADEMIC PRESS, INC. (LONDON) LTD.
24/28 Oval Road, London NW1

Library of Congress Catalog Card Number: 65-28633

PRINTED IN THE UNITED STATES OF AMERICA

Dedicated to
Beruria and Edgar Stroke

PREFACE TO THE SECOND EDITION

To keep pace with the considerable number of important new ramifications in holography since this book first appeared in 1966, it seemed appropriate to include an updating appendix.

The same thought seems to have occurred to the Russian editors and translators of this book. The Russian translation appeared in 1967, published by MIR, under the editorship of Dr. L. M. Soroko, with translations by Drs. I. P. Nalimov and V. I. Kolesnikova. Dr. Nalimov kindly agreed to allow his excellent updating appendix to be translated into English for inclusion in this second edition, and Dr. Soroko graciously updated the Russian foreword for this edition. The kind assistance of Irving I. Goldmacher with the translation is gratefully acknowledged.

The author has further added a 16-figure Photographic Appendix together with extensive figure captions, primarily to describe and illustrate some of the currently most recent advances and applications of holography, including examples of the first pulsed-laser holographic images of a living person (L. D. Siebert), and several examples illustrating the new holographic methods of image restoration (image "deblurring"), now under development, as a part of the now increasingly apparent powerful, simple, and speedy new methods of "optical computing" using holography. Several photographic illustrations have also been added to the translation of the Russian appendix, particularly to illustrate some of the most advanced forms of holographic interferometry (R.M. Grant) and new extensions of the advantages of the author's "lensless Fourier-transform" holography arrangement to imaging through "distorting" media in the case of imaging of 3-D objects. The author wishes to especially thank the colleagues, as mentioned, who have so kindly contributed the several previously unpublished illustrations of their work.

In preparing this new edition of the book the author took advantage of a number of kind comments and suggestions to correct some errors and minor misprints which remained from the transcription of the original Michigan "lecture notes." The author wishes to acknowledge with particular pleasure the extensive comments and suggestions of Reuven Meidan which resulted in several important revisions found throughout the body of the text. The author also wishes to also acknowledge a kind letter of Professor G. Toraldo di Francia who showed that the resolution "gain" of the "lensless Fourier transform holograms" (in comparison with Fresnel-transform holograms) appreciably exceeded the already considerable original estimate, notably in high-resolution microscopy applications. The continued kind interest and encouragement

of Professor Dennis Gabor, together with his most fruitful kind comments
and suggestions have proven to be uniquely fruitful in this work and are
acknowledged with the deepest gratitude and appreciation. The author
will be particularly grateful for further comments and suggestions.

It may be in order to note that it is still not possible, even at this
time, to include extensive treatments of all aspects of the new developments in holography which are of importance. As one example, we may
mention the most recent forms of the increasingly fruitful analogies
between the presumed functioning of the human brain and several holographic processes, particularly as described in two forms [*Nature*, 1968]
by Professor Longuet-Higgins and by Professor Dennis Gabor. Other
examples include detailed discussion of holographic interferometry (based
on the pioneering work of R. L. Powell and K. Stetson, of J. Burch, of
R. E. Brooks, L. O. Heflinger, and R. F. Wuerker, of R. J. Collier,
E. T. Doherty, and K. S. Pennington, of H. Nassenstein, and of many
others), of the different forms of the increasingly important methods of
underwater ultrasonic holographic imaging, and indeed of the remarkable
holographic nature of microwave imaging with synthetic-aperture side-
looking radar (of L. Cutrona, E. N. Leith, C. Palermo. L. J. Porcello,
and others), as well as of the newest forms of holographic microscopy
(notably that based on the method of "white-light color reflection holography" using the Lippmann-Bragg diffraction effect, of G. W. Stroke and
A. E. Labeyrie). These and many other new applications of holography
are, however, presented with various detail in the text and in the appendix.
Experience has indeed shown that a basic treatment of fundamentals,
such as the treatment in this book, illustrated by selected examples, is
capable of adequately dealing with the new developments as they come
about. As an example, the application of the author's new methods of
holographic image restoration (deblurring) to the solution of the newly
proposed possibilities of "aperture synthesis" (imaging with "synthetic
apertures") is given in the *Photographic Appendix* (also p. 226), for the
case of incoherent x-ray radiation imaging, using a random-array scatter-
hole camera of novel design. Further reading and research may also be
helped by the new extensive additions to the bibliography in the various
parts of the text and particularly in the appendices.

The author again wishes to express his sincere thanks for the kind
initiative of Dr. L.M. Soroko and of Dr. I. P. Nalimov in translating the
original version into Russian, and for so generously contributing a
book-review foreword and a valuable appendix.

Stony Brook, New York   GEORGE W. STROKE
October, 1968

PREFACE TO THE FIRST EDITION

This book is based on a series of lectures and courses presented during the last two years, mostly at The University of Michigan. It contains much original scientific material, notably in the field of holography and coherent image-processing, obtained as a part of the author's recent investigations in his laboratory at the University of Michigan.

To some extent, the presentation of original scientific material in a textbook may be compared to a precedent set by A. A. Michelson in his famous small book entitled "Studies in Optics," which was first published by the University of Chicago Press in 1927. In his "Preface," Michelson observed that "In order to give the subject a semblance of continuity, it will be desirable to introduce considerable matter which will be found in any of the standard works on optics—but even here it may be of interest to present these investigations from my own point of view and to convey my own impressions in such a way as to emphasize the ideas of the founders of the science which have made the deepest impression on my own mind." The present author feels, in all humility, that he could not have conveyed his own feelings and approach in any better way.

In view of the great interest in these new advances in optics, and in an effort to make this material more widely and more rapidly available than is possible in lectures, so as to possibly stimulate further advances, much material from the original lecture notes is presented here substantially as it was first published at The University of Michigan in two editions (May, 1964 and March, 1965) under the title "An Introduction to Optics of Coherent and Non-Coherent Electromagnetic Radiations." The author therefore apologizes for a perhaps sometimes more telegraphic style than would be appropriate in a polished text, as well as for any possible omissions of appropriate references in various parts of the text.

Because of the uniquely important place held by Professor D. Gabor's three original "wavefront-reconstruction" papers, which laid the groundwork for the principles of wavefront reconstruction imaging in optics, and more particularly for the retrievable recording of phase and amplitude information in imaging light fields, the three original Gabor papers are reprinted, in their entirety in an Appendix. The author is particularly grateful to Professor Gabor (and to the publishers) for permission to reprint these papers, as well as for continuous encouragement and numerous fruitful conversations and suggestions in connection with the holographic aspects of this work. He is also indebted to a great many other people for continuous inspiration in many forms, in connection with many aspects of this work, and especially to Professors M. Francon, G. R. Harrison, P. Jacquinot, A. Maréchal, O. C. Mohler and E. L. O'Neill.

Finally, the author is greatly indebted to Professor Henry H. Stroke of New York University for many kind comments and suggestions, as well as for invaluable editorial assistance. Whatever improvements have resulted are to his credit, but the final responsibility for the text, and for any possible errors in its contents, is the author's. The kind comments, assistance, and stimulation of the author's students and assistants, both in class and in the laboratory, are also acknowledged with much gratitude. Parts of the experimental results which we report were obtained through the kind assistance of the National Science Foundation and the Office of Naval Research.

Ann Arbor, Michigan   GEORGE W. STROKE
November, 1965

FROM THE FOREWORD TO THE RUSSIAN EDITION

This is the first monograph on holography; it was based on the course of lectures read by the author at the University of Michigan.

In great detail, and for many areas, Stroke shows in his book the tempestuous development of optical holography. The completeness of this monograph is quite natural, since its author is an active participant in the majority of interference and holographic investigations. It is he who performed the classical experiments which form the basis of modern holography. Thus, for instance, working with D. Gabor, the discoverer of holography, Stroke performed the experiment on optical interference of two light beams which are time-nonsuperimposed.

Stroke's book remains the major work on holography despite certain shortcomings which are already noted in the literature. The book also contains Gabor's first work in this field. That this book was published twice within one year of its appearance shows better than any review the usefulness and completeness of the work.

By developing the methods of construction and quality control of diffraction gratings with extremely high resolving powers Stroke was closer to holography than any other optical scientist at a time when the invention of the laser caused its rapid development. Optical phenomena which form the basis of holography and of diffraction grating optics turned out to be very closely connected. After completing in 1959-1961 his investigations on the optics of imperfect diffraction gratings, Stroke began to work on holography.

Holography was created in 1948 when the English physicist Gabor introduced for the first time the notion of a hologram, i.e., a system of complete recording of the spatial structure of a light wave (amplitude and phase), through observation of interference between a diffracted wave emanating from the object and a homogeneous coherent background. Gabor showed that this system of recording possesses the property of *reversibility* which allows one to reconstruct the image in the second stage. The use of an auxiliary coherent background, or a coherent illumination, is the distinguishing trait of any scheme used in holography. The idea of a hologram did not come to Gabor accidentally: From intuitive assumptions, he was deeply convinced of the possibility of extracting information about the object from a diffraction picture which always contains this information in a coded form. Gabor himself obtained the first holograms, and using those, he was able to reconstruct the original object.

However, observations very soon showed that the original scheme suffered from certain shortcomings which did not allow Gabor to com-

pletely solve the posed problem. The real and imaginary images reconstructed at the second stage by means of Gabor's hologram were superimposed on each other and therefore caused mutual interference. Another
shortcoming of Gabor's scheme was the fact that the intensive coherent
background had to pass through the sample and this narrowed considerably the class of objects for which one might apply this method. †

Many scientists attempted to eliminate these shortcomings; however,
twelve years passed before Gabor's idea could be revived. An important
stimulus was provided by the work of two radio-physicists, Emmett
Leith and Juris Upatnieks, who achieved a synthesis of communications
theory and optics. Using notions and principles from communications
technology, Leith and Upatnieks introduced an inclined beam to create
the coherent background, removing the shortcomings in Gabor's original
arrangement.

† There has long appeared to exist a general misunderstanding with regard to
Gabor's very first scheme of holography, the principle of which was first given
in his article "A New Microscopic Principle" [Nature 161, 777-778 (1948)],
which has been reproduced, in extenso, on pages 263 - 265 of this work. It may
be readily seen in Fig. 1 of Gabor's article that the coherent background
("primary wavefront") in Gabor's original scheme does in fact *not* pass
through the object, *but* clearly *on the side* of it. A discussion of the matter can
be found in the paper by G. W. Stroke *et al.*, "On the Absence of Phase Recording
or 'Twin-Image' Separation Problems In Gabor (In-Line) Holography" [*Brit. J.
Appl. Phys.* 17, 497-500, 2 tables (1966)] That paper (see also the "holography
schemes" figure reproduced from that paper on p. 104 of this work) contains
verification of the fact that no phase recording imperfections nor any twin-image
separation problems exist in holography performed according to Gabor's original
scheme. A part of the misunderstanding of Gabor's original work may have been
due to the comparatively imperfect quality of the first holographic imaging and
reconstructions, obtained in very feebly coherent, ordinary mercury light, rather
than with the highly coherent, intense laser light which became available in
optics in 1960-1961, notably thanks to the pioneering work of A.T. Schawlow
and C.H. Townes, A.T. Maiman, A. Javan, and others. Another part of the misunderstanding of Gabor's work may have been due, as pointed out so well by Dr.
Soroko, to the generally unexpected appearance of the very surprising new concepts introduced into optics by Professor Gabor, under the name of "wavefront-
reconstruction imaging," and to apparent difficulties in immediately appreciat-
ing their wide scope, fruitful generality, and deep implications. The importance
of the clarification with regard to Gabor's original scheme of "in-line" holography, aside from academic and historical interest, resides also in the fact
that this scheme, because of its comparatively low coherence-length requirements, permits one to record holograms in situations where very large field
depths must be recorded with comparatively low-coherence length lasers. An
outstanding example of such an application of "in-line" holography is found in
the remarkable results obtained by B.J. Thompson *et al.* since 1963 [see for
example, B.J. Thompson *et al.*, "Application of Hologram Techniques for Particle Size Analysis," *Appl. Opt.* 6, 519-526 (1967)]

The closest anyone came to the solution of the problem of spatial separation of the real and the imaginary images was A. Lohmann who in 1956 proposed one of the possible schemes of single side-band filtering. Almost simultaneously with the brilliant experiments of Leith and Upatnieks in 1961, Lohmann developed an interferometer system which utilized diffraction gratings. This system, due to its achromatism, enabled one in principle to obtain high quality holograms.

A major role was played by the laser which by 1962 had become a very widespread source of intensive coherent light beams. A year later, Leith and Upatnieks showed high-quality three-dimensional images reconstructed from two-beam holograms, in fulfillment of one of Gabor's predictions.

Quite recently, Leith and Upatnieks set up an achromatic holographic system, thereby making another important contribution to holography. Such achromatic systems allow one to obtain large numbers of interference fringes in the holograms, while using a comparatively broad band of the spectrum.

A hologram possesses the property of being able to actively reconstruct the light wave emanating from the object, and thereby it enables one to make the object itself visible. This property of the hologram is a consequence of the fact that when the wave scattered by the object is registered, none of its characteristics are lost. The wave is registered completely by the hologram: Simultaneously and quite distinctly, the amplitude and phase information are registered, i.e., a complete optical wave record is obtained. In classical photography, however, only the intensity of the scattered wave is registered while the distribution of the phase shifts of electromagnetic oscillations in space is completely lost.

Holography, which enables one to obtain a complete optical wave record, presents the experimenter with new and unusual possibilities, which force one to review many of the methods of physical optics and the techniques of physical experiment. The most interesting possibility consists in the following: The observer may correct the optical properties of the object used in the experiment after the experiment is completely finished, i.e., a posteriori. Thus, for instance, a three-dimensional scene may be brought into sharp focus over an arbitrary depth. It is also possible to translate the observation point, to perform optical filtration of the spatial structure of the object and, in particular, to remove the aberrations of the optical image-forming system. However, the most astounding property of holography is that it allows one to perform interference between two light beams which are not superimposed either in time or space.

Holography should not be considered as merely an alternative to photography. Using the complete recording of light with the retention not only of amplitude but also of phase, it is possible now, using holography, to perform a wide variety of mathematical operations on complex functions, when these are given in the form of spatial distributions of amplitudes and phases of light waves.

Holography created a new optics which, according to its properties, is related to preholographic optics as the theory of complex variables is related to real variable theory. This comparison is not simply a pretty analogy; it best reflects the new "mathematical possibilities" of holographic optics. Using the principle of holography, it is already possible today to perform the following operations over complex functions: addition and subtraction, multiplication, differentiation, and certain classes of integral operations.

A brilliant example of the mathematical possibilities of holography can be found in the successful experiments with artificial holograms and two-dimensional translators which were performed by A. Lohmann and D. Paris. No less astounding is the system for the differentiation of optical signals created in the spirit of the method of optical filtration of A. Maréchal and P. Croce by S. Lowenthal and Y. Belvaux. In so far as the operations of addition and subtraction of complex-valued functions are concerned, these were first proposed and experimentally performed by D. Gabor and G.W. Stroke, and also by a number of other authors (R. Powell and K. Stetson).

The advantages of holographic methods of information processing lie in the fact that in holography the initial information is processed in its entirety and almost simultaneously throughout the entire field. Such operations as scanning or spreading the image into lines, which are necessary in electron systems, or the separation of the real and imaginary part of the complex function in different channels, are completely eliminated in the coherent optical system.

The first optical scientist in the Soviet Union who turned his attention to Gabor's holography and began independent experiments for the development of more sophisticated holographic systems was Yu.N. Denisyuk. By his experiments with Lippmann emulsions in 1962, he established a completely new orientation in the field of holography, completely different from Gabor's and related schemes. This method was widely recognized at a later date. The hologram invented by Denisyuk is a three-dimensional interferogram. As a consequence of the fact that in Denisyuk's holography the wave front interferes with the coherent background throughout the thickness of the emulsion, this scheme, from the very beginning, does not yield a superposition of the real and imaginary images. The thick-layered Denisyuk hologram reconstructs only one image, and the information which may be introduced into the thick layered hologram turns out to be much richer than the information contained in the usual Gabor hologram. †

Yu. Denisyuk's latest contribution to holography is a new system for obtaining holograms using light waves of two different frequencies [*Dokl. Akad. Nauk* 176, 1274 (1967)]. Experiments involving a running interference picture which is observed in analogous conditions were described by F. Fischer [*Z. Physik* 199, 541 (1967)].

Summing up, one might say that the following developments were the milestones in the development of holography: Gabor's investigations, which were the first to propose the idea of holography and to realize the first holographic system; Leith's and Upatnieks' investigations, which revived holography using ideas taken over from the theory of communications and founded one of the new areas of radio-optics, modern laser holography; Denisyuk's investigations which proposed and, for the first time, presented, a three-dimensional ("volume") hologram, and, finally, Stroke's investigations which accomplished a cycle of exhaustive experiments in holography and which established the contemporary principles as well as the most effective schemes of holography.

It should not be forgotten, however, that the authors of the above-mentioned four discoveries drew support from the accomplishments of their predecessors and contemporaries, who conducted investigations in optics, radio-communications, and the theory and techniques of information, as well as in spectroscopy.

It is interesting to note that soon after the discovery of holography, approximately since 1951, parallel to the idea of the replacement of the image by its diffraction image (diffraction-transform), one saw a rapid development of Fourier spectroscopy. Instead of the line spectrum or continuous spectrum, one registers a Fourier image (Fourier transform).

---

†An additional advantage of Denisyuk's scheme was first recognized by G.W. Stroke and A.E. Labeyrie in 1965, when they predicted that holographic images could be reconstructed with incoherent white light illumination rather than only with laser light, which had appeared as necessary in all holographic schemes proposed up to that time. The first "white-light reflection holograms" were obtained by Stroke and Labeyrie in 1965 by slightly modifying Denisyuk's scheme, that is, by introducing the reference beam from the back of the emulsion (rather than through the emulsion, as in Denisyuk's original work). This achievement, first reported by G.W. Stroke and A.E. Labeyrie in *Phys. Letters* 20, 368-370 (1966) under the title "White-Light Reconstruction of Holographic Images using the Lippmann-Bragg Diffraction Effect," was reprinted in extenso in the Russian translation of this book. More recently, a complete theory of these types of holograms, based on Denisyuk's original scheme, is given in the paper by D. Gabor and G.W. Stroke, "The Theory of Deep Holograms," *Proc. Roy. Soc. (London)* A304, 275–289 (1968). (Note by GWS.)

An important contribution to Fourier spectroscopy was made by P. Felgett and P. Jacquinot, as well as by Janine and Pierre Connes.

The last stages of Fourier-spectroscopy evolution are marked by significant successes in the technique of fast Fourier transform. The newest algorithm of J.W. Cooley and J.W. Tukey not only expedites the analysis of the interferograms registered in the Fourier spectrometer, but also opens up new possibilities in the technique of obtaining artificial holograms and two-dimensional translators. It should be noted that the concept of holography as a system for total recording based on the phenomenon of light interference is, at present, fundamental in all other fields of physical optics and interferometry.

January, 1968

L. M. SOROKO

CONTENTS

IV. Coherence Characteristics of Light (Experimental Characterization)

V. Image Formation in Coherent Light

VI. Theoretical and Experimental Foundations of Optical Holography (Wavefront-Reconstruction Imaging)

VII. Fourier Transforms, Convolutions, Correlations, Spectral Analysis, and the Theory of Distributions

Reprinted Papers

I. INTRODUCTION

1. Emergence of Modern Optics as a Branch of Electrical Engineering

Many of the most dramatic advances in the field of optics in the last decade or two have been directly stimulated by or originated through progress in electrical engineering and its branches of communication sciences, microwave electronics, and radioastronomy. Among the most noteworthy similarities are the operational Fourier-transform treatment of optical-image-forming processes and of spectroscopy, the introduction of resonant structures and of optical feedback control (for example in lasers, fiber optics, and in interferometric control of machines). The remarkable simplicity of optical computing and of coherent-background (heterodyne) detection used in communications systems are other examples. We find further similarities between electrical engineering and optics in the exploitation of the statistical and coherence properties of electromagnetic signals and radiation, and of polarization in interferometry and astronomy, as well as important developments in light amplification and control in optical masers and, more recently, the new achievements in "lensless" photography and "automatic" character recognition. Nonlinear optics presents another important example of the close similarities of basically comparable theory and techniques throughout broad ranges of the electromagnetic domain. The basic similarities encompass such fields as astronomy, radioastronomy, physics, and electrical engineering. Skillful recognition and exploitation of these basic similarities in pursuits throughout the entire electromagnetic domain is proving most fruitful in pin-pointing new areas of research and of industrial application in what may be called the new field of *electro-optical science and engineering*.

2. Mathematical Character of Electro-Optical Engineering

Perhaps the single most important element in the rapid development of electro-optical sciences is the great simplicity that results from the deliberate use of a sophisticated but powerful mathematical formulation.

To paraphrase Townes,[1] one might say that the recent dramatic developments in electro-optical science, including the maser, "epitomize the great change that has recently come over the character of technological frontiers." The maser, nonlinear optics, optical computers, interferometric gratings, lensless photography, optical filters, and automatic reading systems, to mention only a few, were predicted and worked out "almost entirely on the basis of theoretical ideas of a rather complex and abstract nature." These are not inventions or developments "which

1

could grow out of a basement workshop, or solely from the Edisonian approach of intuitive trial and error.'' They are creatures of our present scientific age which have come almost entirely from modern theory in physics, communication sciences, and, indeed, in electro-optical engineering.

3. Mathematical Methods of Modern Optics [†]

Except for the solution of boundary-value problems, and problems in nonlinear optics, where electromagnetic theory is basic, and of course in the study of the basic physics of radiation, where quantum theory and statistical theories are appropriate, a dominant part of electro-optical engineering is based on exploiting the simplicity of an operational, generally Fourier-transform treatment of the problems. Fourier-transform formulations, already used by Lord Rayleigh and A. A. Michelson near the turn of the century, and more recently the *theory of distribution*, based on the work of Laurent Schwartz (1950–1951), appear as uniquely powerful tools, not only in the *analysis of* more-or-less classical image-forming and communications systems but in the prediction and *synthesis* of new devices and systems. The *matrix formulation* of image formulation by lenses and mirrors has reduced lens computation to very great mathematical simplicity, especially when used with electronic computers. Indeed, *optical analogue correlators and computers*, born out of the new mathematical formulations, have started to complement the sometimes much more complex digital computers. Several examples of optical multiplication and addition of complex signals are given in Chapter V. Generally, in optics, the "signal" is an intensity (or complex amplitude) distribution in *two, spatial* dimensions (for example, x and y, in the focal plane or in the aperture of a lens). Optical computers are therefore marked by simplicity and compactness, and have particular success because of their multichannel, two-dimensional spatial capability (as compared to the single (time) dimension in purely electronic computers); indeed, the light intensity in any Δy channel can be varied along the x direction, and many Δy channels can be stacked up, side by side, along y, and operated on simultaneously (rather than sequentially).

4.   Some Limitations of Operational Formulation of Optical Image Formation and the Need for Boundary-Value Solutions in the Study of Diffraction of Electromagnetic Waves in Optics

The emphasis placed on operational treatment of image-formation and

[†]A list of the readily available mathematical references is given at the end of this chapter. See also Chapter VII.

communications systems in optics also calls for a clear understanding of its limitations, and also of domains where it is inapplicable.

For instance, the recent advances in attaining unprecedented high efficiencies in *optical diffraction gratings*[2] were based on recognizing that polarization, and consequently electromagnetic theory (rather than extensions of elementary scalar "diffraction theory"), are basic in determining the energy distribution among the waves of various diffracted orders (or modes). Indeed, a complete, exact solution of the electromagnetic boundary-value problem of optical gratings (added to the very few electromagnetic boundary-value solutions in existence) has recently been obtained in a series of papers by Stroke, Bousquet, Petit, and Hadni,[2] based on the method of solution given by Stroke in 1960.[3]

The most important limitation in the use of Huygens' principle, as given by its Fourier-transform expression, is generally not serious in most cases. The Fourier-transform relation between the complex amplitude distribution in the wavefront and the complex amplitude at any given point in the image applies only to the near vicinity of the center of the (aperture-limited) quasi-spherical image-forming wavefront, for instance near the focus of a lens, although regardless of whether this be a principal or a secondary focus. The domain of applicability of Huygens' principle and of the Fourier-transform expression of image formation can be made quite obvious when the Fourier-transform expression is derived from Maxwell's equations.[4,5]

However, it is not true, as sometimes assumed, that the scalar formulation of diffraction will always provide at least a qualitative description of optical diffraction. For instance, in the case of the grating illustrated in Fig. 1a, it is found on the basis of electromagnetic the-

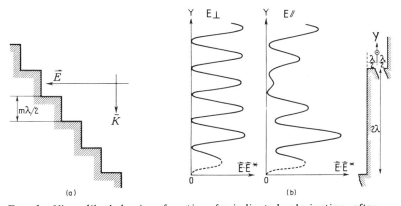

FIG. 1. Mirror-like behavior of grating, for indicated polarization, after Marechal and Stroke.[3] Actual experimental recording of Hertz-like standing waves in 3-cm microwave domain are shown in (b).

ory[3] that the grating illustrated will behave like a *perfect mirror*, reflecting only a single wave back along the $\overline{k}$ direction, showing none of the "classical" diffraction into any other side orders, regardless of the narrowness of the horizontal facet (parallel to $\overline{E}$). Experiments with 3-cm microwaves in the $E_\perp$ polarization, shown in Fig. 1b, confirm this classically unexplainable, but electromagnetically easily foreseen prediction, made by Maréchal and Stroke in 1959.[3]

The basically *electromagnetic nature of diffraction* should be constantly borne in mind when systematic inconsistencies between experiment and theory are encountered in the use of the otherwise very powerful Fourier-transform formulation of diffraction and image formation in optics.

In the most general sense, diffraction of electromagnetic waves results simply from the requirement that easily specified *boundary conditions* must be satisfied at the boundary of the diffracting object *by the total field*, incident plus diffracted.

Frequently the great simplicity in the actual solution of a diffraction problem will result from writing the diffracted field as an integral *sum of plane waves*, with the direction cosines as parameters for the waves of different amplitudes. An important example is given in Section 5 for the case of optical gratings.

The theory of gratings and interferometry is basic to much of modern optical-image-formation theory.[4,5]

5. Grating Equation as an Example of Boundary-Value Solution of a Diffraction Problem[5]

Perhaps the best equation describing the diffraction of light by gratings is the so-called *grating equation*,

$$\sin i + \sin i' = \frac{m\lambda}{a} \tag{1}$$

where i and i' are the angles formed by the incident and diffracted wavefronts with the mean grating surface, a is the grating spacing constant, and m is an integer.

The derivation usually found in textbooks with the help of Huygens' principle is straightforward. However, the Huygens' solution makes it appear, incorrectly, that the plane-diffracted waves are "built up," as it were, by the "envelope" of little "wavelets."

More generally, it has been the practice to *assume* that, when a plane wave is incident on a plane grating, the diffracted waves are also plane and exist only in a discrete set. In fact, the existence of plane diffracted waves is merely the result of the periodic nature of the grating.

The *proof* of the existence of a discrete set of plane diffracted waves, satisfying the grating equation (1) when produced by the incidence of a plane wave on a grating, follows.

Ruled gratings are essentially two-dimensional structures. As such, their surface can be described by a function $S = f(x, y)$ (Fig. 2) which

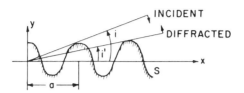

FIG. 2. Parameters used in the grating equation. Sin i + sin $i'= m\lambda/a$.

is independent of one of the coordinates, say z, but is periodic as a function of x,

$$S_{x+a} = S_x \tag{2}$$

The groove length is along z. One next recalls the existence of the class of *two-dimensional* problems (here, independent of z). Two-dimensional problems are essentially scalar in nature. However, this does not mean that they can be identified with nonelectromagnetic problems (for example, with acoustical problems). Rather, two-dimensional problems can be expressed in terms of only a single dependent electromagnetic-field variable (for example, E_z or H_z). It can be further shown that E_z or H_z, in these problems, satisfies the wave equation (written for E_z, as an example)

$$\frac{\partial^2 E_z}{\partial x^2} + \frac{\partial^2 E_z}{\partial y^2} + k^2 E_z = 0 \tag{3}$$

where $k = 2\pi/\lambda$. The factor exp $(-i\omega t)$ is implicit.

A fundamental elementary solution of (3) can be written

$$\exp[ik(x \sin \theta + y \cos \theta)] \tag{4}$$

which is the equation of a plane wave, where θ is now the angle of diffraction (not yet restricted to be i' of the grating equation). When θ is imaginary (4) represents an inhomogeneous or evanescent wave. [5,6]

A complete solution of Eq. (3) is formed by an angular spectrum of plane waves and can be represented by a Fourier integral,

$$\int_{-\infty}^{+\infty} E(\theta) \exp[ik(x \sin \theta + y \cos \theta)] \, d\theta \tag{5}$$

where $E(\theta)$ are the amplitudes of the various waves. The coefficients $E(\theta)$ are generally complex, and are to be determined to solve any special problem described by (5).

In the case of optical gratings, it is necessary to investigate the diffraction of polarized waves, and either $\overline{E}$ or $\overline{H}$ in the incident plane wave are chosen to be parallel to the groove length, along z. Let E_z^i or H_z^i be the components of the incident field for the two cases. One has

$$E_z^i = \exp\left[-ik(x \sin i + y \cos i)\right] \tag{6}$$

(The i in front of the k is clearly equal to $\sqrt{-1}$.) Let E_z^d be the diffracted field. Since the total field, $E_z = E_z^i + E_z^d$ satisfies the wave equation, and since the incident field satisfies the same equation, the diffracted field E_z^d must also satisfy the wave equation. In its most general form, the diffracted field E_z^d can be represented by its *angular spectrum of plane waves*:

$$E_z^d = \int_{-\infty}^{+\infty} E_z^d(i') \exp[ik(x \sin i' + y \cos i')]\, di' \tag{7}$$

Here $E_z^d(i')$ are the amplitudes of the plane diffracted waves corresponding to the angles i'. According to (7) there exists an infinity of diffracted waves, distributed in continuous angular directions.

We shall next show that the periodic nature of the grating boundary, and the fact that the boundary condition on the surface of the grating must be satisfied by the total field E_z (incident plus diffracted), restrict the continuous angular distribution of diffracted waves to only a *discrete set* of waves, satisfying the grating equation (1).

Indeed, one must have †

$$\left|(E_z)_{x+a}\right| = \left|E_z(x)\right| \tag{8}$$

on the grating surface. Equations (6) and (7), when introduced in (8), give

$$\exp\left[-ik(x \sin i + y \cos i)\right] \exp(-ika \sin i)$$

$$+ \int_{-\infty}^{+\infty} E_z^d(i') \exp\left[ik(x \sin i' + y \cos i')\right] \exp(ika \sin i')\, di'$$

$$= \exp\left[-ik(x \sin i + y \cos i)\right]$$

$$+ \int_{-\infty}^{+\infty} E_z^d(i') \exp\left[ik(x \sin i' + y \cos i')\right] di' \tag{9}$$

†See mathematical references at the end of the chapter, notably Ramo *et al.*, p. 478; also L. Brillouin, especially in relation to Floquet's theorem.

After division by $\exp[-ik(x \sin i + y \cos i)]$ and factoring of $\exp(-ika \sin i)$, Eq. (9) gives

$$| \exp(-ika \sin i) \left\{ 1 + \int_{-\infty}^{+\infty} E_z^d (i') \exp[ik(x + pa)(\sin i + \sin i')] \right.$$

$$\left. \times \exp[iky(\cos i + \cos i')] \right\} di' | = | 1$$

$$+ \int_{-\infty}^{+\infty} E_z^d (i') \exp[ikx(\sin i + \sin i')] \exp[iky(\cos i + \cos i')] di' | \quad (10)$$

For any given k, a, and any given angle of incidence i, the factor $|\exp(-ika \sin i)|$ is equal to unity. Therefore, for (6) to (10) to be satisfied, one must have

$$ka (\sin i + \sin i') = m 2\pi \quad (m = \text{integer}) \quad (11)$$

that is,

$$k (\sin i + \sin i') = m \frac{2\pi}{a} \quad (12)$$

Recalling that $k = 2\pi/\lambda$, (12) gives finally

$$\sin i + \sin i' = \frac{m\lambda}{a} \quad (1)$$

which is indeed the grating equation. We have just shown that a plane wave

$$E_z^i = \exp[-ik(x \sin i + y \cos i)] \quad (13)$$

incident on a periodic surface, having a period a, gives rise to a diffracted field E_z^d formed of a discrete set of plane waves, such that

$$E_z^d = \sum_m (E_z^d)_m \exp[ik(x \sin i'_m + y \cos i'_m)] \quad (14)$$

It is noted that it is truly the boundary condition and the periodicity of the grating which are *at the origin* of the diffraction of light by gratings. It is remarkable that the exact nature of the boundary condition does not enter the part of the solution required to demonstrate the *existence of plane diffracted waves, satisfying the grating equation.* Clearly, though, the amplitudes $E_z^d(i'_m)$ of the diffracted waves do depend on the exact nature of the boundary, and more particularly on its material nature (dielectric, conductor, etc.), as well as on the groove shape (see Section 4).

6. Gratings as Information Carriers in Optics (Application to Wavefront-Reconstruction Imaging)[†]

Just as a sinusoidally time-varying wave is the basic carrier for time-varying signals in radio communications, so a spatially varying grating (for instance, the sinusodial grating produced on a photographic plate by interference of two plane waves) is a basic carrier for spatially varying images in optics (Fig. 3). In both cases, when either space or time are the parameters, the signal (or image) involves variations of both ampli-

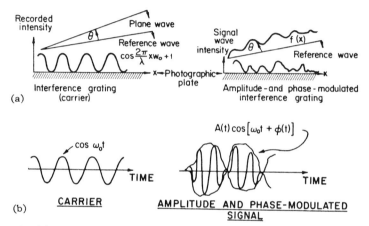

FIG. 3. (a) Interference grating as carrier of complex amplitude (amplitude and phase) imaging information in optics. (b) Electrical carrier analogue of interference grating.

tude and phase. Single or double side-band modulation can both be achieved, and the use of interference gratings as information carriers can be immediately understood with the help of elementary image-formation theory.

Consider first the case of two mirrors M_1 and M_2 illuminated by a plane wave S_1 (Fig. 4). The two mirrors are inclined with respect to each other, so that the two waves Σ_1 and Σ_2 incident on a distant photographic plate form angles θ_1 and θ_2 with the plate (that is, the two waves form an angle $\theta_1 - \theta_2$ with each other). Let $A_1(x) \exp[i\phi_1(x)]$ and $A_2(x) \exp[i\phi_2(x)]$ be the corresponding spatial component of the electric field vectors on the photographic plate. For the case of plane waves $A_1(x)$ and $A_2(x)$ are constants. As we show below $\phi_1(x)$ and $\phi_2(x)$ are linear functions of x. Clearly the intensity recorded on the photographic plate is nothing but a sinusoidal interference grating, with

[†]A more extensive treatment of the principles of optical holography (wavefront-reconstruction imaging) is given in Chapter VI.

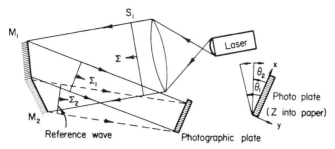

FIG 4. Recording of hologram of a plane mirror used as object.

straight-line fringes parallel to z. One has

$$I(x) = [A_1 \exp(i\phi_1) + A_2 \exp(i\phi_2)][A_1^* \exp(-i\phi_1) + A_2^* \exp(-i\phi_2)]$$

$$= |A_1|^2 + |A_2|^2 + |A_1||A_2| \exp[-i(\phi_1 - \phi_2)$$

$$+ |A_1||A_2| \exp[+i(\phi_1 - \phi_2)] \quad (15)$$

where one recognizes

$$I(x) = |A_1|^2 + |A_2|^2 + 2|A_1||A_2| \cos(\phi_1 - \phi_2) \quad (16)$$

In this case one also notes that, for small angles (otherwise use $\sin \theta_1$, $\sin \theta_2$),

$$\phi_1 = \frac{2\pi}{\lambda} x \theta_1 \quad (17a)$$

$$\phi_2 = \frac{2\pi}{\lambda} x \theta_2 \quad (17b)$$

The period of the fringes is $a = \lambda/(\theta_1 - \theta_2)$. When the interference grating is illuminated by a plane wave, as shown in Fig. 5, (15) shows immediately that three waves will emerge from the plate: one wave on the axis, and two waves forming angles $+(\theta_1 - \theta_2)$ and $-(\theta_1 - \theta_2)$ with the plate.

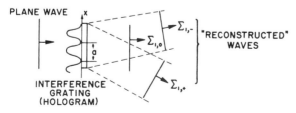

FIG. 5. Reconstruction of image-forming waves from hologram recorded in Fig. 4.

It will appear (see Chapter VI) that it is possible to take suitable precautions (e.g., relatively high reference-beam intensity) in the recording
of the hologram, so that it is indeed possible to suppress all but the
first orders (one first order on each side of the zero order).

Equation (15) can be written

$$I(x) = (|A_1|^2 + |A_2|^2) + |A_1||A_2| \exp [-i(2\pi/\lambda) x (\theta_1 - \theta_2)]$$
$$+ |A_1||A_2| \exp [+ i (2\pi/\lambda) x (\theta_1 - \theta_2)] \quad (18)$$

where (for this case) $[|A_1|^2 + |A_2|^2]$ and $|A_1||A_2|$ are constants, within
the aperture of the plate.

One has from Fourier-transform theory the following equation:

If $f(x)$ $\longrightarrow$ $F(\omega)$

then $f(x) \exp (2\pi i \theta_1 x)$ $\longrightarrow$ $F(\omega - \theta_1)$

and $f(x) \exp (-2\pi i \theta_1 x)$ $\longrightarrow$ $F(\omega + \theta_1)$ (19)

where $\longrightarrow$ indicates "by Fourier transformation."

It follows that the Fraunhofer diffraction pattern at infinity consists
of three images, one on the axis, and one each at $\pm (\theta_1 - \theta_2)$. [For a
plate of finite width, the image "points" have a $(\sin \theta)/\theta$ distribution.]
To each of the image "points" at infinity corresponds a plane wave
at the emergence from the plate. Let us now consider one of the original waves reflected by the mirrors, and forming the interference
grating, to be the reference wave. Let Σ_2 be the reference wave.
One can then say that the two waves $\Sigma_{1,-}$ and $\Sigma_{1,+}$ obtained by
illuminating the interference grating with a plane wave are the reconstruction of the original unknown or modulating wave.

The importance of this analysis will become clear when one examines
how an arbitrary wave Σ_1, reflected by an object of arbitrary amplitude
and phase, is recorded on the interference grating, used as carrier.

For example, let Σ_1 be formed by the spherical waves originating by
scattering at the various points P of the object. In other words, Σ_1 can
be taken as being equal to the resultant complex amplitude produced in
the plane forming an angle θ_1 with the photographic plate, when the
plate is illuminated by light scattered (or transmitted) by the object
(see Fig. 6). The intensity recorded is now given by

$$I (x) = (A_1 A_m)(A_1 A_m)^* + A_2 A_2^*$$
$$+ [A_m^* \exp (-i \phi_m)] [A_1^* A_2 \exp [-i (\phi_1 - \phi_2)]$$
$$+ [A_m \exp (i \phi_m)] [A_1 A_2^* \exp [+i (\phi_1 - \phi_2)] \quad (20)$$

where Σ_1 has been written as a modulated plane wave,

$$\Sigma_1 = A_m (x) \exp [i \phi_m (x)] A_1 \exp (2\pi/\lambda) x \theta_1] \quad (21)$$

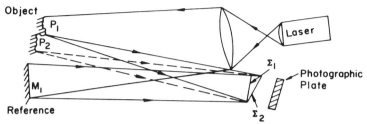

FIG. 6. Recording of hologram of three-dimensional object after Stroke[7] and Leith and Upatnieks.[9]

As before, $|A_1|$ = constant, $|A_2|$ = constant, and

$$\Sigma_2 = A_2 \exp \left[i (2\pi/\lambda) \, x \, \theta_2 \right] \tag{22}$$

It is important to note that both $A_m(x)$ and $\phi_m(x)$ are now functions of x.

By comparing (20) with (15) and by noting (19), one immediately sees that the modulating wave $A_m(x) \exp \left[i \phi_m(x) \right]$ can now be considered as "carried" by the interference grating of the case described by (15). Therefore, by illuminating the plate with a plane wave, as before, one again obtains three waves—a zero-angle (or zero-order) wave, carrying no information, and two side-band waves, modulated by $A_m \exp(i \phi_m)$ and by $A_m^* \exp(-i \phi_m)$, respectively, and leaving the plate at angles $\pm(\theta_1 - \theta_2)$, respectively. Clearly, the side-band waves are complete reconstructions of the Σ_1 wave, which, in turn, was formed by scattering of light by the original object. The two reconstructed waves will form a real and virtual image, identical to the original object, as will be shown in Chapter VI.

The preceding scheme of wavefront reconstruction with linearly spaced interference gratings is, of course, applicable to tridimensional objects. It has now been successfully verified in a number of laboratories[†] and was suggested by the author and his associates as a practical basis for truly image-forming x-ray microscopy.[10, 26] Despite statements sometimes found in connection with the preceding "off-axis" holographic recording scheme, Stroke (with students)[42] has recently again verified that perfectly separated images may indeed be obtained in the original "in-line" Gabor recording scheme, notably in high-resolution imaging and with diffused illumination.

Some possibilities for use of optical interference gratings as carriers were explicity considered at least as early as 1944 by Duffieux in his classical treatise[16] and again in 1958.[17] More recently, Lohmann[18] has again explicitly suggested the use of the optical equivalents of single side-band modulation in the "lensless" hologram photography, originated

[†]See Chapter VI, as well as Refs. 10 and 30.

by Gabor in 1948 (see also Refs. 19 and 8). A more extensive discussion of the theoretical and experimental foundations of holographic imaging can be found in Ref. 10, as well as in Chapter VI.

There is of course, no need to place the reference mirror next to the object, in actual practice, when recording the hologram, nor is it necessary to illuminate the object with a plane wave. For instance, illumination with a wave scattered by a ground glass is perfectly adequate. Diffused or scattered light illumination in holography was first suggested by the author in 1964[20] and was subsequently incorporated with other aspects of a collaborative effort with Leith, in Ref. 9. The only obvious requirement in the recording is that there should exist the possibility of recording an interference grating, in the case that the object and the reference beam both give rise to plane waves. It will be shown (Chapter VI) that spherical as well as plane wavefronts can be used with various advantages as reference wavefronts in the recording of holograms. As such (see Fig. 7), plane and spherical wavefronts, respectively, can be considered as building-block wavefronts in holography.

Extensive work on wavefront reconstruction, following the initial work by Gabor, was carried out by many authors in particular by El-Sum[14,15,22,23] and others.[8-20,22-27] In particular, the zone-plate lens-

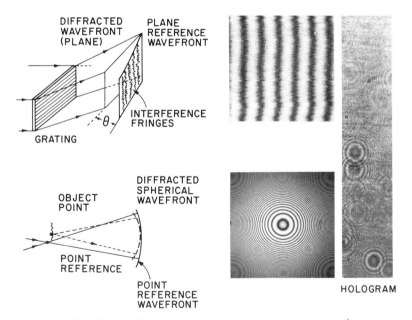

FIG. 7. Building blocks for holograms, after Stroke.[4]

like image-forming properties of holograms were particularly noted by Rogers in 1950.[24]

The surprisingly large magnifications (in excess of 1 million) attainable in holographic two-step imaging in going from 1-Å x-ray construction to 6328-Å laser-light reconstructions were stressed by Gabor in 1948.[11-13, 24] However, it had generally been considered that these magnifications would—in holography—be "empty,"[25] with resolutions limited to only 5000 to 10,000 Å, until Stroke and Falconer demonstrated[26] in 1964 that high resolutions on the order of 1 Å should be attainable in holographic imaging by new Fourier-transforming construction and reconstruction schemes,[27] with a generalization of the remarkable results obtained by the x-ray microscopy methods using digital computers (Kendrew[28]) or optical reconstruction (Buerger[29]). We shall return to these considerations in detail in Chapter VI (Section 4).

7. Optics and Communication Theory. Historical Background

Obvious similarities between modulation and demodulation in the use of interference gratings as carriers and heterodyne methods used in communications are easily recognized. Quite basic similarities between the methods described and the phase-contrast (coherent background) methods introduced into optics by Zernike in 1934[30-32] should be recognized. The Nobel Prize in Physics was awarded to Zernike for this work in 1953. Indeed, many of the beginnings of this work may be traced back at least as far as Abbe (1873)[33] and Töpler (1867).[34]

It has been stressed by Duffieux,[16] in reviving interest in Fourier-transform treatment of optical processes, that there is little doubt that the modern foundations of operational methods in optical communications, image processing, and spectroscopy can be found in the work of Michelson[35] and Lord Rayleigh.[36] Attention to the use of Fourier methods in English-speaking countries is largely due to a paper on "Optics and Communication Theory" published by Elias[37] in 1953 and to lecture notes on "Selected Topics in Optics and Communication Theory," first published by O'Neill in 1956.[38] Extensive references and credit to the considerable work that preceded their own papers is given by both Elias and O'Neill. Much of the credit for stimulating modern developments in image-formation theory and filtering is due to Maréchal,[39] in optical communication theory to Blanc-Lapierre,[40] Duffieux,[16] and O'Neill,[38] and in spectroscopy to Jacquinot.[41]

The foundations for the dramatic achievements in optical computing, filtering character recognition, and more generally for the new sophisticated optical materialization of communication-theory principles at The

University of Michigan have resulted in large part from the work of
Cutrona and his associates.[19] More recently work by the author on interferometry and diffraction-grating principles has played a major role in
these advances.[5,10,20]

References

1. C. H. Townes, *in* "The Age of Electronics" (C. F. J. Overhage, ed.), p.
166. McGraw-Hill, New York, 1962.
2. See, for example, G. W. Stroke, *J. Opt. Soc. Am.* **54**, 846 (1964).
3. G. W. Stroke, *Rev. Optique* **39**, 291–298 (1960).
4. A. Maréchal and M. Françon, "Diffraction." Revue d'Optique, Paris, 1960.
5. G.W. Stroke, "Diffraction Gratings," *in* "Handbuch der Physik" (S. Flügge,
ed.), Vol. 29, pp. 426–754. Springer, Berlin, 1967.
6. For general background, see, for example, P. C. Clemmow, *in* "Principles of
Optics" (M. Born and E. Wolf, eds.), pp. 553–588. Pergamon Press, New
York, 1959.
7. G. W. Stroke, private communication to E. N. Leith (1963).
8. E. N. Leith and J. Upatnieks, *J. Opt. Soc. Am.* **52**, 1123 (1962).
9. J. Upatnieks and E. Leith, *J. Opt. Soc. Am.* **54**, 1295 (1964).
10. G. W. Stroke, *in* "Optical Information Processing" (J. T. Tippett, L. C.
Clapp, D. Berkowitz and C. J. Koester, eds.). M.I.T. Press, Cambridge,
Massachusetts, 1965; see also G. W. Stroke and E. N. Leith, private
communication to the National Science Foundation (Dec. 6, 1963).
11. D. Gabor, *Nature* **161**, 777 (1948).
12. D. Gabor, *Proc. Roy. Soc. (London)* **A197**, 454–487 (1949).
13. D. Gabor, *Proc. Phys. Soc. (London)* **B64**, 449–469 (1951).
14. See, for instance, H. M. A. El-Sum, "Reconstructed Wavefront Microscopy,"
Ph.D. thesis, Stanford Univ., Stanford, California, November 1952; available
from University Microfilms, Inc., Ann Arbor, Michigan (Dissertation Abstracts
4663, 1953).
15. P. Kirkpatrick and H. M. A. El-Sum, *J. Opt. Soc. Am.* **46**, 825 (1956).
16. P. M. Duffieux, "L'Intégrale de Fourier et ses applications à l'optique."
Faculte des Sciences, Besançon, 1946.
17. P. M. Duffieux, *Rev. Optique* **37**, 441–457 (1958).
18. A. Lohmann, *Opt. Acta*, **3**, 97–99 (1956).
19. L. J. Cutrona, E. N. Leith, C. J. Palermo, and L. J. Porcello, *IRE Trans.
Inform. Theory* **6**, (3), 386–400 (1960).
20. G. W. Stroke, "An Introduction to Optics of Coherent and Noncoherent Electromagnetic Radiations," Univ. Michigan Engineering Summer Conf. on
Lasers, Lecture Notes, May 1964, pp. 1–77.
21. G. W. Stroke, *Intern. Sci. Tech.* **41**, 52 (1965).
22. A. V. Baez and H. M. A. El-Sum, *in* "X-Ray Microscopy and Microradiography" (V. E. Cosslett, A. Engström, and H. H. Pattee, Jr. eds.), pp. 347–
366. Academic Press, New York, 1957.
23. See also U.S. Patent 3,083,615.
24. G. Rogers, *Nature* **166**, 273 (1950).
25. A. Baez, *J. Opt. Soc. Am.* **42**, 756 (1952).
26. G. W. Stroke and D. G. Falconer, *Phys. Letters* **13**, 306 (1964).
27a. G. W. Stroke and D. G. Falconer, *J. Opt. Soc. Am.* **55**, 595 (1965).

27b. G. W. Stroke and D. G. Falconer, *Phys. Letters* **15**, 238 (1965).
28. J. C. Kendrew, G. Bodo, H. M. Dinitz, R. G. Parrish, H. Wyckoff, and D. C. Phillips, *Nature* **181**, 662 (1958).
29. M. J. Buerger, *J. Appl. Phys.* **21**, 909–917 (1950).
30. F. Zernike, *Physica* **1**, 43 (1934).
31. F. Zernike, *Physik Z.* **36**, 848 (1935).
32. F. Zernike, *Z. Tech. Phys.* **16**, 454 (1935).
33. E. Abbe, *Arch. Mikrosk. Anat.* **9** (1873).
34. See H. Wolter, *in* "Handbuch der Physik" (S. Flügge, ed.), Vol. 24, pp. 555–645. Springer, Berlin, 1956.
35. A. A. Michelson, *Phil. Mag.* **34**, 280 (1892).
36. Lord Rayleigh, *Phil. Mag.* **34**, 407 (1892).
37. P. Elias, *J. Opt. Soc. Am.* **43**, 229–232 (1953).
38. E. L. O'Neill, Optical Research Laboratory, Boston Univ., 1956.
39. A. Maréchal and P. Croce, *Compt. Rend.* **237**, 607 (1953).
40. A. Blanc-Lapierre, *Ann. Inst. Henri Poincaré* **13**, 275 (1953).
41. P. Jacquinot, *Rept. Progr. Phys.* **23**, 267–312 (1960).
42. G. W. Stroke, D. Brumm, A. Funkhouser, A. Labeyrie, and R. C. Restrick, *Brit. J. Appl. Phys.* **17**, 497–500 (April 1967).

Mathematical References

J. Arsac, " Transformation de Fourier et Théorie des Distributions." Dunod, Paris, 1960.

M. Bouix, "Les Fonctions Généralisées ou Distributions," Masson, Paris, 1964.

A. Erdélyi, "Operational Calculus and Generalized Functions." Holt, New York, 1962.

E. A. Guillemin, "The Mathematics of Circuit Analysis," Wiley, New York, 1951.

R. C. Jennison, "Fourier Transforms and Convolutions for the Experimentalist." Pergamon Press, New York, 1961.

A. A. Kharkevich, "Spectra and Analysis" (translated from Russian).Consultants Bureau, New York, 1960.

M. J. Lighthill, "An Introduction to Fourier Analysis and Generalized Functions." Cambridge Univ. Press, 1958.

A. Papoulis, "The Fourier Integral and Its Applications." McGraw-Hill, New York, 1962.

L. Schwartz, "Méthodes Mathématiques pour les Sciences Physiques." Hermann, Paris, 1965.

S. Ramo, J.R. Whinnery, and T. Van Duzer, "Fields and Waves in Communication Electronics." Wiley, New York, 1965.

L. Brillouin, "Wave Propagation in Periodic Structures." Dover, New York, 1953.

R. Bracewell, "The Fourier Transform and its Applications." McGraw-Hill, New York, 1965.

II. DIFFRACTION THEORY (QUALITATIVE INTRODUCTION)

1. The Two Aspects of the Diffraction of Light

Two characters of the diffraction of light have been recognized ever since the early investigations of the properties of light in the eighteenth and nineteenth centuries (Fraunhofer, Fresnel, Young, Airy, etc.):

(1) The departures from straightness in the directions of propagation of waves when interacting with boundaries.

(2) The formation of diffraction patterns in image planes when waves of finite aperture are brought into focus with the help of mirrors, lenses, or other means.

As a result of work by Maxwell, Lord Rayleigh, Sommerfeld, Marechal, and others, the two characters of diffraction of electromagnetic waves that need to be distinguished can be categorized as follows:

(1) An *electromagnetic character* which is fundamental to the diffraction of light by boundaries, and with the direction, polarization, and amplitudes of the various waves.

(2) A *scalar aspect*, which is important in the image-forming properties of the various diffracted waves, when they are limited in size by the aperture of the optics and limited in quality by inherent or manufacturing imperfections.

Moreover, the instrumental applications of optical elements (mirrors, lenses, prisms, diffraction gratings, interferometers, and so on) for the formation and transformation of images, and in general for the analysis and processing of light, involve certain *geometric* characteristics of optics (magnification, image-object distance equations, spectral dispersion in spectroscopic instruments, resolution, etc.)

2. Theoretical Calculation of Energy Distribution in Diffraction and of Spectral Diffraction Patterns

2.1. Electromagnetic Boundary-Value Solutions

Maxwell's electromagnetic equations and the boundary conditions that must be satisfied by the total field (incident plus diffracted) on the surfaces of boundaries in the electromagnetic field are sufficient to determine any electromagnetic diffraction problem in a homogeneous dielectric medium. In fact, if rigorous solutions of Maxwell's equations were easy to obtain in practice, all optical diffraction and image-formation problems would be attacked and solved with the help of these equations.

If the incident electric field vector is described by $\overline{E}_i$ and the scattered field vector by $\overline{E}_s$, then one simply has for the scattered field,

$$\overline{E}_s = \overline{E} - \overline{E}_i \tag{1}$$

where the total field vector

$$\overline{E} = \overline{E}_i + \overline{E}_s \tag{2}$$

has to satisfy appropriate boundary conditions on the diffracting surfaces or apertures and boundaries. Of course, one simple type of boundary is the infinitely extended perfectly conducting plane acting as a reflector. In this case, one of the boundary conditions is that the tangential component of the total field $\overline{E}$ must vanish on the surface of the conductor. From this condition and Maxwell's equations, the laws of reflection of electromagnetic waves are immediately obtained.

Maxwell's equations written in differential form for free space are

$$\begin{aligned}
\text{curl } \overline{E} &= -\mu_0 \frac{\partial \overline{H}}{\partial t} \\
\text{curl } \overline{H} &= \varepsilon_0 \frac{\partial \overline{E}}{\partial t} \\
\text{div } \overline{E} &= 0 \\
\text{div } \overline{H} &= 0
\end{aligned} \tag{3}$$

To these equations one must always associate appropriate boundary conditions which are, for the normal components $\overline{E}$ and $\overline{H}$ (see Fig. 1),

$$\begin{aligned}
\overline{n} \cdot \mu_0 (\overline{H}_1 - \overline{H}_2) &= 0 \\
\overline{n} \cdot \varepsilon_0 (\overline{E}_1 - \overline{E}_2) &= \sigma
\end{aligned} \tag{4}$$

where σ is the surface charge, and for the tangential components,

$$\begin{aligned}
\overline{n} \times (\overline{H}_1 - \overline{H}_2) &= \overline{K} \\
\overline{n} \times (\overline{E}_1 - \overline{E}_2) &= 0
\end{aligned} \tag{5}$$

where $\overline{K}$ is the surface current density.

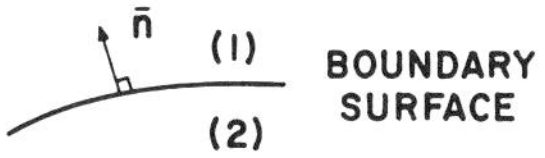

FIG. 1. Boundary-condition parameters.

When Maxwell's equations are written in integral form, the boundary conditions are implicit. One has (see Fig. 2)

$$\oint_c \overline{E} \cdot \overline{ds} = -\frac{d}{dt} \int_S \mu_0 \overline{H} \cdot \overline{n} \ da$$

$$\oint_c \overline{H} \cdot \overline{ds} = \frac{d}{dt} \int_S \varepsilon_0 \overline{E} \cdot \overline{n} \ da + \int_S \overline{J} \cdot \overline{n} \ da$$

(6)

where $\overline{J}$ is the current density. Maxwell's equations in the differential form and the associated boundary conditions are particularly suitable for the solution of the diffraction of electromagnetic waves (light) by boundaries of simple geometries, which result in separable solutions of the differential equations. This is the case of isolated cylinders, or

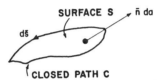

FIG. 2.   Boundary-condition parameters.

developed cylindrical surfaces of infinite extent (for example, diffraction gratings), spheres, ellipsoids, arrays of cylinders, spheres, and so on. The solutions become particularly straightforward for two-dimensional surfaces which are a function of two coordinates only (cylinders, gratings).

However, even when they appear straightforward, rigorous solutions of electromagnetic boundary-value problems have so far been obtained in only a very small number of cases (edge, slit, wedge, sphere, gratings of sinusoidal profile,[†] etc.). In reality, the mathematical complexity of electromagnetic boundary-value problems is generally rather formidable, and comparable to that of boundary-value problems in other domains of physics (quantum theory, for example).

In general, one is interested in the diffraction of waves having a simple harmonic frequency of the form

$$\overline{E}_i = (\overline{E}_i)_0 \ \exp \ (-i\omega t)$$

(7)

[†]Recently, rigorous solutions for gratings of triangular profile have been obtained, based on the solution for the sinusoidal grating given by Stroke in 1960 (see Ref. 1).

or of a superposition of such waves, where $(\overline{E}_i)_o$ is the time-independent part of the vector. In such a case, the $(\overline{E}_i)_o$ and $(\overline{H}_i)_o$ vectors are known to satisfy wave equations of the form

$$\nabla^2 \overline{E} + k^2 \overline{E} = 0$$
$$\nabla^2 \overline{H} + k^2 \overline{H} = 0$$

(8)

where $k = 2\pi/\lambda$ and $\omega = 2\pi f$. The frequency f and the wavelength λ are related by the equation

$$c = f\lambda$$

(9)

where c is defined as the velocity of light. For the case of mono-chromatic (single-frequency) waves, one has

$$\text{curl } \overline{E} = i\omega\mu_0\overline{H}$$
$$\text{curl } \overline{H} = -i\omega\varepsilon_0\overline{E}$$

(10)

and at boundaries, in the second medium (having a conductivity γ) we have

$$\overline{E} = \frac{1}{\gamma - i\omega\varepsilon_0} \text{ curl } \overline{H}$$
$$\overline{H} = \frac{1}{i\omega\mu} \text{ curl } \overline{E}$$

(11)

Clearly, at a perfect conductor, for which $\gamma \longrightarrow \infty$ and $\overline{E}_2 \longrightarrow 0$, $\overline{H} \longrightarrow 0$, inside the conductor, one has from (5), $\overline{n} \times (\overline{E}_1 - \overline{E}_2) = 0$, the result

$$\overline{n} \times \overline{E}_2 = \overline{n} \times \overline{E}_1 = 0$$

(12a)

which is the previously stated boundary condition

$$\overline{E}_{\text{tangential}} = 0$$

(12b)

on the surface of the conductor.

The boundary condition for the $\overline{H}$ field is particularly simple to obtain for two-dimensional surfaces which are independent of one coordinate. The method is used to illustrate some of the steps used in dealing with diffraction problems. Consider a plane-polarized wave H_z^i incident along the $-y$ direction in vacuo (medium 1) to a surface of equation $y = f(x)$ independent of z. Let H_z^i be parallel to z. By symmetry, the diffracted field will also only have a single magnetic component, H_z^d. Let the total magnetic field component in the medium 1 be $H_z = H_z^i + H_z^d$. One has for the total field in medium 1,

$$\text{curl } \overline{H} = -i\omega\varepsilon_0 E$$

(13)

and on the surface

$$\bar{n} \times (\nabla \times \bar{H}) = -i\omega \mathcal{E}_0 \bar{n} \times \bar{E} = 0 \tag{14}$$

from (12). By expanding $\bar{n} \times (\nabla \times \bar{H})$ and noting that $H_x = H_y = 0$ one obtains

$$n_x \frac{\partial H_z}{\partial x} + n_y \frac{\partial H_z}{\partial y} = 0 \tag{15}$$

applicable to the total field. But

$$\frac{\partial H_z}{\partial n} = \bar{n} \cdot \nabla H_z = n_x \frac{\partial H_z}{\partial x} + n_y \frac{\partial H_z}{\partial y} + n_z \frac{\partial H_z}{\partial z} \tag{16}$$

which gives the boundary condition

$$\frac{\partial H_z}{\partial n} = 0 \tag{17}$$

We find that the boundary condition for the $\bar{H}$ field is that the normal derivative of the component of the total magnetic field parallel to the surface coordinate z must be zero, in the case when the perfectly conducting surface is independent of the coordinate z to which the incident magnetic-field vector is parallel. More generally, it can be shown that all normal derivatives of the covariant components of the magnetic field must vanish on the surface of a perfect conductor, for the class of surfaces for which the conductor surface coincided with the coordinate surface of a curvilinear orthogonal coordinate system.

In addition to Maxwell's equations and the boundary conditions, other conditions need to be considered and satisfied when attempting to solve electromagnetic boundary-value diffraction problems. One of these is the Sommerfeld "radiation condition at infinity," which has to do with the requirement that the amount of energy flowing from sources in a finite domain through an infinitely extended boundary at an infinite distance from the sources must tend to zero. (In fact, the condition is somewhat stronger and states that the sources must be sources and not sinks of energy.) Another condition results from the use of energy-conservation laws and Poynting's theorem. Still another arises when the fields are expanded as Fourier-transform superpositions of plane waves, in which case all the waves, not only with real but also with imaginary propagation constants, must be included. Waves with imaginary propagation constants, or evanescent waves, and more generally inhomogeneous waves with complex propagation constants, have surfaces of constant phase and surfaces of constant amplitude which do not coincide. In a two-dimensional case, for example, in an ordinary cylindrical lens, for which the glass-thickness variation results in a variation of absorption across the lens surface, the surfaces of constant phase and constant amplitude are orthogonal to each other. Inhomogeneous waves form the most general and most frequently encountered types of waves in optics.

2.2. Image-Formation Solutions Using Huygens' Principle

Diffraction phenomena in optics have traditionally been described in the Helmholtz-Kirchhoff integral form, often called a formulation of Huygens' principle. Basically, it can be shown that the various Huygens'-principle solutions used in optics (for example, its "rigorous" vector form of its Fourier-transform formulation) can be obtained from Maxwell's electro-magnetic equations provided that certain important approximations are made. One of the fundamental approximations is that the Huygens'-principle formulation applies only to the vicinity of the center of quasi-spherical image-forming wavefronts. It is essential to be aware of the approximate nature of the Huygens'-principle expressions in their application to the solution of diffraction and image-formation problems. The apparent simplicity and even reasonable form of Huygens'-principle expressions, when evaluated heuristically on the basis of superposition principles and plane-wave spectra, should not be mistakenly used as the basis for "more rigorous" solutions by the inclusion of higher-order terms obtained from the initially approximate formulation. When correctly applied, however, Huygens' principle, in its Fourier-transform expression, forms a most powerful tool for dealing with image-formation problems. In particular, it is being used for the calculation of the distribution of energy in diffraction patterns formed by wavefronts of finite size produced by reflection, transmission, and diffraction from optical elements (mirrors, lenses, prisms, and gratings).

A particularly useful approximate form of the Helmholtz-Kirchhoff integral formulation of Huygens' principle which may be derived directly from Maxwell's equations is [†]

$$\overline{E}_P(k\alpha, k\beta) = \frac{i}{\lambda} \frac{\exp(-ikR)}{R} \iint_{\text{aperture}} \overline{E}_0 \exp[ik\Delta(x, y)]$$

$$\times \exp[-ik(\alpha x + \beta y)]\, dx\, dy \quad (18)$$

where $\overline{E}_P$ is the complex amplitude of the electric-field vector at a point P in the image plane in the vicinity of the center of a quasi-spherical wavefront of radius R. α and β are the direction cosines defining the position of P as seen from the aperture centered on the point 0 on the quasi-spherical wavefront, x and y the coordinates defining the aperture, $\Delta(x, y)$ the aberration of the wavefront from sphericity, $k = 2\pi/\lambda$,

[†]For a more complete discussion of this very important subject, see, for example, Refs. 2–6.

$i = \sqrt{-1}$, and $\overline{E}_0$ the amplitude $\overline{E}_0(x, y)$ in the wavefront. The direction cosines α and β, and in fact the coordinates x and y, refer to the quasi-plane wavefront side of a perfect-focusing system (mirror or lens). A perfect-focusing system is defined as having the property of transforming a perfectly plane wavefront into a perfectly spherical wavefront. Consequently, it transforms a quasi-plane wavefront into a quasi-spherical wavefront. Moreover, the aberrations from the plane wavefront are identical to the aberrations from the sperical image-forming wavefront within the approximations used.

Equation (18) has the form of a Fourier transformation and can be read as follows: The complex amplitude of the electric-field vector at a point in the image plane is equal to the Fourier transform of the distribution of complex amplitude of the electric field within the image-forming aperture. Clearly here the field vectors in the aperture, and the field vectors in the image plane, are parallel to the image plane. One Fourier transformation needs to be carried out for each point in the diffraction pattern.

For example, for a perfectly uniform plane wavefront within a rectangular aperture of width A along the x axis, $|\overline{E}_0| = 1$, $\Delta = 0$,

$$\overline{E}_P(k\alpha) = \frac{i}{\lambda} \frac{\exp(-ikR)}{R} \int_{-A/2}^{+A/2} \exp(-ik\alpha x)\, dx \qquad (19)$$

which immediately gives upon integration,

$$\overline{E}_P(k\alpha) = \left[\frac{i}{\lambda} \frac{\exp(-ikR)}{R} \frac{A}{2}\right] \left[\frac{\sin k\,\alpha(A/2)}{k\alpha(A/2)}\right] \qquad (20)$$

The complex amplitude $\overline{E}_P(k\alpha)$ thus varies according to the well-known $\sin \S'/\S'$ function along $\S' \cong R\alpha = f\alpha$ in the image plane and has a first zero for

$$\alpha_0 = \frac{\lambda}{A} \qquad \text{radians} \qquad (21)$$

or

$$\S'_0 = R\frac{\lambda}{A} = f\frac{\lambda}{A} \qquad \text{(with } R = f = \text{focal length of focusing system)} \qquad (22)$$

In general only the time-average of the instantaneous power (see Chapter IV), in optics called the intensity

$$I_P = \overline{\overline{E}_P \cdot \overline{E}_P^{\,*}} \qquad (23)$$

is detectable (with the help of photoelectric, photographic, or other receivers). (The symbol $*$ denotes "complex conjugate," and the $\overline{}$ indicates a time average.) For the case of Eq. (20) we simply have $I_P = |\overline{E}_P|^2$.

Hence I_p varies as $(\sin \S'/\S')^2$ and also has, of course, a first minimum at $(\lambda/A) \cdot f$ from the central maximum. I_p is generally called the *diffraction pattern* corresponding to the aperture A.

Equation (18), which expresses the diffraction at infinity by an aperture, can be made heuristically plausible on the basis of superposition and Huygens' principle as it is generally understood. Consider a pupil in the xy plane and a point M centered on an element of area $dx\,dy$ in that plane. The element of area centered on M emits an elementary wave $f(x,y)\,dx\,dy$, where $f(x,y)$ describes the scalar component of the $\overline{E}$ vector in the pupil. In the direction defined by the direction cosines α, β, γ, the wave from M is dephased by $(2\pi/\lambda)(\alpha x + \beta y)$ with respect to an elementary wave emitted from the center of the pupil O. By superposition, we have in the direction α, β, γ the sum of these waves,

$$\iint_{\text{pupil}} f(x, y) \exp\left[i(2\pi/\lambda)(\alpha x + \beta y)\right] dx\,dy \tag{24}$$

All elementary waves emitted by the pupil (or transmitted by the pupil) in the direction α, β, γ come to focus at a single point $P'(\xi', \eta')$ in the focal plane of a perfect-focusing system, such that

$$\xi' = \frac{f\alpha}{\gamma} \qquad \eta' = \frac{f\beta}{\gamma} \tag{25}$$

Inasmuch as $\gamma \cong 1$ and $\alpha^2 + \beta^2 + \gamma^2 = 1$, we have

$$\xi' = f\alpha \qquad \eta' = f\beta \tag{26}$$

for the coordinates in the image, in the focal plane of the focusing system (focal length $= f$) and

$$\alpha = \frac{\xi'}{f} \qquad \beta = \frac{\eta'}{f} \tag{27}$$

One obtains for the diffraction pattern the equations

$$F(k\alpha, k\beta) = \iint_{\text{pupil}} f(x,y) \exp\left[\frac{2\pi i}{\lambda}(\alpha x + \beta y)\right] dx\,dy \tag{28}$$

and

$$F\left(\frac{k}{f}\xi', \frac{k}{f}\eta'\right) = \iint_{\text{pupil}} f(x,y) \exp\left[\frac{2\pi i}{\lambda}\left(\frac{\xi'}{f}x + \frac{\eta'}{f}y\right)\right] dx\,dy \tag{29}$$

which are expressions for the diffraction pattern identical to that ob-

tained from Maxwell's equations in (18), by using the there-stated approximations. (Recall that $k = 2\pi/\lambda$).

In summary, a knowledge of the distribution of complex amplitude
(amplitude and phase) of the electromagnetic field (or more exactly of a
component of the electric field) within an aperture, no matter how this
field is created in the aperture, permits us to compute the diffraction
patterns corresponding to the aperture and light distribution. A unique
relation between the field distribution in the aperture and the light distribution in the diffraction patterns exists, within the stated approximations, and takes the form of a Fourier transformation. The powerful
techniques of operational calculus have been extensively applied to
optical-image-formation problems with very fruitful results. In particular,
a basic similarity has been recognized between problems in electrical
engineering and problems of optical image formation and spectroscopy,
whenever superposition and operational methods are appropriate. One
way by which a complex amplitude distribution $f(x,y)$ may be created in
the aperture of a lens is by placing a *hologram* into the aperture, illuminated by a plane or spherical wave, for example.

3. Image-Formation Theory and Optical Signal Processing in Fourier-Transform Formulation

The well-known operational theory of signal processing used when
dealing with electrical and electronic systems can be immediately translated into optics, provided that the parameter "time" used in electrical
engineering is translated as "space" when used in optics.

The diffraction patterns (intensity) can be considered as the impulse
response of an optical system. The intensity distribution in the image
as a function of spatial coordinates in the image space is simply given
by the *convolution* integral of the intensity distribution in the object (as
geometrically imaged in the image plane) with the intensity distribution
in the diffraction pattern (as it appears in the image plane). Also the
Fourier transform of the image intensity distribution is equal to the product of the transform of the object intensity distribution by the transform
of the diffraction pattern. The transform of the diffraction pattern is also
called the *frequency-response function* of the optical system, since it
gives the distribution of light in the image of a spatially periodic intensity distribution in the object. Finally, the frequency response (function) of an optical system can be immediately shown to be equal to the
spatial convolution of the light distribution (complex amplitude) in the
aperture with itself. For example, for the uniformly illuminated rectangular aperture considered before, the frequency-response function is im-

mediately seen to have a triangular shape as a function of *spatial* frequency, that is, as a function of (distance)$^{-1}$ (see Fig. 5, Chapter III).

Clearly, optical systems, when used as elements of a signal-processing system for electromagnetic waves, for example, in television, radio astronomy, and similar systems, present a degree of two-dimensional flexibility unavailable in electrical systems. Moreover, the parameter "time" can be reintroduced and used to advantage.

Skillful use of the relations between object, aperture, and image space permits us to considerably simplify measurements of the desired distribution of light (complex amplitude and intensity) in any of these domains in practical applications.

References

1. G. W. Stroke, *J. Opt. Soc. Am.* **54**, 846 (1964).
2. A. Maréchal and M. Françon, "Diffraction." Revue d' Optique, Paris, 1960.
3. A. Maréchal, "Imagerie Géométrique." Revue d'Optique, Paris, 1952.
4. G.W. Stroke, "Diffraction Gratings," *in* "Handbuch der Physik" (S. Flügge, ed.),Vol. 29, pp. 426-754. Springer, Berlin, 1967.
5. G. Toraldo di Francia, "La Diffrazione della Luce." Edizione Scientifice Einaudi (Publ. Paolo Boringhieri), Torino, 1958.
6. M. Born and E. Wolf, "Principles of Optics," 2nd revised ed. Pergamon Press, Oxford, 1964.

III. IMAGE FORMATION IN NONCOHERENT LIGHT (ELEMENTS AND DEFINITIONS)†

1. Image of Point Source

Under the conditions where the Fourier-transform relation between the complex amplitude distribution in the pupil $f(x, y)$ and the complex amplitude distribution at a point in the focal plane $F_P[k(\xi'/f), k(\eta/f)]$ holds (Fig. 1), it was shown [(28) and (29), Chapter II] that

$$F_{P'}\left(k\frac{\xi'}{f}, k\frac{\eta'}{f}\right) = \iint_{\text{pupil}} f(x, y) \exp\left[\frac{2\pi i}{\lambda}\left(\frac{\xi'}{f}x + \frac{\eta'}{f}y\right)\right] dx\,dy \quad (1a)$$

or, equivalently,

$$F_{P'}(k\alpha, k\beta) = \iint_{\text{pupil}} f(x, y) \exp\left[\frac{2\pi i}{\lambda}\left((\alpha x + \beta y)\right)\right] dx\,dy \quad (1b)$$

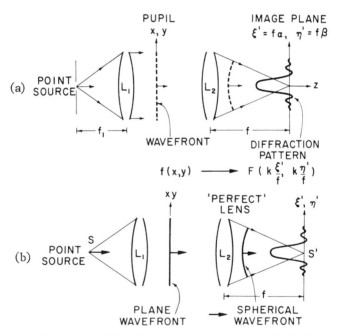

FIG. 1. (a) Point-image diffraction-pattern parameters. (b) Systems definition of *perfect lens system*. Note: The point source is at a distance f_1 from L_1 not necessarily equal to the focal length of L_1. Similarly, the "diffraction pattern" (of which the magnitude squared is called the "spread function") is at a distance f_2 from L_2, not necessarily equal to the focal length of L_2.

†For a more detailed analysis, see Refs. 1-3. In this text we use the term "noncoherent light" in preference to "incoherent light," according to a kind suggestion by F. B. Llewellyn and W. G. Dow, thus also avoiding possible uncertainties between the expressions "in coherent" and "incoherent."

$F_{P'}$ is the electric field vector amplitude at the point P'. We recall that

$$\xi' = f\alpha \qquad \eta' = f\beta \tag{2}$$

and

$$k = \frac{2\pi}{\lambda} \tag{3}$$

With T = "Fourier transform of," one can write (1)

$$F_{P'}\left(\frac{k}{f}\xi', \frac{k}{f}\eta'\right) = T\left[f(x,y)\right] \tag{4}$$

or

$$f(x,y) \longrightarrow F_{P'}\left(\frac{k}{f}\xi', \frac{k}{f}\eta'\right) \tag{5}$$

to indicate the Fourier-transform relationship between pupil and diffraction image. Note that

$$\frac{k}{f}\xi' = \frac{2\pi}{\lambda}\alpha \tag{6}$$

$$\frac{k}{f}\eta' = \frac{2\pi}{\lambda}\beta \tag{7}$$

From Equation (1) it is seen that *the image of a point source is not a point source*. In general, as a result of the electromagnetic nature of light and the finite dimensions of the pupil, the light from a point source is spread out in a diffraction pattern of which the intensity (magnitude squared) will hence forth be called the *spread function* (after E.L. O'Neill).

Optical receptors are generally sensitive to the electric field $\overline{E}$, or, rather, to its magnitude squared, namely, to $\overline{E} \cdot \overline{E}* = |\overline{E}|^2$. For this reason, we shall use $\overline{E} = F$ or f, when appropriate, in the Fourier-transform treatment of image formation.

A physical receptor (photoelectric cell, photographic plate, eye, etc.) can only detect energy, at best. In optics, only the time average of the electric field is observable, under ordinary conditions. The detectable quantity is then the *intensity*

$$I_{P'} = \overline{E} \cdot \overline{E}^* = |\overline{E}|^2 \tag{8}$$

and *not* the complex amplitude $\overline{E} = |\overline{E}| \exp\left[i\phi(\overline{E})\right]$ of the electric-field vector. The phase $\phi(\overline{E})$ in the electric-field vector *can*, however, be detected, for example, if an interferometric arrangement is used.

2. Summation of Light from Several Source Points Reaching One Image Point

It is very important to observe that light originating from more than one source point may reach any point in the image. Let $\overline{E}_1$ and $\overline{E}_2$ be the light-amplitude vectors corresponding to the light reaching a point P' from two different source points. Two extreme situations may arise. They are illustrated for two source points.

(1) *The two source points radiate coherently:* $\overline{E}_1$ and $\overline{E}_2$ may interfere, and the detected intensity is

$$[I_{P'}]_{\text{coherent}} = (\overline{E}_1 + \overline{E}_2) \cdot (\overline{E}_1 + \overline{E}_2)^* = |\overline{E}_1 + \overline{E}_2|^2 \tag{9}$$

(2) *The two source points radiate completely noncoherently:* $\overline{E}_1$ and $\overline{E}_2$ cannot interfere, and the detected intensity is

$$[I_{P'}]_{\text{noncoherent}} = (\overline{E}_1\overline{E}_1{}^*) + (\overline{E}_2\overline{E}_2{}^*) = |\overline{E}_1|^2 + |\overline{E}_2|^2 = I_1 + I_2 \tag{10}$$

The distinction between the detection according to (9) and (10) is all-important. It is basic for the further discussion of image formation.

Equations (8), (9), and (10) describe ideal situations. In general, only suitable *time averages* of the indicated intensities are detected. Clearly also, in case of the light from several source points reaching *one* image point, one has

$$[I_{P'}]_{\text{coherent}} = \left| \sum_i \overline{E}_i \right|^2 \tag{11}$$

and

$$[I_{P'}]_{\text{noncoherent}} = \sum_i |\overline{E}_i|^2 = \sum_i I_i \tag{12}$$

We have therefore the following well-known "prescriptions":

In *coherent light*: *Sum amplitudes and take square of magnitude*
In *noncoherent light*: *Sum intensities* (i.e., *sum squares of magnitudes*)

It will be shown, when dealing with the mathematical formulation of coherence, partial coherence, and incoherence, that the absence of interference is indeed sufficient to characterize noncoherence as described by the above.

Note: It is very important to note in this "wave" treatment of optics that the transition from complex amplitude $\overline{E}$ to the detected intensity $I = \overline{E} \cdot \overline{E}^*$ occurs only at a detector. An important support for this fact was first obtained in experiments by Forrester *et al.*[4] (see also Mandel[5] and Chapter 4).

3. Spread Function

Note: In noncoherent light, a point source will produce *not a* point image but an intensity distribution. For the case of any point source (within the applicable approximations), the intensity distribution in its image may be written in the form

$$I_{P'}(k\alpha, k\beta) = \overline{F_{P'}(k\alpha, k\beta)} \cdot \overline{F_{P'}(k\alpha, k\beta)}^* = |F_{P'}(k\alpha, k\beta)|^2 \tag{13}$$

The intensity distribution in the image of a point source is defined as the *spread function*

$$s(k\alpha, k\beta) \equiv |F_{P'}(k\alpha, k\beta)|^2 \qquad (14a)$$

or better,

$$s(k\alpha, k\beta) \equiv \overline{F_{P'}(k\alpha, k\beta)} \cdot \overline{F_{P'}(k\alpha, k\beta)^*} \qquad (14b)$$

In many practical cases in optics the spread functions can be made to be locally spatially invariant, i.e., the same except for translation parameters.

4. Image of Extended Source in Noncoherent Light

An extended source can be considered as formed by *individual elementary radiators* (dipole, atomic, molecular, etc.). This is true for self-radiating or illuminated surfaces, gases, lasers, etc. (see Fig. 2).

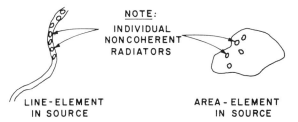

FIG. 2. Line-element and source-element concepts.

For the purposes of image-formation theory, we shall concern ourselves at first with sources in areas at right angles to the optical axis of a lens system. Two conclusions reached previously are basic to what follows:

(1) *The light from a point source is spread out into a spread function.* As a corollary, it is clear that light from two neighboring points in the source will spill over from one of the geometric-optics images to the other, and vice versa.

(2) *In noncoherent light*, the contribution to the *light intensity at any image point* is obtained by a suitable *summation of the intensity contributions* to the particular image point of the light "spilled" over or centered there, as a result of its origin in the discrete object points.

5. Method of Computing Image Intensities and Transfer Functions

The light intensity at an image point in the presence of spreading can be computed by first considering the *image formed according to "perfect"* 'Gaussian' *geometric optics* (called *geometric-optics image* henceforth) and, next, by computing the weighted contribution by starting from the geometric-optics image rather than from the object.

Let $O(k\xi'/f, k\eta'/f)$ be the image according to geometric optics, and let $s(k\xi'/f, k\eta'/f)$ be the spread function according to (14). $s(k\xi'/f, k\eta'/f)$ is

the intensity distribution in the image of a point source (located at infinity). To simplify the notation, it is frequently convenient, at this
stage, to introduce so-called *reduced coordinates* u' and v', in place of
$k\xi'/f$ and $k\eta'/f$, that is, in place of $k\alpha$ and $k\beta$. Let us define the reduced
coordinates:

$$u' \equiv \frac{k}{f}\,\xi' = k\alpha \qquad (6a)$$

$$v' \equiv \frac{k}{f}\,\eta' = k\beta \qquad (7a)$$

We need to compute the intensity $I(u_0', v_0')$ at a point (u_0', v_0') in the
image (Fig. 3). Clearly, the *contribution of the image of a source element*

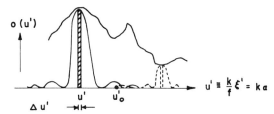

F IG. 3. Image parameters, for convolution analysis of image formation.

of width $\Delta u'$ *and having an intensity* $O(u')$ when centered at the point u'
is given by

$$\Delta I(u_0') = s(u_0' - u')\,O(u')\,\Delta u' \qquad (15)$$

under the suitable assumptions of spatial invariance over the appropriate image region.

Equation (15) is nothing but the diffraction pattern (or spread function)
weighted by the intensity $O(u')$ at the point u' where it is centered.
Since even an elementary radiator (atom, molecule, etc.) has some finite
width $\Delta u'$, the spread-function maximum intensity must be weighted by
$O(u')\Delta u'$. If several source (image) points contribute to the intensity at
u_0', one has

$$I(u_0') = \sum_{j} O(u')_j\, s(u_0' - u')\, du' \qquad (16)$$

and in the limit of a great number of source points,

$$I(u_0') = \int_{-\infty}^{+\infty} O(u')s(u_0' - u')\, du' \qquad (17a)$$

Equation (17a) is immediately recognized as having the form of a convolution integral, that is (see Chapter VII),

$$I(u_0') = O(u') \otimes s(u') \qquad (17b)$$

The several important image-formation equations established so far can be written in terms of the reduced coordinates [(6a) and (6b)] as:

Diffraction pattern (complex amplitude distribution):

$$F(u', v') = \iint f(x, y) \exp \left[i(u'x + v'y) \right] dx \, dy = T \left[f(x, y) \right] \qquad (18)$$

Spread function (intensity distribution):

$$s(u', v') = T \left[f(x, y) \right] \cdot T^* \left[f(x, y) \right] \qquad (19)$$

Image (intensity distribution):

$$I(u', v') = O(u', v') \otimes s(u', v') \qquad (20)$$

where T is the Fourier transform, and $\otimes$ is the convolution integral; that is,

Spatial-domain representation of image:

$$I(u_0', v_0') = O(u', v') \otimes s(u', v')$$

$$= \iint_{-\infty}^{+\infty} O(u', v') s(u_0' - u', v_0' - v') \, du' \, dv' \qquad (20a)$$

Equation (20a) is the two-dimensional equivalent of (17).

6. Analysis of Image Formation in Terms of Spatial Fourier Space

Some important advantages are obtained by analyzing the image-formation problem in the *spatial Fourier space* (also known as the *spatial frequency domain*).

The concepts of spatial Fourier space can be introduced, in analogy with electrical engineering, by taking the Fourier transforms of the equations [(18) to (20)] which express the image-formation problem in the spatial domain.

Consider

$$I(u_0', v_0') = O(u', v') \otimes s(u', v') \qquad (20b)$$

We recall that

$$u' = \frac{2\pi}{\lambda f} \xi' = \frac{\text{direction cosine}}{\text{wavelength}} = (\text{length})^{-1} \qquad (6a)$$

In other words, u' has the dimensions of an angle(radians) per unit length (wavelength), in the image domain. We now take the Fourier transform of both sides of (20), which will bring (20) back into the pupil (i.e., wavefront) domain, where the coordinates are x and y. We have

$$T\left[I\left(u',\,v'\right)\right] = \iint_{-\infty}^{+\infty} I\left(u',\,v'\right)\exp\left[-2\pi i\left(u'x + v'y\right)\right]du'\,dv' \quad (21a)$$

and

$$T\left[O\left(u',\,v'\right)\otimes s\left(u',\,v'\right)\right]$$

$$= \iint_{-\infty}^{+\infty} \left[O\left(u',\,v'\right)\otimes s\left(u',\,v'\right)\right]\exp\left[-2\pi i\left(u'x + v'y\right)\right]du'\,dv' \quad (21b)$$

We may therefore write, by Fourier transformation of (20), that

$$T\left[I\left(u_0',\,v_0'\right)\right] = T\left[O\left(u',\,v'\right)\otimes s\left(u',\,v'\right)\right] \quad (21c)$$

According to the convolution theorem, (VII.34b), (21) becomes

Fourier-domain representation of image formation:

$$T\left[I\left(u_0',\,v_0'\right)\right] = T\left[O\left(u',\,v'\right)\right]\cdot T\left[s\left(u',\,v'\right)\right] \quad (22)$$

Equation (22) states the important conclusion that the Fourier transform of the image-intensity distribution is given by the *product* of the Fourier transform of the object-intensity distribution with the Fourier transform of the spread function.

The importance of (22) becomes additionally clear when the spread function $s\left(u',\,v'\right)$ is expressed in terms of the complex amplitude distribution in the wavefront $f(x,\,y)$, to which it is related according to

$$s\left(u',\,v'\right) = T\left[f(x,\,y)\right]\cdot T^*\left[f(x,\,y)\right] \quad (19)$$

It follows from the second form of the convolution theorem, (VII.41), that

$$T\left[s\left(u',\,v'\right)\right] = f(x,\,y)\otimes f^*\left(-x,\,-y\right) \quad (23)$$

which gives another important conclusion, stating that the Fourier transform of the spread function is given by the convolution of the pupil function with its complex conjugate (folded over!). The operation in (23) (very similar to the convolution encountered so far) is known as a *Faltungintegral* (from the German word "Faltung," meaning "folding"). For a real, symmetric pupil function $f(x,\,y)$, (23) becomes

$$T\left[s\left(u',\,v'\right)\right] = f(x,\,y)\otimes f(x,\,y) \quad \text{(real, symmetric } f(x,\,y)) \quad (23a)$$

which is a convolution in the sense used heretofore. We may now write (22) in perhaps its most useful form:

Fourier representation of image formation:

$$T\left[I\left(u',\,v'\right)\right] = T\left[O\left(u',\,v'\right)\right]\cdot\left[f(x,\,y)\otimes f^*\left(-x,\,-y\right)\right] \quad (24)$$

$$T[I(u', v')] = T[O(u', v')] \cdot [f(x, y) \circledast f(x, y)] \qquad (24a)$$

for the, in practice, very interesting case of real, symmetric $f(x, y)$.

Because of the analogy with circuit theory, and transfer functions used in electrical engineering, the function

$$T[s(u', v')] = f(x, y) \circledast f^*(-x, -y) \equiv \tau(x, y) \qquad (23b)$$

may be described as the *transfer function* of the image-forming instrument, and identified by the symbol $\tau(x, y)$. It is a complex function.

We shall show in Section 7 that the transfer function $\tau(x, y)$ may be considered also as the frequency-response function of the optical instrument. When the instrument is used to form images of sinusoidally periodic intensity objects, then the frequency-response function $\tau(x, y)$ gives a measure of the attenuation in the contrast of the images of sinusoidal intensity objects of different spatial frequencies, as well as of the shift of the sinusoidal images with respect to the geometric images of the sinusoidal objects. Strictly speaking, the attenuation in the contrast is given by the magnitude of $\tau(x, y)$, and the shift of the images by the phase of $\tau(x, y)$.

We may finally note that it follows from (VII.64) that another, perhaps simpler, way of writing (23) is

$$\tau(x, y) \equiv T[s(u', v')] = [f_1 \star f_1^*]_{-x, -y} \qquad (23c)$$

which states that the transfer function $\tau(x, y)$ is equal to the autocorrelation function of the pupil function, evaluated at $(-x, -y)$. In terms of Eq. (23b) we may finally write Eq. (22) in the form

$$T[I(u_0', v_0')] = T[O(u', v')] \cdot \tau(x, y) \qquad (23d)$$

The normalized form of these equations is discussed by O'Neill.[2]

7. Physical Significance of Spread Function and of Fourier-Space Analysis of Image Formation

We continue to examine the image-formation problem in terms of the geometrical image of the object, rather than in terms of the actual object. We implicitly also avoid the need for introducing magnification parameters at this time. (In many cases, for instance, in the case of astronomical instruments, the result of image-formation calculations can, at best, only yield the geometric image. The geometric image is the image which would be formed by a perfect optical instrument, without any limits set by diffraction and instrument imperfections. Accordingly, the use of "geometric image" of the object, rather than "object" is well justified.)

We now consider the image-formation problem for two particularly important objects, the point object (delta function) and the sinusoidal intensity object. The qualitative role of these two objects, in the image-formation analysis, is symbolized by the arrow $\longrightarrow$.

7.1. Point Source (Dirac Delta Function) $\longrightarrow$ Diffraction Pattern = Spread Function

Consider a point source with a geometric image

$$\delta\ (0) = \lim_{\Delta \to 0} \int_{-\Delta/2}^{+\Delta/2} \frac{1}{\Delta}\ du' \tag{25}$$

According to (20), we have

$$I(u_0') = \delta(u_0') \circledast s(u') = s(u_0') \tag{26}$$

and we may conclude that the spread function

$$s(u') = \text{image of a point source} \tag{27}$$

Implicitly, (27) was assumed in establishing (20) in the first place. Note that $s(u')$ does not in any way have to be the "ordinary" diffraction pattern of the simplest type [for example, the $(\sin u'/u')^2$ distribution, or some equivalent]. In its most general form, $s(u')$ and $s(u', v')$ are the *images* of an "infinitely narrow" slit, or of a point source, respectively. In fact, $s(u', v')$ or $s(u')$ do not have to be in the focal plane, as far as (20) is concerned.

7.2. Sinusoidal Intensity Objects $\longrightarrow$ Spatial Frequency-Response Function = Fourier Transform of Diffraction Pattern

In analogy with a simple harmonic time function

$$\exp\left[2\pi\, i\ \frac{t}{T} \right]$$

with a period T, consider a sinusoidal intensity object

$$O(u', v') = \exp\left[2\pi\, i \left(\frac{u'}{U_0'} + \frac{v'}{V_0'} \right) \right] \tag{28}$$

where U_0' and V_0' are the periods of the object in the (u', v') domain. (Note, in this analysis, that the real parts of the complex representation of the object and image, respectively, must be taken in the end of the computation.) In view of the transformation of $O(u', v')$ into the (x, y) domain, we may choose to describe the object as

$$O(u', v') = \exp\left[2\pi\, i(u'x_0 + v'y_0) \right] \tag{29}$$

where $x_0 = 1/U_0'$ and $y_0 = 1/V_0'$ are now "frequencies" in the (x, y) domain.

We shall now show that the image of a sinusoidal object is also sinusoidal, with the same frequency (i.e., period), but attenuated in contrast, and shifted in phase, according to the transfer function $\tau(x, y)$.

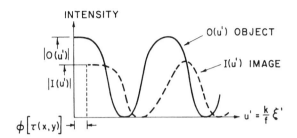

FIG. 4. Frequency-response-function parameters (noncoherent light).

Consider (20),

$$I(u_0{'}, v_0{'}) = O(u', v') \circledast s(u', v') \tag{20}$$

According to Section 2 of Chapter VII, the convolution integral is commutative, and we have

$$I(u_0{'}, v_0{'}) = s(u', v') \circledast O(u', v') \tag{20a}$$

that is, here

$$I(u_0{'}, v_0{'}) = \iint_{-\infty}^{+\infty} s(u', v') \exp \{2\pi i[(u_0{'} - u')x_0 + (v_0{'} - v')y_0]\} \, du' \, dv' \tag{30}$$

which can be written

$$I(u_0{'}, v_0{'}) = \iint_{-\infty}^{+\infty} s(u', v') \exp [2\pi i(u_0{'}x_0 + v_0{'}y_0)] \times \exp [-2\pi i(u'x_0 + v'y_0)] \, du' \, dv' \tag{31}$$

By noting that the integration in (31) is with respect to u' and v', we have

$$I(u_0{'}, v_0{'}) = \exp [2\pi i(u_0{'}x_0 + v_0{'}y_0)] \iint_{-\infty}^{+\infty} s(u', v') \times \exp [-2\pi i(u'x_0 + v'y_0)] \, du' \, dv' \tag{32}$$

Finally, we recognize [see (23)] that the integral in (32) is nothing but the transfer function $\tau(x, y)$. We may therefore finally write that the image $I(u_0{'}, v_0{'})$ of the sinusoidal object is given by the equation

$$I(u_0{}', v_0{}') = \exp\left[2\pi\,i\,(u_0{}'x_0 + v_0{}'\,y_0)\right]\,\tau(x_0, y_0) \qquad (33)$$

that is,

$$I(u_0{}', v_0{}') = O(u_0{}', v_0{}')\,\tau(x_0, y_0) \qquad (34)$$

We may also write

$$\tau(x_0, y_0) = \frac{I(u_0{}', v_0{}')}{O(u_0{}', v_0{}')} \qquad (35)$$

which states that the transfer function, for the case of a sinusoidal intensity object, is simply equal to the ratio of the sinusoidal image to the sinusoidal object. In reality, since we are dealing with complex amplitude objects, and images, we must write (35) in the form

$$|\,\tau(x_0, y_0)\,| = \frac{|\,I(u_0{}', v_0{}')\,|}{|\,O(u_0{}', v_0{}')\,|} \qquad (35a)$$

and

$$\text{phase of } \tau(x_0, y_0) = [\text{phase of } I(u', v')] - [\text{phase of } O(u', v')]$$

that is,

$$\phi\,[\tau(x_0, y_0)] = \phi\,[I(u', v')] - \phi\,[O(u', v')] \qquad (35b)$$

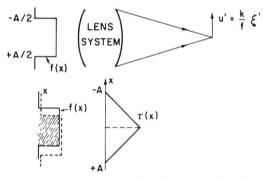

FIG. 5. Frequency-response function for $f(x)$ = rect. function (noncoherent light).

The graphical interpretation of (35) for the case of the sinusoidal object is given in Fig. 4. For the case of an object which can be analyzed in terms of a Fourier spectrum (series or integral) of sinusoidal intensity components, each spatial frequency x_0, y_0 gives one point on the transfer-function graph for the instrument. Of course, to each point on the transfer-function representation, corresponds an amplitude $|\tau(x, y)|$ and a phase $\phi\,[\tau(x, y)]$. We may finally recall that the transfer function $\tau(x, y)$ may be obtained from the wavefront by a convolution process, according to (23),

$$\tau(x,\ y) = f(x,\ y) \circledast f^*(-x,\ -y) \tag{23b}$$

An example of $\tau(x,\ y)$ for the case of a real function $f(x,\ y) = 1$, within a rectangular aperature of width A, is shown in Fig. 5.

8. Application of "Fourier-Space" Representation to "Synthesis" of Object-Functions

The Fourier-space representation of the image-formation problem appears particularly useful in "matched filtering" problems, such as those arising in spectroscopy, as first recognized by Golay, Girard, and others.[6-9] The problem here is to synthesize a very narrow slit by means of a materialization (in the spatial domain) of the frequency representation of the slit, as it appears in the "spatial-frequency domain". The object of the synthesis is to obtain a slit-scanning function (the result of the scanning of the entrance-slit image, in monochromatic light, past a similar exit slit) with the same spatial (i.e. wavelength) resolution as that of the fine slits, but with a greatly increased flux transmission. Since a family of sinusoidal functions is required to "synthesize" the fine slit, and since it is necessary to use sufficiently long "sinusoids" in order to obtain an adequate representation of each sinusoid, it is clear that the fine slit can be synthesized by a rather large entrance (and exit) slit area, in which sinusoidally transmitting functions are stacked up along the slit height; the increase in luminosity then corresponds to the ratio of this area to the area of the narrow slit.

Many other examples of image synthesis are conceivable, both in the case of noncoherent illumination (this chapter) and in the case of coherent illumination.

References

1. A. Maréchal and M. Françon, "Diffraction." Revue d' Optique, Paris, 1960
2. E. L. O'Neill, "An Introduction to Statistical Optics." Addison-Wesley, Reading, Massachusetts, 1963.
3. R. C. Jennison, "Fourier Transforms and Convolutions for the Experimentalist." Pergamon Press, New York, 1961.
4. A. T. Forrester, R. A. Gudmundsen, and P. O. Johnson, *Phys. Rev.* **99**, 1691 (1955).
5. L. Mandel, *in* "Progress in Optics" (E. Wolf, ed.), Vol. II. North-Holland Publ., Amsterdam, 1963.
6. M. J. E. Golay, *J. Opt. Soc. Am.* **41**, 468 (1951).
7. A. Girard, *Opt. Acta* **1**, 81 (1960); *Appl. Opt.* **2**, 79 (1963).
8. G.W. Stroke, *in* "Handbuch der Physik" (S. Flügge, ed.), Vol. 29 "Diffraction Gratings," pp. 426-754. Springer, Berlin, 1967.
9. P. Jacquinot and B. Roizen-Dossier, *in* "Progress in Optics" (E. Wolf, ed.), Vol. III. North-Holland, Amsterdam, 1964.

IV. COHERENCE CHARACTERISTICS OF LIGHT
(EXPERIMENTAL CHARACTERIZATION)

1. Introduction

We may distinguish at least the following types of coherence:

(1) Spectral (temporal or phase) coherence (short-term temporal coherence).

(2) Spatial coherence.

(3) Amplitude and frequency stability in time (long-term temporal coherence).

In general, light must be described by probability functions. In general also, light from ordinary sources is either unpolarized or partially polarized, and has a finite frequency spread and an amplitude which varies in time. In cw lasers, the light is generally polarized when Brewster-angle windows are used.

The narrowest definition for light will describe it as a single-frequency, polarized electromagnetic wave of constant amplitude. This ideal condition may not be readily approached in practice, except for lasers. However it is of great practical interest to consider the case of quasi-monochromatic (narrow-band radiation). Let $\widetilde{E}(x,y,z,t)$ describe the instantaneous value of the light vector for such a case.

Over short periods of time one has (see Fig. 1a) $\widetilde{E}(t) = \widetilde{E}_0(t) e^{i\omega_0 t}$

where ω_0 is the center frequency and where $\widetilde{E}_0(t)$ is a time-varying vector which varies slowly in comparison with $(1/\omega_0)$. Over long periods of time the mean of $\widetilde{E}(t)$, defined by

$$\lim_{\Delta T \to \infty} \frac{1}{\Delta T} \int_0^{\Delta T} \widetilde{E}(t) \ dt \equiv \ <\widetilde{E}(t)> \ = \overline{\widetilde{E}(t)}$$

is equal to zero. The angle brackets or overbars indicate "time average." However the mean square $\overline{\widetilde{E}(t)\,\widetilde{E}(t)^*}$ is equal to the power (intensity $= I$) and is different from zero:

$$I = \overline{\widetilde{E}(t)\,\widetilde{E}(t)^*} = \lim_{\Delta T \to \infty} \frac{1}{\Delta T} \int_0^{\Delta T} \widetilde{E}(t)\,\widetilde{E}* \ dt \neq 0$$

that is, $I = \ <\widetilde{E}(t)\,\widetilde{E}(t)^*> \ \neq 0$.

2. Characterization of Coherence

2.1. Correlation Method

Consider a light source S and an observation point P (see Fig. 2). It is assumed that the light arriving at P via the two paths shown has collinear polarizations.

FIG. 1. Characterization of sources at optical frequencies, after Maréchal and Françon.[17]

The resultant field at P is

$$\widetilde{\overline{E}}_R(t) = \widetilde{\overline{E}}_1(t) + \widetilde{\overline{E}}_2(t) = \widetilde{\overline{E}}(t) + \widetilde{\overline{E}}(t - \tau) \qquad (1)$$

Assuming quadratic detection with long-time integration, the observable quantity is

$$I_P(t,\tau) = \lim_{\Delta T \to \infty} \frac{1}{\Delta T} \int_0^{\Delta T} \widetilde{\overline{E}}_R(t)\, \widetilde{\overline{E}}_R(t)^* \, dt$$

$$= \overline{\widetilde{\overline{E}}_R(t)\, \widetilde{\overline{E}}_R(t)^*} \neq 0 \qquad (2)$$

If one further assumes that the source is stationary, $I_P(t, \tau)$ becomes $I_P(\tau)$, and we may write

$$<\widetilde{\overline{E}}(t)\, \widetilde{\overline{E}}(t-\tau)^*> = \overline{\widetilde{\overline{E}}(t)\, \widetilde{\overline{E}}(t-\tau)^*} = \varphi(\tau) \qquad (3)$$

where φ is a function only of τ, and where τ is the time difference between the two beams. One has (with Re{ } indicating "real part of")

$$I_P(\tau) = \overline{|\widetilde{\overline{E}}(t) + \widetilde{\overline{E}}(t-\tau)|^2}$$

$$= \overline{|\widetilde{\overline{E}}(t)|^2} + \overline{|\widetilde{\overline{E}}(t-\tau)|^2} + 2 \operatorname{Re} \overline{\{\widetilde{\overline{E}}(t)\, \widetilde{\overline{E}}(t-\tau)^*\}} \qquad (4)$$

The three terms in (4) are recognized as autocorrelation functions $\varphi(\tau)$. One also has

$$\overline{|\widetilde{\overline{E}}(t)|^2} = \overline{|\widetilde{\overline{E}}(t-\tau)|^2} \equiv \varphi(0) \qquad (5)$$

since the source was assumed stationary. One finally has for the observed intensity,

$$I_P(\tau) = 2\varphi(0) + 2 \operatorname{Re}\{\varphi(\tau)\} \qquad (6)$$

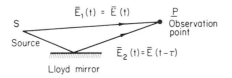

FIG. 2. Lloyd's mirror interferometer, for discussion of coherence.

where the varying quantity of interest is $\varphi(\tau)$, the autocorrelation function of the electric field in the light beam. Equation (6) will appear particularly interesting when it is compared to the expression for $I(\tau)$ obtained from interference theory.[†] It might be useful to note (see Section 7) that two different sources, emitting two different frequencies, are capable of interference, provided that they are *stationary* in the sense of (3)!

2.2   Alternate Way of Looking at Time Coherence (Interference Theory) [‡]

a.   Monochromatic (single-frequency) light

Consider any ordinary two-beam interferometer: For example, the Lloyd mirror illustrated in Fig. 2 or a Michelson-Twyman-Green interferometer, illustrated in Fig. 3.

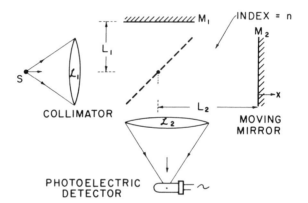

FIG. 3.   Michelson-Twyman-Green interferometer, with photoelectric detector.

Consider the $I(\tau)$ obtained with a finite delay τ at a given frequency ω, as one of the mirrors M_2 is made to move at constant velocity, in the x direction. One has

$$I(\tau) = 1 + 1 + 2\cos 2\pi \ \frac{\tau}{T_\omega} = 2(1 + \cos 2\pi\omega\tau) \qquad (7)$$

with $T_\omega = 1/\omega$. This is shown in Fig. 4. In an actual interferometer, one has: path difference $\Delta = n(L_2 - L_1)$, phase difference $\phi = 2\pi \Delta/\lambda$, time difference $\tau = \Delta/c$, c = velocity of light in vacuo, and n = refractive index.

[†]See, for example, Born and Wolf.[1]

[‡]For symbolical simplicity, the frequency f is described by ω in Sections 2.2-2.5.

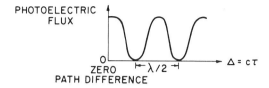

FIG. 4. Interference fringe signal with single-frequency radiation, in Michelson-Twyman-Green interferometer of Fig. 3, as a function of the path difference $\Delta = nL_1 - nL_2$.

b. Polychromatic light (noncoherent source)

Let $\Phi(\omega)$ be the spectral energy distribution in the light source per unit frequency interval $\Delta\omega$ (Fig. 5). It is assumed that a noncoherent source is such that the different spectral frequencies in the source are statistically uncorrelated, so that no interference in the ordinary sense is possible between waves of different frequencies. (The case of beat frequencies between waves of different frequencies is analyzed below.) Under this assumption, one has for

$$\omega_1: I_1(\tau) = \Phi(\omega_1)\Delta\omega \; 2(1 + \cos 2\pi\omega_1\tau)$$
$$\omega_2: I_2(\tau) = \Phi(\omega_2)\Delta\omega \; 2(1 + \cos 2\pi\omega_2\tau) \qquad \text{etc.}$$

and since the different frequencies are *noncoherent* with each other, the intensity observed is (Fig. 6)

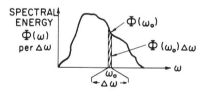

FIG. 5. Spectral-energy distribution in noncoherent (thermal) polychromatic light source.

FIG. 6. Fringe signal in photoelectric interferometer with polychromatic light source.

$$I(\tau) = 2 \int_{\omega} (1 + \cos 2\pi\omega\tau)\Phi(\omega)\,d\omega \tag{8}$$

where $\displaystyle\int_{\omega} = \int_{0}^{\infty}$

2.3. Comparison of the Correlation and Interferometric Methods

Equation (6) is

$$I(\tau) = 2\varphi(0) + 2\,\mathrm{Re}\{\,\varphi(\tau)\,\} \tag{6a}$$

Equation (8) is

$$I(\tau) = 2 \int_{\omega} \Phi(\omega)\,d\omega + 2 \int_{\omega} \Phi(\omega)\cos(2\pi\omega\tau)\,d\omega \tag{8a}$$

One immediately recognizes the physical meaning of the autocorrelation functions. One has

$$\varphi(0) = \int_{\omega} \Phi(\omega)\,d\omega \tag{9}$$

and

$$\mathrm{Re}\{\varphi(\tau)\} = \int_{\omega} \Phi(\omega)\cos(2\pi\omega\tau)\,d\omega \tag{10}$$

Equation (10) shows that the *real part of the autocorrelation function of the electric field in the light beam is equal to the Fourier cosine transform of the energy distribution in the spectrum.*

Equation (10) has found important applications in Fourier-transform spectroscopic analysis of light, and also in radioastronomy.

2.4. Narrow Spectrum (Flat-Top Line)

This special case is particularly important for laser analysis. Consider the idealized line shown in Fig. 7. One has for the intensity,

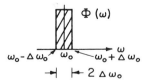

FIG. 7. Special example of light-intensity distribution in a polychromatic light source.

$$I(\tau) = 2 \int_{\omega_0 - \Delta\omega_0}^{\omega_0 + \Delta\omega_0} d\omega + 2 \int_{\omega_0 - \Delta\omega_0}^{\omega_0 + \Delta\omega_0} \cos\left(2\pi\omega\,\tau\right) d\omega \qquad (11)$$

that is,

$$I(\tau) = 4\Delta\omega_0 + \frac{2}{2\pi\tau} \left[\sin 2\pi\omega\,\tau\right]_{\omega_0 - \Delta\omega_0}^{\omega_0 + \Delta\omega_0} \qquad (12)$$

and finally,

$$I(\tau) = 4\Delta\omega_0 \left(1 + \frac{\sin 2\pi\Delta\omega_0\,\tau}{2\pi\Delta\omega_0\,\tau} \cos 2\pi\omega_0\,\tau\right) \qquad (13)$$

The $I(\tau)$ described in (13) is shown in Fig. 8. The first minimum of the sin x/x envelope occurs for $2\pi\Delta\omega_0\,\tau_0 = \pi$. The sin x/x envelope is some-

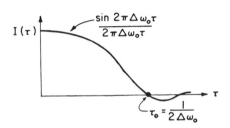

FIG. 8. Envelope of fringe signal with source of Fig. 7.

times described as the fringe-visibility curve. It is seen that the decrease of fringe visibility with path difference is inversely proportional to the width of the spectral line.

2.5. Photoelectric Interferometry with Gaussian Line Shapes, in Interferometers with Moving Mirrors

We now consider the case of the· photoelectric interferometer of Fig. 3 when it is illuminated with a light source having a Gaussian spectral energy distribution. This case is particularly interesting, because most thermal light sources (not lasers) have line shapes which are essentially Gaussian under usual conditions. We show that, for a Gaussian line-shape function, the Fourier transform is also Gaussian.

Indeed, let

$$g(\omega) = \exp\left(-a^2\omega^2\right) \qquad (14)$$

where a is a constant. By Fourier transformation,

$$g(\omega) = \exp\left(-a^2\omega^2\right) \longrightarrow G(x) = \int_{-\infty}^{+\infty} \exp\left(-a^2\omega^2\right) \exp\left(2\pi i\omega x\right) d\omega \qquad (15)$$

Equation (15) can be separated into

$$G(x) = \int_{-\infty}^{+\infty} \exp(-a^2\omega^2) \, \cos(2\pi\omega x) \, d\omega$$

$$+ i \int_{-\infty}^{+\infty} \exp(-a^2\omega^2) \, \sin(2\pi\omega x) \, d\omega \quad (16)$$

The second term is zero because $g(\omega)$ is even, and therefore,

$$G(x) = 2 \int_0^{\infty} \exp(-a^2\omega^2) \, \cos(2\pi\omega x) \, d\omega = \frac{\sqrt{\pi}}{a} \exp\left(-\frac{\pi^2 x^2}{a^2}\right) \quad (17)$$

In other words,

$$g(\omega) = \exp[-(a\omega)^2] \longrightarrow G(x) = \frac{\sqrt{\pi}}{a} \exp\left[-\left(\frac{\pi x}{a}\right)^2\right] \quad (18)$$

Equation (18) shows the remarkable feature that the Fourier transform of of a Gaussian function is also a Gaussian function, with a width that is inversely proportional to the width of the original function.

More generally, the photoelectric recording must also take into consideration the effects of the source aperture (see also Section 2 of Chapter II). The complete expression[2] for the photoelectric flux F_G, for the case of a line of Gaussian line shape

$$\Phi_\sigma = \Phi_0 \exp[-(mc^2/2kT)(\sigma - \sigma_0)^2/\sigma^2] \quad (19)$$

is

$$F_G(x) = 4\pi^{3/2} R_s^2 I_s \sigma \left\{ 1 + \exp[-(4\pi^2 x^2 \sigma^2 n^2)] \frac{\sin(\pi x \alpha_s^2 n/\lambda)}{(\pi x \alpha_s^2 n/\lambda)} \right.$$

$$\left. \times \cos\left[\frac{2\pi x(1 - \alpha_s^2/4)}{(\lambda/n)/2}\right] \right\} \quad (20)$$

where account has now been taken not only of the line shape, but also of the effect of finite-source aperture α_s. In (20) (see Fig. 9), R_s is the

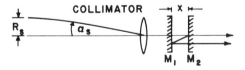

FIG. 9. Parameters in photoelectric interferometers.

source radius, I_s is the intensity in the interferometer beam, $\sigma = 1/\lambda$ (with λ the wavelength in centimeters, x is the mirror separation, n is the refractive index inside the interferometer, and α_s is the angle (in radians) of the source aperture, as seen from the collimator (in the case of small angles). The line-shape effect is contained in the $\exp[-4\pi^2 x^2 \sigma^2 n^2]$ factor, while the effect of the finite-source aperture is contained in the $\sin x/x$ factor. [Note that this factor results here from a different cause than the factor in (13).] With a finite-source aperture, beams go through the interferometer not only normally, but also at angles within a finite cone (see Fig. 9). In practice, the geometric $\sin x/x$ factor can be made negligibly small compared to the line-shape factor for distances up to the order of meters. For instance, with $\alpha_s \cong 6 \times 10^{-4}$ radians, the first zero of the $\sin x/x_0$ factor appears at $x \cong 1.5$ meters. However, as we have shown,[2] a nonnegligible length correction needs to be made in the use of interferometers with finite-source apertures. For instance, with the same $\alpha_s \cong 6 \times 10^{-4}$, the $\alpha_s^2/4 \cong 1 \times 10^{-7}$ in the cosine term in (20), which is far from negligible in practice (an error of 0.05 $\lambda/2$ would result in a mirror motion of $x = 136$ millimeters, if no account were taken of this correction).

2.6. Physical Significance of Power Spectra

a. Case of a single-frequency signal

We shall consider the analysis from its electrical-circuit analogue (see Fig. 10). Consider the circuit shown. Let

$$i(t) = I_0 \cos \omega_0 t \tag{21}$$

and, therefore,

$$v(t) = RI_0 \cos \omega t \tag{22}$$

It follows that the instantaneous power is

$$w(t) = RI_0^2 \cos^2 \omega_0 t$$
$$= \tfrac{1}{2} RI_0^2 + \tfrac{1}{2} RI_0^2 \cos 2\omega_0 t \tag{23}$$

In the frequency domain one has

$$I(\omega) = T[i(t)] = \tfrac{1}{2} I_0 \delta(\omega - \omega_0) + \tfrac{1}{2} I_0 \delta(\omega + \omega_0) \tag{24}$$

FIG. 10. Circuit analogue, for discussion of power spectra.

where the delta function is now defined by the scalar products (see also Section 3 of Chapter VII)

$$\int_{-\infty}^{+\infty} f(x)\delta(x)\,dx = f(0) \tag{25}$$

and

$$\int_{-\infty}^{+\infty} f(x)\delta(x-a)\,dx = f(a) \tag{26}$$

The spectral representation $I(\omega)$ of the current, according to (24) is shown in Fig. 11.

$$I(\omega) \qquad \tfrac{1}{2}I_o \Big| \qquad\qquad\qquad \Big| \tfrac{1}{2}I_o$$
$$\underline{\qquad +\omega_o \qquad 0 \qquad +\omega_o \qquad}\;\omega$$

FIG. 11. Spectral representation of $I(\omega)$ for single-frequency signal.

The frequency-domain representation $W(\omega)$ of the power is

$$W(\omega) = T\,[w(t)] = \tfrac{1}{2}RI_0{}^2\,\delta(0)$$
$$+ \tfrac{1}{4}RI_0{}^2\,\delta(\omega - 2\omega_0) + \tfrac{1}{4}RI_0{}^2\,\delta(\omega + 2\omega_0) \tag{27}$$

It is shown in Fig. 12.

One notes that the spectral representation $W(\omega)$ of the *power* $w(t)$ as described in (23) shows energy at frequencies where there is no current! One concludes that it is necessary to seek a more suitable representation of the power spectrum.

Consider the average power over the time interval T. Examine

$$\frac{1}{2T}\int_{-T}^{+T} i^2(t)\,dt = \tfrac{1}{2}I_0{}^2\left[\frac{\sin 2\omega_0 T}{2\omega_0 T} + 1\right] \tag{28}$$

W(ω)

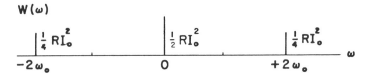

$$\Big| \tfrac{1}{4}RI_o^2 \qquad\qquad \Big| \tfrac{1}{2}RI_o^2 \qquad\qquad \Big| \tfrac{1}{4}RI_o^2$$
$$\underline{-2\omega_o \qquad\qquad 0 \qquad\qquad +2\omega_o \qquad}\;\omega$$

FIG. 12. Spectral representation of $W(\omega)$ for single-frequency signal.

In the limit, one has

$$\lim_{T \to \infty} \frac{1}{2T} \int_{-T}^{+T} i^2(t) \, dt = \tfrac{1}{2} I_0^2 \tag{29}$$

and, therefore, an autocorrelation function $\varphi(\tau)$ exists:

$$\varphi(\tau) = \lim_{T \to \infty} \frac{1}{2T} \int_{-T}^{+T} (I_0 \cos \omega_0 t)[I_0 \cos \omega_0(t - \tau)] \, dt \tag{30}$$

that is,

$$\varphi(\tau) = \tfrac{1}{2} I_0^2 \cos \omega_0 \tau \tag{31}$$

The Fourier transform $\Phi(\omega)$ of the autocorrelation function is

$$\Phi(\omega) = T[\varphi(\tau)] = \tfrac{1}{4} I_0^2 [\delta(\omega - \omega_0) + \delta(\omega + \omega_0)] \tag{32}$$

$\Phi(\omega)$ is defined as the *power spectrum* or the *spectral energy distribution* of the current. It is readily seen that $\Phi(\omega)$ has power at the same frequencies at which the current exists (Fig. 13). One is led to conclude

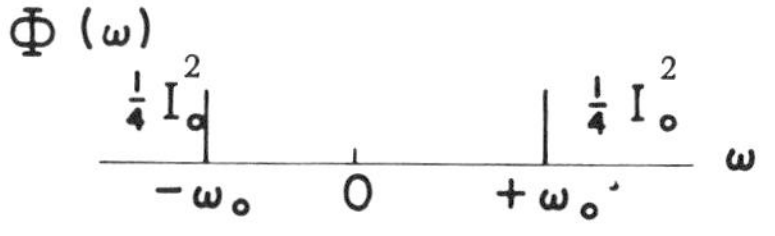

FIG. 13. Spectral representation of $\Phi(\omega)$ for single-frequency signal.

that the *power spectrum* (equal to the Fourier transform of the autocorrelation function of the signal) is a meaningful representation of the spectral energy distribution in the signal (here, the current).

b. Case of a multiple-frequency signal

Let

$$f(t) = \sum_{-\infty}^{+\infty} a_k \exp(i \omega_k t) \tag{33}$$

be the signal. One has

$$F(\omega) = \sum_{-\infty}^{+\infty} a_k \delta(\omega - \omega_k) \tag{34}$$

and the power spectrum is

$$\Phi(\omega) = T[\varphi(\tau)] = \sum_{-\infty}^{+\infty} |a_k|^2 \delta(\omega - \omega_k) \tag{35}$$

according to Fourier-transform theory, and the theory of distribution (see Section 3 of Chapter VII).

2.7. Heterodyne Analysis of Signals, Beat Frequencies, etc.

a. Beat frequencies between coherent waves and relativistic effect for light reflected from moving mirrors

The widths of the spectral lines emitted by optical masers are generally too narrow for an analysis by high-resolution optical spectrometers, using gratings or even Fabry-Perot interferometers. Monochromaticities $\Delta \lambda / \lambda = 10^{-14}$ at the 1.153-micron wavelength of the Javan He-Ne laser were reported as early as 1961 by Javan et al.[3] These values were obtained by heterodyning the light emitted from two similar lasers (Fig. 14). The heterodyning experiments are essentially inter-

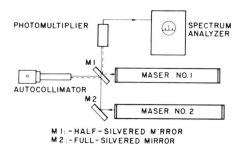

FIG. 14. Beat-frequency heterodyning of light signals from two cw He-Ne optical masers, after Javan et al.[3]

ference experiments, very similar, in fact, to the experiments using photoelectric interferometers with moving mirrors,[4] such as that described in Section 5. Because of the relativistic frequency shift in the light reflected from the moving mirror M_2 (Fig. 3), relative to the light reflected from the stationary mirror M_1, the analysis of the photoelectric signal detected must take into account the *interference between* waves of *different frequencies*, rather than be conducted along the lines of Section 5.

Because of the similarities between the Doppler-effect analysis of interferometers with moving mirrors (Fig. 3) and the beat-frequency analysis of spectral line shapes in heterodyning experiments, we shall first give the Doppler-effect analysis of the photoelectric interferometer, using moving mirrors.

According to the theory of relativity, and for motions at velocities v small compared to the velocity of light c, the frequency f_s of the source appears to the observer O to have the frequency f_0, when it is reflected at angle θ_0 from a mirror moving at the velocity v (in the x direction,

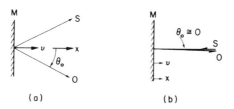

FIG. 15. Parameters for discussion of relativistic effects, in the reflection of light from moving mirrors.

Fig. 15). We have

$$f_0 \cong \frac{f_s}{1 - 2(v/c)\cos\theta_0} \cong f_s\left(1 + 2\frac{v}{c}\cos\theta_0\right) \tag{36}$$

for the case when both the photoelectric detector and the source S are in the same (stationary) framework, and when the angle θ_0 is also measured in the observer frame.[5] For the case of normal incidence on the mirror (Fig. 3), when $\theta_0 = 0$, (36) is

$$f_0 \cong f_s\left(1 + 2\frac{v}{c}\right) \tag{37}$$

More generally, when no small-velocity approximation is made, (36) becomes

$$f_0 = \frac{f_s\,[1 - (v^2/c^2)]^{1/2}}{1 - 2(v/c)\cos\theta_0} \tag{36a}$$

showing that there appears a frequency shift, even in the case when the light is reflected from a reflector moving at an angle $\theta_0 = 90°$ from the line of sight. (For the case of light reflected from satellites, and other objects in space, this remark might be of importance.)

The effect of the moving mirror in the photoelectric interferometer can at first be analyzed in a very qualitative manner as follows. According to (37), the relative frequency shift (and the corresponding relative wavelength shift) are given by

$$\left|\frac{\Delta f}{f}\right| = \left|\frac{\Delta\lambda}{\lambda}\right| \cong \left|\frac{2v}{c}\right| \tag{38}$$

where

$$x = vt \tag{39}$$

and

$$c = f\lambda \tag{40}$$

where $f = f_s$. In a time interval t, there will appear in the interferometer,

$$n\lambda = ft \tag{41}$$

cycles at the frequency $f = f_s$, and

$$n_{\lambda + \Delta \lambda} \cong f \left(1 + \frac{2v}{c} \right) t \qquad (42)$$

cycles at the frequency $f + \Delta f$. There will therefore appear *one fringe cycle* (n) *coincidence*, or *beat*, whenever

$$n_{\lambda + \Delta \lambda} - n_\lambda = p \qquad (p = \text{integer}) \qquad (43)$$

that is, for

$$ft + \frac{2vtf}{c} - ft = \frac{2xf}{c} = \frac{2x}{\lambda} = p \qquad (44)$$

for

$$x = p \, \frac{\lambda}{2} \qquad (45)$$

in agreement with the ordinary theory given in Section 5. We may therefore conclude that optical waves of two different frequencies are capable of interference with each other, provided that the two frequencies are coherent with each other in the stationary sense given by (3).

A somewhat more complete beat-frequency analysis, applicable both to the laser-heterodyning experiments and to interferometers with moving mirrors, can be given as follows.

[For the case of laser-heterodyning, it is important to recall that the two lasers can be considered as emitting stationary frequencies during the observation time, and therefore that the lasers are also coherent relative to each other in the stationary sense given by (3), and in a sense comparable to the coherence of the two frequencies f_0 and f_s in the interferometer with moving mirrors, which have been demonstrated to be capable of interference; see Ref. 2.]

Let E_1 and E_2 be the complex amplitudes of the electric fields of two plane (or spherical waves) emitted by the two sources S_1 and S_2 and propagating in the *same* direction, x. (Tolerances on the parallelism of the two waves are given below.) We may write

$$\begin{aligned} E_1 &= \cos \left[\left(k + \frac{\Delta k}{2} \right) x - \left(\omega + \frac{\Delta \omega}{2} \right) t \right] \\ E_2 &= \cos \left[\left(k - \frac{\Delta k}{2} \right) x - \left(\omega - \frac{\Delta \omega}{2} \right) t \right] \end{aligned} \qquad (46)$$

where $k = 2\pi/\lambda$ and $\omega = 2\pi f$, and where Δk and $\Delta \omega$ are measures of the wavelength and frequency differences between E_1 and E_2, respectively.

The resultant field E is given by the equation

$$E = E_1 + E_2 = 2 \cos (kx - \omega t) \cos \left(\frac{\Delta k}{2} x + \frac{\Delta \omega}{2} t \right) \qquad (47)$$

in which $\cos [(\Delta k/2)x + (\Delta \omega/2)t]$ is the beat wave. The $\cos (kx - \omega t)$ wave is recognized as the wave (or fringe signal) which would be obtained with a perfectly monochromatic source at the frequency f, in the case of the photoelectric interferometer.

For the case of the photoelectric interferometer, the beat wave will have maxima for

$$p \, 2\pi = \frac{\Delta k}{2} x + \frac{\Delta \omega}{2} t = 2\pi \left(\frac{\Delta f}{2c} x + \frac{\Delta f}{2} t \right) \qquad (48)$$

that is, for

$$\Delta f \frac{x}{c} = p \qquad (49)$$

or

$$\frac{2vt}{\lambda} = p \qquad (50)$$

In other words, one fringe cycle will be obtained for a mirror motion

$$x = p \frac{\lambda}{2} \qquad (51)$$

in agreement with our previous analysis.

For the case of heterodyning, beat-frequency maxima will therefore appear at the frequency Δf. In the case of spectral lines having a finite width, it follows from (8) or from a convolution analysis (see, for example, Section 2 of Chapter VII) that the essential nature of the interference is not changed, but the beat frequencies become smeared out by the width of the spectral lines, rather than being infinitely narrow, as they would be in the case of spectral delta functions.

b. Parallelism tolerances for the two interfering waves in photoelectric interferometry and heterodyning experiments

Because of the very close analogy between photoelectric interferometers with moving mirrors and photoelectric heterodyning of waves emitted by two (and more) lasers, we may give the tolerances for the parallelism of the interfering waves in terms of the analysis which we have first given in Ref. 2.

Basically (see Fig. 3) we may say that the photoelectric tube receives the integral of the light-intensity distribution within the aperture of the lens L_2, which is used to focus the interfering beams onto its sensitive surface. (The argument does not change in the case where no lens is used, but does when the photosensitive surface is exposed to some extended portions of the interfering waves.) If the two interfering waves form an angle with each other, the interference field within the aperture will not be uniformly bright, but rather its intensity will vary sinusoidally, because of the interference fringe system formed in the aperture, in the case of two plane waves, for example. The effect of having a lack of parallelism between the two waves will therefore appear as a decrease in the fringe contrast (or fringe signal amplitude) in the case of the interferometer with moving mirrors. Similarly, the lack of parallelism will reduce the apparent amplitude of the beat-frequency signal in heterodyning experiments. According to our analysis in Ref. 2, we may examine the ratio of the photoelectric flux for two phase positions ϕ between the two interfering waves, such that $\Delta\phi = \pi/2$. In the case of a square aperture, we have (see Fig. 16)

$$\frac{\text{flux at } \phi = 0}{\text{flux at } \phi = \pi/2} = \frac{\int_{-\delta/2}^{+\delta/2} (1 + \cos \phi)\, d\phi}{\int_{(\pi/2)-\delta}^{(\pi/2)+\delta} (1 + \cos \phi)\, d\phi} = R \tag{52}$$

which readily gives

$$R = 1 + \frac{\sin \delta}{\delta} \tag{53}$$

The fringe signal amplitude (i.e., the apparent amplitude of the beat-frequency signal) is then

$$M = R - 1 \tag{54}$$

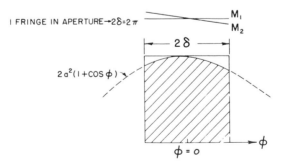

FIG. 16. Parameters for discussion of effect of tilted wavefronts in heterodyning of light from two lasers, and in photoelectric interferometers with moving mirrors.

or

$$M = \frac{\sin \delta}{\delta} \tag{55}$$

An experimentally important parameter, in terms of which M can be represented, is the number of fringes within the interferometer aperture. A graph of M in terms of the tolerable number of fringes within the aperture is given in Fig. 17. It is seen that a one-half fringe within the aperture still maintains about 62 per cent of the maximum amplitude, but that one

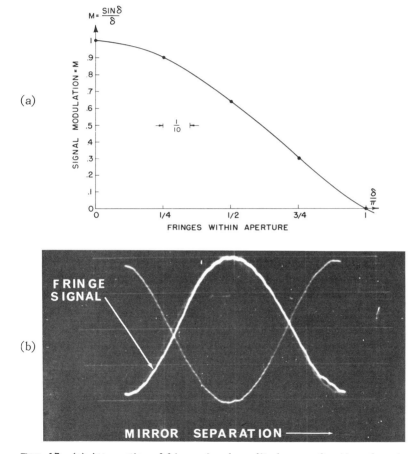

(a)

(b)

FIG. 17. (a) Attenuation of fringe-signal amplitude, as a function of number of fringes within aperture (the attenuation is described as M = signal modulation here, after Stroke.[2]   (b) Photoelectric fringe signal, obtained with mercury isotope source, Hg-198, at a 200-mm path difference, on M.I.T. grating ruling engine, after Harrison and Stroke.[6]

entire fringe within the aperture will reduce the signal to zero! This can be readily understood, qualitatively, in noting that one fringe in the aperture fills the aperture with one-half cycle of darkness and one-half cycle of brightness; any phase change will merely change the distribution of brightness within the aperture, so that the integrated amount, in this case, will remain zero. The parallelism tolerance values given above are in agreement with those subsequently developed, with specific application to heterodyning,[7] as well as earlier,[8] in connection with photoelectric mixing of *incoherent* light.

c. Photoelectric mixing of incoherent light

This is the problem which has been dealt with in the work by Forrester *et al.*[8] Strictly speaking, we may consider the heterodyning and other experiments, performed with lasers, to belong to the class of interferometric experiments in *coherent* light, dealt with in the preceding sections. Accordingly, extensive treatment of this subject falls outside of the scope of this chapter.

Some insight into the problems arising in the mixing of incoherent light may be gained in examining the following problem, which may have some possible applications.

d. Photoelectric mixing of coherent laser light with noncoherent thermal light[†]

This case may arise in an experiment where one may wish to lock the frequency of the light emitted by an optical maser to some mean frequency of the light emitted by an external gas discharge, for instance to a Kr-86 or Hg-198 lamp or to an atomic-beam light source. A conceivable hypothetical arrangement for locking a laser to a gas discharge is shown in Fig. 18.

The feedback system is designed to maintain a constant frequency difference between the center of the spectral line from the thermal source and the spectral line from the laser. The two light waves are mixed by means of a photomultiplier, and the resultant beat frequency is used to produce an error signal, to control the length of the laser cavity.

Even though the Doppler width of the thermal line is much broader than the width of the laser line, it will appear that the width of the thermal line should be much narrower than the Doppler width of the gas in the laser. (This would not be the case for the isotope sources mentioned above, when compared to the neon lines. However, an atomic-beam light source, such as the Hg-198 atomic-beam source described by Kessler

---

[†]This section is the result of a collaborative effort between the author and Professor Henry H. Stroke, with assistance from F. B. Rotz.

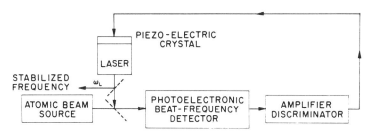

FIG. 18. Frequency stabilization of laser by locking to atomic-beam source.

might be suitable.) Our analysis may also apply to the laser frequency
stabilization method.

Let $s(t)$ be the complex amplitude of the light from the atomic-beam
light source, and let exp $[i\omega_L t]$ be the amplitude of the light from the
laser (we assume single-frequency operation), and let $i(t)^2$ be the
square of the output current of the photoelectric detector (see Section
6). We have

$$i(t)^2 = [\, s(t) + e^{i\omega_L t}\,]\,[\, s(t) + e^{i\omega_L t}\,]* \tag{56}$$

that is

$$i(t)^2 = 1 + |s(t)|^2 + s(t)e^{-i\omega_L t} + s(t)*e^{i\omega_L t} \tag{57}$$

According to Eq. (57) the output (photocurrent)^2 is seen to consist of
four terms. The first term contributes to the "dc." The second term
also contributes to the "dc," while also contributing a second-harmonic
component. It is the third and fourth "signal" terms which are of interest in the stabilization.

Next we wish to obtain a spectral representation of these components
of the current $i(t)$, along the lines developed in Section 6. The current-
component functions, of which we wish to obtain the spectral representa-
tions, are: $s(t)$, $|s(t)|^2$, and

$$y(t) = s(t)e^{-i\omega_L t} + s(t)*e^{i\omega_L t} \tag{58}$$

The spectral representations $\Phi_{ss}(\omega)$ and $\Phi_{s^2 s^2}(\omega)$ of $s(t)$ and $s^2(t)$ immediately follow from the analysis given in Section 6. They are shown
in Fig. 19. It was assumed that $s(t)$ is a Gaussian function centered at
the frequency ω_c. Similarly, because of the multiplication of $s(t)$ and
$s(t)*$ with exp $(i\omega_L t)$ in (58), the spectral representation $\Phi_{yy}(\omega)$ of $y(t)$ is
given in Fig. 20. The spectral energy of interest for servostabilization
is located in the two central lobes (centered at $\omega_L - \omega_c$ and at $\omega_c - \omega_L$,
respectively). In comparing Fig. 20 with Fig. 19b, it is seen that the two
central lobes of interest may be smeared out by overlap of $\Phi_{s^2 s^2}(\omega)$ if the

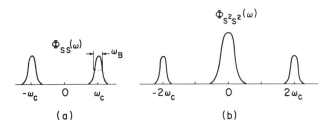

FIG. 19. Spectral representations of $\Phi_{ss}(\omega)$ and of $\Phi_{s^2s^2}(\omega)$, in connection with laser stabilization according to Fig. 18.

Gaussian line $s(t)$ is too broad. The overlap will not be serious if the frequency difference

$$\omega_c - \omega_L > \tfrac{3}{2}\,\omega_B \tag{59}$$

where ω_B is the bandwidth of $s(t)$ (see Fig. 19a).

In practice, stabilization of the laser may involve passing the light from the laser through the atomic beam, and adjusting the laser frequency to obtain maximum absorption. By splitting the laser beam twice, and bringing three laser beams through the atomic beam at suitable angles (one as close to normal as possible), three absorbtion filters may be gen-

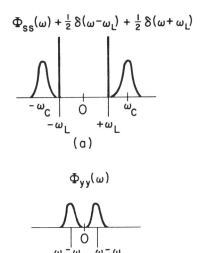

FIG. 20. Spectral representation in connection with laser stabilization according to Fig. 18.

erated, and their outputs combined in a suitable discriminator, for the purpose of avoiding the difficulty of determining the very accurate normal direction necessary for the accurate location of the frequency ω_c. Also, in a practical case, a signal-to-noise-ratio analysis, and assessment of the effects of amplitude differences between the laser and the thermal source, would be necessary to determine the possibilities of experiments, such as that indicated above, for the purpose of illustrating the analysis.

2.8. Spatial Coherence

The spatial coherence in a light beam generally has to do with the coherence between two points in the field illuminated by one or more light sources. In its elementary sense, the degree of coherence between the two points simply describes the contrast of the interference fringes that are obtained when the two points are taken as secondary sources.

The meaning of spatial coherence can best be understood with the help of Young's two-slit experiment, or some equivalent. The important, and generally unexpected, conclusion to be demonstrated is that even an extended source, composed of millions of statistically incoherent oscillators, can produce a coherent field, provided only that the angular diameter of the source, as seen from the plane of the two slits, is small compared to the slit separation, and that the radiation from the source be "quasi-monochromatic" (narrow band).

We shall now derive the equation defining the *partial coherence factor*. Let S_1 and S_2 be the two slits illuminated by an extended, noncoherent source (Fig. 21). Let $E_1(t)$ and $E_2(t)$ be the electromagnetic field ampli-

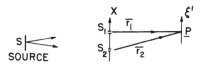

FIG. 21. Young's interferometer, for discussion of coherence.

tudes at S_1 and S_2. The *amplitude* at P is then

$$E_R(t) = E_1(t) \exp\left(i\,\frac{\phi}{2}\right) + E_2(t) \exp\left(-i\,\frac{\phi}{2}\right) \tag{60}$$

where

$$\phi = \frac{2\pi}{\lambda}\,(|\bar{r}_1| - |\bar{r}_2|) \tag{61}$$

The observable quantity at P is

$$\overline{E_R(t)E_R(t)^*} = \frac{1}{2T}\int_{-T}^{+T} E_R(t)E_R(t)^*\,dt \tag{62}$$

where $\overline{}$ denotes time average and $T \gg T_0$ (period of oscillation). One has

$$\overline{E_R(t)E_R(t)^*} = \overline{(E_1 e^{i\,\phi/2} + E_2 e^{-i\phi/2})(E_1^* e^{-i\phi/2} + E_2^* e^{i\phi/2})}$$

$$= \overline{E_1 E_1^*} + \overline{E_2 E_2^*} + \overline{E_1 E_2^* e^{i\phi}} + \overline{E_1^* E_2 e^{-i\phi}} \qquad (63)$$

Let

$$\overline{E_1 \overline{E}_2}^* = |\overline{E_1 \overline{E}_2}^*| e^{i\theta} \qquad (64)$$

Then

$$\overline{E_1}^* \overline{E}_2 = |\overline{E_1}^* \overline{E}_2| e^{-i\theta} \qquad (64a)$$

and

$$\overline{\overline{E}_R \overline{E}_R}^* = \overline{\overline{E}_1 \overline{E}_1}^* + \overline{\overline{E}_2 \overline{E}_2}^* + |\overline{E_1 \overline{E}_2}^*| e^{i\,\phi} e^{i\theta} + |\overline{E_1}^* \overline{E}_2| e^{-i\phi} e^{-i\theta} \qquad (65)$$

since

$$(|\overline{E_1}^* \overline{E}_2| e^{-i\phi} e^{-i\theta}) = (|\overline{\dot{E}_1 \overline{E}_2}^*| e^{+i\phi} e^{+i\theta})^* \qquad (66)$$

it follows that

$$\overline{E_R(t)E_R(t)^*} = \overline{E_1 E_1^*} + \overline{E_2 E_2^*} + 2|\overline{E_1 \overline{E}_2}^*| \cos(\phi + \theta) \qquad (67)$$

It is now appropriate to compare (67) with the interference fringe equation, obtained in Young's experiment with two perfectly coherent sources, namely,

$$I = |E_1|^2 + |E_2|^2 + 2(E_1 E_1^*)^{1/2} (E_2 E_2^*)^{1/2} \cos\phi \qquad (68)$$

or

$$I = I_1 + I_2 + 2(I_1)^{1/2} (I_2)^{1/2} \cos\phi \qquad (68a)$$

It is seen from (68) that one can write (67)

$$\overline{E_R(t)E_R(t)^*} = \overline{E_1 E_1^*} + \overline{E_2 E_2^*}$$
$$+ 2|\gamma_{12}| \overline{(E_1 E_1^*)}^{1/2} \overline{(E_2 E_2^*)}^{1/2} \cos(\phi + \theta) \qquad (69)$$

where we have defined (see Refs. 1, 9, 14, 17) a

Partial coherence factor:

$$\gamma_{12} = \frac{\overline{E_1 E_2^*}}{(E_1 E_1^*)^{1/2}(E_2 E_2^*)^{1/2}} = \frac{|\overline{E_1 E_2^*}| \exp(i\theta)}{(E_1 E_1^*)^{1/2}(E_2 E_3^*)^{1/2}} \qquad (70)$$

One can also write (69)

$$\overline{E_R(t)E_R(t)^*} = I_1 + I_2 + 2|\gamma_{12}|(I_1)^{1/2} (I_2)^{1/2} \cos(\phi + \theta) \qquad (70a)$$

The meaning of the partial coherence factor γ_{12} becomes immediately apparent by comparing (69) with (68). It is seen that the *magnitude* $|\gamma_{12}|$ of the partial coherence factor is a measure of the *contrast* in the two slit interference fringe systems formed by S_1 and S_2 when they are illuminated by the source S. The phase θ of the partial coherence factor is a measure of the phase shift of the fringe maxima, compared to the phase obtained by perfectly coherent illuminated slits. For instance, θ measures the position shift along ξ' in the observation plane (Fig. 22).

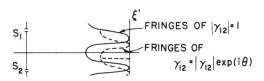

FIG. 22. Parameters defining the partial coherence factor γ_{12}, in connection with Young's interferometer of Fig. 21.

It is clear that

$$\begin{aligned} |\gamma_{12}| &= 1 \longrightarrow \text{perfect coherence} \\ |\gamma_{12}| &= 0 \longrightarrow \text{perfect incoherence} \dagger \\ 0 < |\gamma_{12}| &< 1 \longrightarrow \text{partial coherence} \end{aligned} \qquad (71)$$

The meaning of θ can be illustrated for a few special cases. For example,

$$\phi + \theta = 2m\pi \longrightarrow \cos(\phi + \theta) = \text{max} = +1$$

and

$$\overline{E_R E_R{}^*} = \overline{(E_R E_R{}^*)}_{max} = I_1 + I_2 + 2|\gamma_{12}|(I_1)^{1/2}(I_2)^{1/2}$$

i.e., maximum illumination,

$$\phi + \theta = (2m + 1)\pi \longrightarrow \cos(\phi + \theta) = \text{min} = -1$$

and

$$\overline{E_R E_R{}^*} = \overline{(E_R E_R{}^*)}_{min} = I_1 + I_2 - 2|\gamma_{12}|(I_1)^{1/2}(I_2)^{1/2}$$

i.e., minimum illumination.

The meaning of $|\gamma_{12}|$ can be illustrated by comparison with the case of perfect coherence, described by (68) and (68a): Assuming $I_1 = I_2$ one has

$$I_c = 2I_1(1 + \cos\phi) \qquad (72)$$

and, therefore,

$$I_{c,max} = 4I_1$$

†I.e., noncoherence (see footnote on page 26).

and

$$I_{c,\min} = 0$$

This is illustrated in Fig. 23a.

For the case of $|\gamma_{12}| < 1$, the partially coherent fringes can be described [for the case $(I_1)^{\frac{1}{2}} = (I_2)^{\frac{1}{2}}$] by

$$I_{pc} = 2I_1 |\gamma_{12}| [1 + \cos(\phi + \theta)] \tag{73}$$

The fringe contrast is then smaller by $|\gamma_{12}|$, as shown in Fig. 23b.

For completely incoherent sources, $|\gamma_{12}| = 0$, and

$$\overline{(E_R E_R{}^*)}_{\max} = \overline{(E_R E_R{}^*)}_{\min}$$

Consequently, the fringe contrast is zero, as shown in Fig. 23c.

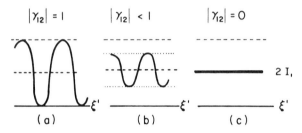

FIG. 23. Interference-fringe amplitude, in Young's interferometer, for three values of the partial coherence factor γ_{12}.

In general, it should be noted that the mere *observation of fringes*, by placing two slits in the field (for example at the exit of a laser, only shows that there is *partial coherence* in the field. To truly describe the degree of coherence in the field, it is necessary to *measure* the fringe contrast.

2.9. Partial Coherence with Extended Noncoherent Source

Let a source S illuminate the two slits S_1, S_2, as shown in Fig. 24. The source is centered on the z axis, normal to the xy plane at one of the slits, say S_1. The source is perfectly noncoherent, according to the preceding section. That is to say, no interference fringes can be obtained by placing two slits in the plane of the source. It will be shown, however, that if the two slits are placed far enough away from the noncoherent source, interference fringes of good and even excellent contrast can be obtained, again assuming "quasi-monochromatic" radiation.

Consider the two wavefronts Σ_1 and Σ_2, originating from an oscillator

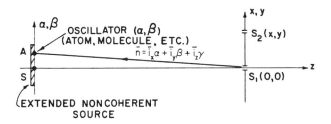

FIG. 24. Attainment of spatial coherence in the xy plane (plane of two slits), illuminated by extended, spatially *noncoherent* source: definition of parameters.

in the direction $\bar{n}$ as seen from S_1 (Fig. 25). One has

$$\bar{n} = \bar{i}_x\, \alpha + \bar{i}_y\, \beta + \bar{i}_z\, \gamma \tag{74}$$

where α and β are the direction cosines and $\bar{i}_x$ and $\bar{i}_y$ the unit vectors. Let

$$\bar{\rho} = \bar{i}_x x + \bar{i}_y y \tag{75}$$

describe the coordinates of $S_2\,(x,\,y)$ with respect to $S_1\,(0,\,0)$. The path difference between the two wavefronts Σ_1 and Σ_2 from A is

$$\Delta = \bar{n}\cdot\bar{\rho} = \alpha\, x + \beta y \tag{76}$$

The electric fields at S_1 and S_2, resulting from the oscillator A (Fig. 24), are therefore

at S_1: $E_{s1}(t) = E(t)$

at S_2: $E_{s2}(t) = E(t) \exp[-ik(\alpha\, x + \beta y)]$ $\qquad(77)$

where $k = 2\pi/\lambda$.

In the case of many atoms A_i, one has

at S_1: $E_{s1}(t) = \Sigma\, E_i(t)$

at S_2: $E_{s2}(t) = \Sigma E_i(t) \exp[-ik(\alpha_i x + \beta_i y)]$ $\qquad(78)$

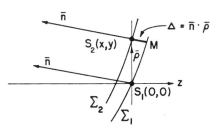

FIG. 25. Attainment of spatial coherence, in the xy plane of Fig. 24.: definition of parameters.

Monochromatic light is assumed. In practice, only reasonable monochromaticity is necessary for the observation of fringes in two-slit
measurements, when the fringes are observed with reasonably small total
time difference τ between the source and the observation plane along
the two paths. (Filters are used in the case of polychromatic sources.)

To evaluate the fringe contrast it is necessary to obtain an expression
for

$$\overline{E_{s1}(t)\,E_{s2}(t)^*} \tag{70a}$$

One has

$$E_{s1}(t) = E_1(t) + E_2(t) + \cdots$$

$$E_{s2}(t) = E_1(t)\exp(-i\phi_1) + E_2(t)\exp(-i\phi_2) + \cdots \tag{79}$$

where $\phi_1 = 2\pi\Delta_1/\lambda$, $\phi_2 = 2\pi\Delta_2/\lambda$ [Δ_1 and Δ_2 are optical path lengths
as defined in (76)]. It follows that

$$\overline{E_{s1}(t)\,E_{s2}(t)^*} = \overline{[E_1 + E_2 + \cdots][E_1{}^*\exp(+i\phi_1) + E_2{}^*\exp(+i\phi_2) + \cdots]}$$

that is,

$$\overline{E_{s1}(t)\,E_{s2}(t)^*} = \overline{E_1 E_1{}^*\exp(i\phi_1)} + \overline{E_2 E_2{}^*\exp(i\phi_2)} + \overline{[E_2 E_1{}^*\exp(i\phi_1)}$$

$$+ \overline{E_3 E_1{}^*\exp(i\phi_1)} + \text{other cross terms}] \tag{80}$$

However, all the cross terms are equal to zero, as a result of the noncoherence of the source.

The only terms remaining in (80) are

$$\overline{E_{s1}(t)\,E_{s2}(t)^*} = \overline{E_1 E_1{}^*\exp(i\phi_1)} + \overline{E_2 E_2{}^*\exp(i\phi_2)} + \cdots$$

that is,   (81)

$$\overline{E_{s1}(t)\,E_{s2}(t)^*} = \overline{\Sigma\,E_i(t)E_i(t)^*\exp[+ik(\alpha_i x + \beta_i y)]}$$

In the limit of an infinitely great number of oscillators, (81) becomes,
to within a constant of normalization,

$$\overline{E_{s1}(t)E_{s2}(t)^*} = \iint_{source} I_s(\alpha,\beta)\exp[ik(\alpha x + \beta y)]\,d\alpha\,d\beta \tag{82}$$

where

$$I_s(\alpha,\beta) = \frac{\overline{E_i E_i{}^*}}{d\alpha\,d\beta}$$

$$= \text{energy per unit angular area of the source at given } \alpha,\beta \tag{83}$$

In other words, $I_s(\alpha,\beta)$ is the apparent intensity distribution in the
source as seen from S_1.

It follows immediately from (82) and (83) that the partial coherence factor γ, describing the coherence between S_1 $(0, 0)$ and S_2 (x, y) when illuminated by the extended noncoherent source is given by

$$\gamma = \frac{\iint_{source} I_s(\alpha, \beta)\, \exp[ik(\alpha x + \beta y)]\, d\alpha\, d\beta}{\iint_{source} I_s(\alpha, \beta)\, d\alpha\, d\beta} \tag{84}$$

Equation (84) is very remarkable indeed. It shows that *the degree of partial coherence between two points* S_1 *and* S_2 *illuminated by an extended, noncoherent source is given by the Fourier transform of the intensity distribution* $I_s(\alpha, \beta)$ *of the source as seen from the* S_1, S_2, *plane.* It is quite essential to note that the source is centered on S_1. This will be clarified in the following example.

a. Two-slit (Young) interferometer and Michelson's stellar interferometer

These two interferometers are found to be very helpful in clarifying the concept of partial coherence factor, Eq. (70), and the relation between this factor and the source-intensity distribution, through the Fourier-transform relation given in (84). Consider the two-slit interferometer arrangement shown in Fig. 26. Let $I_s(\alpha, \beta) = I_s$, a constant. Experiment shows that an interference fringe system is observed, under suitable circumstances, in the plane P (Fig. 26).

The nature of the fringe system is easily described in Fourier-transform notation [see (III.1a) and (III.18)]. Indeed [see (VII.12)], if

$$f(x) \longrightarrow F(u')$$

then

$$f(x - d) \longrightarrow F(u')\, \exp[2\pi i u'd] \tag{85}$$

and

$$f(x + d) \longrightarrow F(u')\, \exp[-2\pi i u'd]$$

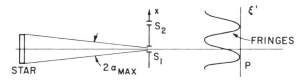

FIG. 26. Interference fringes in a Young's interferometer, illuminated by extended spatially *noncoherent* star.

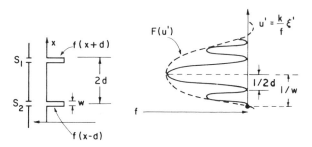

FIG. 27. Interference-fringe formation in stellar interferometer (see Fig. 26.)

Therefore,

$$f(x + d) + f(x - d) \longrightarrow 2\,F(u')\cos 2\pi u'd \qquad (86)$$

and the observed intensity is (Fig. 27)

$$I(u') = 4\,|F(u')|^2\,\cos^2 2\pi\,u'd \qquad (87)$$

that is,

$$I(u') = 2\,|F(u')|^2\,(1 + \cos 4\pi u'd) \qquad (88)$$

where the envelope $F(u')$, which modulates the fringe amplitude, is given by

$$F(u') = T[f(x)]$$

$$= \int_{-W/2}^{+W/2} \exp(2\pi i u'x)\,dx$$

$$= W\,\frac{\sin(\pi i u'W)}{\pi i u W} \qquad (89)$$

and

$$|F(u')|^2 = W^2\left[\frac{\sin(\pi i u'W)}{\pi i u'W}\right]^2 \qquad (90)$$

The contrast in this fringe system is obtained as

$$\gamma(x) = \frac{\displaystyle\int_{-\alpha_{max}}^{\alpha_{max}} I_s\,\exp(-ik\alpha x)\,d\alpha}{\displaystyle\int_{-\alpha_{max}}^{+\alpha_{max}} I_s\,d\alpha}$$

$$= \frac{\sin(k\,\alpha_{max}\,x)}{k\,\alpha_{max}\,x} = \frac{\sin[(2\pi/\lambda)\,\alpha_{max}\,x]}{(2\pi/\lambda)\,\alpha_{max}\,x} \qquad (91)$$

In the case where the two-slit interferometer is used to measure the diameter of a star, the visibility of the fringes varies according to the $\gamma(x)$ curve shown in Fig. 28, as a function of the slit separation. It is very important to note that the $\gamma(x)$ graph in Fig. 28 implies that *one of the slits is centered on the maximum of the* $\gamma(x)$ *graph.* It is clear that the apparent angular diameter of the star can be obtained simply by determining the distance $2d$ of S_2 from S_1 along x such that the fringes disappear.

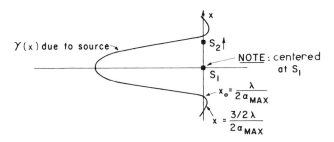

FIG. 28. Spatial coherence factor $\gamma(x)$, in x plane of Figs. 24 and 26, when the plane is illuminated by a distant, spatially noncoherent source.

The remarkable feature is that there is always a separation $S_1 S_2$ at which good fringe contrast is obtained with an extended noncoherent source. This remark may be useful in estimating the conditions under which a spatially coherent spherical (or plane) wavefront may be obtained from an ordinary, thermal light source, for instance, from a small flashlight bulb. We shall now show that a flashlight bulb, used with a suitable monochromatic red filter, may be made to provide a sufficiently coherent wavefront for the observation of a hologram (such as a Fourier-transform hologram), if the flashlight is held at arm's length from the eye (and the Fourier-transform hologram is held in front of the eye).

As a numerical example, consider a 0.1-mm-diameter flashlight bulb held at a distance of 500 mm from the eye. The first minimum in the $\gamma(x)$ graph of Fig. (28) occurs at

$$x_0 = \frac{\lambda}{2\alpha_{max}} = \frac{6/10^4}{\frac{1}{10}/500} = 3 \text{ mm}$$

and we may conclude that good spatial coherence will be obtained over at least 1.5 to 2 mm, which is adequate to cover the pupil of the eye, and hence make the holographic Fourier-transform reconstruction observable!

It is sometimes said that the coherence in light beams increases with distance "by mere propagation." What is happening is simply that with all dimensions (source dimensions and slit separation) remaining constant, the angular separation $S_1 S_2$ becomes increasingly small compared to $\gamma(x)$ as the distance of the slits from the source increases.

2.10. Intensity Correlations in Partially Coherent Fields

Much more can be said about the coherence properties of light than is sufficient for the experimental characterization given above. The reader is referred to Refs. 9–11. Much of the basic theory of coherence can be found in Refs. 12–20.

Only one more of the many aspects of coherence will be examined, to further illustrate the simplicity which results from the use of modern image-formation theory and Fourier-transform methods in dealing with coherence.

2.11 Intensity Interferometers

A great deal of interest in the understanding of the coherence characteristics of light was created by some classically unexpected and not easily understandable results of an experiment reported by Hanbury Brown and Twiss.[21] Basically, the experimental arrangement is quite similar to the two-slit experiments described above. However, instead of observing the interference fringe system formed at some distance behind the two slits S_1 and S_2, photoelectric receivers are placed directly behind the slits, and their outputs are *correlated*. A perfectly equivalent arrangement consists of correlating the outputs of two radioastronomical antennas, following square-law detection. The interpretation of the experiment in terms of image-formation theory is immediate and perfectly straightforward.

Consider the two-slit arrangement shown in Fig. 29, which is perfectly equivalent to the arrangement used by Hanbury Brown and Twiss and shown in Fig. 30. We may call I_A and I_B the intensities recorded at the two slits A and B. The problem is to give an expression for the product $I_A I_B$ in terms of the source intensity distribution $I_S(\alpha, \beta)$ and the separation of the two slits. A very simple method gives the correct interpretation, as we now show.

Consider the schematic arrangement of Fig. 29 which represents the apparatus used by Hanbury Brown and Twiss[4] (see Fig. 30). The instantaneous power recorded by the photoelectric detector S_1 is

$$\bar{E}_{S_1} \, \bar{E}_{S_1}^{*} \tag{92}$$

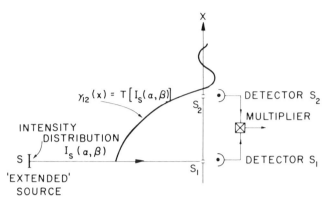

FIG. 29. Intensity interferometer, analogous to correlation apparatus used by Hanbury Brown and Twiss[21] (see Fig. 30).

and that recorded by the photoelectric detector S_2 is

$$\overline{E}_{S2} \ \overline{E}_{S2}^{*} \tag{93}$$

The time average of the instantaneous product obtained by the "multiplier" is

$$\left\langle \overline{E}_{S1} \ \overline{E}_{S1}^{*} \ \overline{E}_{S2} \ \overline{E}_{S2}^{*} \right\rangle \tag{94}$$

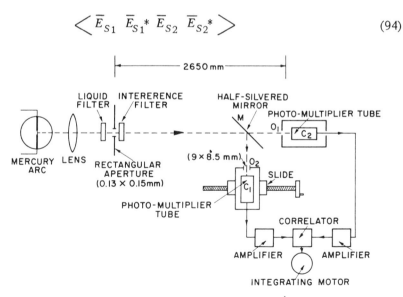

FIG. 30. Apparatus used by Hanbury Brown and Twiss[4] for demonstration of correlation between photoelectric signals.(Equivalent arrangement, used for discussion in the text, is shown in Fig. 29.

We may write (94) in the form

$$\left\langle \overline{E}_{S1} \; \overline{E}_{S2}^{*} \; \overline{E}_{S1}^{*} \; \overline{E}_{S2} \right\rangle \qquad (95)$$

For the case of radiation which is characterized by Gaussian statistics [see, e.g., Ref. 9], (95) may be written as

$$\left\langle \overline{E}_{S1} \; \overline{E}_{S1}^{*} \right\rangle \left\langle \overline{E}_{S2} \; \overline{E}_{S2}^{*} \right\rangle \; [\, 1 + |\, \gamma_{12}(x)\, |^2 \,] \qquad (96)$$

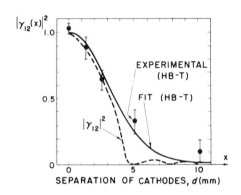

FIG. 31. Intensity (or correlation) interferometer. Experimental values and fit value of product between the two photoelectric signals (Fig. 30), after Hanbury Brown and Twiss,[21] and value of $|\gamma_{12}|^2$, following the equivalent arrangement of Fig. 29.

which shows immediately that the "signal" obtained, upon multiplication and integration in the Hanbury Brown-Twiss intensity interferometer, appears expressed simply as dependent linearly on $|\, \gamma_{12}(x)\, |^2$, that is, on the square of the magnitude of the partial coherence factor as described in the previous sections. It is important to note that the assumption of Gaussian statistics for the characterization of the radiation is a very good approximation for the description of thermal light sources. Fig. 31 shows the very good accord of the signal predicted according to (96) [assuming infinitely narrow slits] with the original experimental results.

It may be of some interest to note the simplicity of the derivation given according to this method, especially if one recalls some difficulties encountered in interpreting the remarkable results obtained by Hanbury Brown and Twiss after they first reported them in 1954. For a more general analysis, the reader may wish to consult the original paper [21] and the references.

References

1. M. Born and E. Wolf, "Principles of Optics," 2nd rev. ed., pp. 491–554. Pergamon Press, New York, 1964.
2. G. W. Stroke, *J. Opt. Soc. Am.* **47**, 1097–1103 (1957).
3. A. Javan, E. A. Ballik, and W. L. Bond, *J. Opt. Soc. Am.* **52**, 96 (1961).
4. G. W. Stroke, Ref. 2, p. 1103.
5. G. W. Stroke, *in* "McGraw-Hill Encyclopedia of Science and Technology," Vol. 4, pp. 264–265. McGraw-Hill, New York, 1960.
6. G. R. Harrison and G. W. Stroke, *J. Opt. Soc. Am.* **45**, 112 (1955).
7. W. S. Read and D. L. Fried, *Proc. IEEE* **51**, 1787 (1963).
8. A. T. Forrester, R. A. Gudmundsen, and P. O. Johnson, *Phys. Rev.* **99**, 1691 (1955).
9. L. Mandel, *in* "Progress in Optics" (E. Wolf, ed.), Vol. II, pp. 183–248. North-Holland Publ., Amsterdam, 1963.
10. "Quantum and Statistical Aspects of Light" (selected reprints). American Institute of Physics, New York, 1963.
11. "Quantum Electronics" (P. Grivet and N. Bloembergen, eds.), Vol. III. Columbia Univ. Press, New York, 1964.
12. P. H. Van Cittert, *Physica* **1**, 201 (1934).
13. F. Zernike, *Physica* **5**, 785 (1938).
14. H. H. Hopkins, *Proc. Roy. Soc. (London)* **A208**, 263 (1951); **A217**, 408 (1953).
15. A. Blanc-Lapierre and P. Dumontet, *Rev. Optique* **34**, 1 (1955).
16. E. Wolf, *Nuovo Cimento* **12**, 884 (1954).
17. A. Maréchal and M. Françon, "Diffraction." Revue d'Optique, Paris, 1960.
18. A. Einstein, *Z. Physik* **10**, 185 (1909); **10**, 817 (1909).
19. A. Einstein, *Ann. Physik* **47**, 879 (1915).
20. A. Einstein and L. Hopf, *Ann. Physik* **33**, 1096 (1910).
21. R. Hanbury Brown and R. Q. Twiss, *Nature* **177**, 28 (1956).
22. M. Françon, "Optical Interferometry." Academic Press, New York, 1967.

V. IMAGE FORMATION IN COHERENT LIGHT

1. Introduction

The basic characteristic of coherent-light image-forming systems is that *complex amplitudes*, rather than intensities in the field, add before recording. Of course, just as in the case of noncoherent image formation, only the intensity $\overline{E}\,\overline{E}*$ of the resultant field can be recorded. In other words, what is finally recorded is always the *resultant field intensity*, that is, $|\overline{E}|^2 = \overline{E}\,\overline{E}*$, the square of the magnitude of the resultant electric-field vector $\overline{E}$. However, it is possible to superimpose on the signal field a coherent background field, by interference, and thus make *both* the amplitude *and* the phase in the signal field *retrievably recordable*.

A basic example of an interferometric heterodyning method, as used in wavefront reconstruction, was described in Section 6 of Chapter I. A more complete discussion of *phase recording* in optics will be given in Chapter VI.

It appears that in many cases coherent-light optical systems permit more flexible image processing than noncoherent systems, for example, in communications systems and holography. We now introduce some of the basic concepts of coherent optical imaging.

2. Coherent Illumination

Coherence, in this chapter, is referred to principally in terms of *spatial* coherence. On the basis of the discussion of spatial coherence, given in Chapter IV, coherent illumination can be best illustrated by the following examples.

2.1. Example 1. Illumination of an Extended Field by a Point Source

Any object placed in the region indicated in Fig. 1 can be considered to be coherently illuminated, even when the point source is itself non-coherent, and has in fact a finite diameter D_S, provided that the lens diameter D_L is

$$D_L < \frac{1}{n}\frac{f\lambda}{D_S} \qquad (n \simeq 5 \text{ to } 10) \tag{1}$$

according to (IV.91).

2.2. Example 2. Equivalence of Coherent Object and Coherently Illuminated Object

A coherent object, such as that represented in Fig. 2, may, for example, be the output mirror of a laser. Another way of illustrating the equiva-

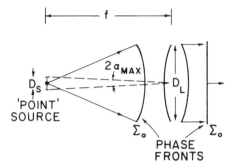

FIG. 1. Coherent illumination parameters, with extended, spatially non-coherent source D_s.

FIG. 2. Equivalence of spatially coherent object and coherently illuminated object (see Fig. 1).

lence between the output of a spatially coherent laser and a coherently illuminated lens is shown in Fig. 3.

A coherently illuminated object, such as a photographic transparency, must conserve uniform phase transmission to a degree which may require, in some cases, immersion of the film into a liquid gate, filled with a liquid of the same refractive index as the film. In most cases, however, we have found that immersion was not necessary with such emulsions as Polaroid P/N film and Kodak 649 F plates when used in spatial filtering and holography of the type described in this text.

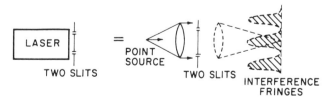

FIG. 3. Equivalence of spatially coherent laser illumination and coherent illumination according to Fig. 1, with noncoherent or coherent source.

3. Image Formation in Coherent Light, Considered as Double Diffraction

In many cases, it is convenient to consider image formation in coherent light as a "double-diffraction" process. The double-diffraction concept may be best illustrated by means of an image-forming arrangement, such as that of Fig. 4. Basically, we may say that the object, with a complex amplitude distribution $T(\xi)$, produces by diffraction a complex amplitude

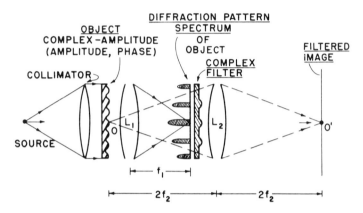

FIG. 4. Spatial filtering arrangement, after Maréchal and Croce[1] and O'Neill.[2]

distribution $f(x)$ in the pupil of the lens L_2, where $f(x)$ is in the focal plane of the lens L_1, if the object itself was illuminated in collimated light (as shown). In this case, the complex amplitude $f(x)$ is equal to the Fourier transform of the complex amplitude $T(\xi)$ in the object. The complex amplitude $f(x)$ is frequently described as the complex diffraction pattern (or diffraction pattern, for short) of the object. Sometimes it is also convenient to recognize $f(x)$ as the spectrum of the object.

The light arriving at the lens L_2 continues on to form an image in the ξ' plane, and we may consider the image $T(\xi')$ to be obtained by a second diffraction, taking place by the field $f(x)$, in the pupil of L_2. It is clear from Fig. 4 that the image amplitude $T(\xi')$ is obtained by a Fourier transformation of the amplitude $f(x)$, and therefore that the image amplitude $T(\xi')$ will be equal to the object amplitude $T(\xi)$, if perfect lenses are used and if no complex filter is used next to $f(x)$, as shown in Fig. 4. (The role of the complex filter will be dealt with in the following sections.)

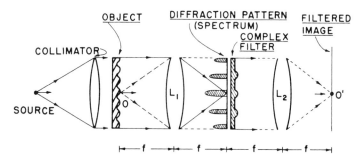

FIG. 5. Spatial filtering arrangement, after Cutrona et al.[3]

The double-diffraction arrangement shown in Fig. 4 is that described by Zernike in 1935 for phase contrast filtering, and it also forms the basis for the spatial filtering arrangements first described by Maréchal and Croce in 1953[1] and by O'Neill in 1954.[2] A comparable arrangement, introduced for spatial filtering in 1960 by Cutrona et al.,[3] is shown in Fig. 5, and is quite generally used for such *correlation filtering* work today. [Because of the *multiplication* of the diffraction pattern $f(x)$ by the filter, we shall indeed show that the final filtered image is equal to the *correlation* of the object with the Fourier transform of the filter.]

Some further clarification of the importance of the double-diffraction concept may be obtained in examining the effect of filtering out spectral orders [produced in the focal plane of L_1 (see Fig. 6) by a periodic object (i.e., a grating)] before the second diffraction takes place. The filtering is accomplished, in this case, with the aid of a mask, arranged to let only the even orders, and the zero order, pass through the lens L_2.

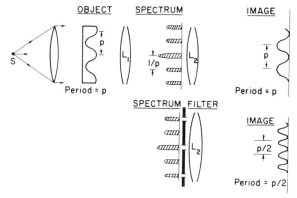

FIG. 6. Double-diffraction principle, after Abbe[4a] and Wood.[4b]

With this filter the filtered field in the pupil L_2 is equivalent to that which would have been obtained with a grating of half the period actually used as an object, and the image obtained by the second diffraction is found (see Abbe,[4a] Wood,[4b] Maréchal, and Françon,[4c] etc.) to be that of a grating of the period $p/2$ (see Fig. 6) rather than the period p, corresponding to the unfiltered image.

As we have noted, the filtering arrangements described so far involve *multiplication* of diffraction patterns, that is, multiplication of the diffraction pattern of the object by a filter which is itself the diffraction pattern of the function, which gives the filtered image by a correlation (or convolution) with the unfiltered image of the object. The filtering processes using *multiplication of diffraction patterns* may therefore be described as *correlation filtering*.

An important advance in optical filtering, and indeed in interferometry, was introduced in 1965 by Gabor and Stroke *et al.*[5] (see Section 9), when they showed the possibility of performing the addition of complex amplitudes in images (or wavefronts), to be formed by interference, by *successively* adding, in the *same* hologram, the *intensities* of the holograms corresponding to the diffraction patterns of the images (or of the wavefronts) of which the complex amplitudes are to be added. In contrast with the methods of correlation filtering, Gabor and Stroke *et al.* also noted that their method could be considered a method of *optical image synthesis* (see Section 9).

4. Abbe and Rayleigh Resolution Criteria

As a further example of the power of the double-diffraction concept, we may discuss the problem of resolution in terms of the qualitative discussion given in Section 3. We shall show that the double-diffraction concept permits one to define resolution in the object space in a very straightforward manner, and that the resolution criterion obtained (Abbe resolution criterion) agrees well with the well-known Rayleigh criterion.

Consider the coherently illuminated periodic amplitude object of period p (Fig. 7), and let the complex amplitude of the object be (except for a constant "bias" term)

$$2F(\xi)\cos\left(2\pi\frac{\xi}{p}\right) \tag{2}$$

producing in the pupil of the lens L a diffraction pattern (spectrum)

$$f\left(x+\frac{1}{p}\right)+f\left(x-\frac{1}{p}\right) \tag{3}$$

according to Fourier-transform theory. The diffraction pattern consists in this case of two separated plane waves which (for the case of a small ob-

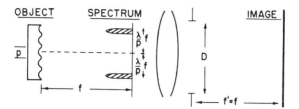

FIG. 7. Abbe resolution criterion (see equivalent Rayleigh criterion in Fig. 8) in *coherent* light.

ject) are symbolized by the two spectral peaks shown. (For convenience, we call the peaks *spectral lines* here.)

Let D be the aperture of the objective lens looking at the object. If the two spectral "lines" at $\pm(\lambda/p)f$ get transmitted through the aperture, the amplitude in the image, according to Fourier-transform theory, is

$$A(\xi') = T\left[f\left(x + \frac{1}{p}\right) + f\left(x - \frac{1}{p}\right)\right]$$

$$= 2F(\xi') \cos\left(2\pi \frac{\xi'}{p}\right) \tag{4}$$

and the intensity in the image is

$$I(\xi') = A(\xi')A(\xi')^*$$

$$= 4|F(\xi')|^2 \cos^2\left(2\pi \frac{\xi'}{p}\right)$$

$$= 2|F(\xi')|^2 \left(1 + \cos 4\pi \frac{\xi'}{p}\right) \tag{5}$$

It is seen that the image has the same periodicity as the amplitude object. In fact, the object is fully resolved because the two waves, which it produced in the first diffraction, both get through the lens to form the image in the second diffraction.

As long as the two spectral lines pass through the aperture, complete resolution will be obtained. It is seen from Fig. 7 that resolution, without loss of contrast, is obtained up to the value of

$$\frac{\lambda}{p} f \leq \frac{D}{2} \tag{6}$$

where f is the focal length of the lens L, that is, up to

Abbe resolution criterion:
$$\boxed{p \geq p_0 = \frac{2\lambda}{D} f} \tag{7}$$

p_0 is the smallest (or finest) period in the object (grating) which can be resolved according to the Abbe criterion of (7).

We may compare the Abbe criterion of (7) with the well-known Rayleigh criterion (see Fig. 8). For the case of coherent-light imaging, we may

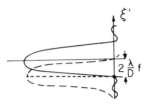

FIG. 8. Rayleigh resolution criterion (see equivalent Abbe criterion in Fig. 7) in *coherent light.*

write the Rayleigh criterion in the form

Rayleigh criterion (coherent-light imaging):

$$\Delta \xi \geq \Delta \xi_0 = \frac{2\lambda}{D} f \tag{8}$$

where $\Delta\xi_0$ is the limit of resolution in the object space. In comparing (7) and (8) we may conclude that the Abbe criterion and the Rayleigh criterion are indeed equivalent. In many cases, however, the Abbe criterion may provide a better insight into the experimental factors which could limit ultimate resolution. This is the case, for example, in holographic imaging.

5. Transfer Functions in Coherent and in Noncoherent Light

According to Fig. 7 and Section 4, we can represent the frequency-transfer function $\tau(x)_{coh}$ for coherent light by the graph in Fig. 9, which shows that transmission without loss of contrast is obtained from $x = 0$ to $x = D/2 = (\lambda/p_0)\, f$, and then the contrast is zero.

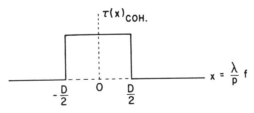

FIG. 9. Transfer function of perfect lens in *coherent* light (square aperture).

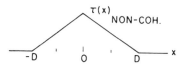

FIG. 10. Transfer function of perfect lens in *noncoherent* light (square aperture).

We recall (Section 6 of Chapter III), for comparison, the frequency response for noncoherent light (Fig. 10).

Finally, in terms of the double-diffraction concept, it is easy to understand why the resolution appears to be doubled in coherent light with the use of oblique illumination. This is illustrated in Fig. 11.

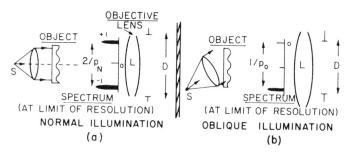

FIG. 11. Double-diffraction concept: increase of limit of resolution by oblique illumination (b) compared to normal illumination (a), for a periodic object. (Oblique illumination may result in an increase of resolution, but may involve loss in the fidelity of the image.)

In carrying out an analysis comparable to that given in Section 4, we find that the image formed, in the case of oblique illumination, contains not only the sinusoidal object terms, but also new terms, which may be considered as noise. We may conclude, that the single side-band imaging obtained with oblique illumination does not represent faithful imaging. However, there is no doubt that increased resolution is obtained.

6. Phase-Contrast Filtering

The concepts of double diffraction permit a very simple interpretation of the phase-contrast filtering used in microscopy, in order to transform unobservable small phase variations into easily observable intensity variations. We recall that Zernike received the Nobel Prize for this work in 1953.

Consider an object $f(x, y)$ which is a pure phase object, that is,

$$f(x, y) = \exp\left[i\,\phi(x, y)\right] \tag{9}$$

Let the phase $\phi(x, y)$ have sufficiently small values, so that one can write

$$f(x, y) \simeq 1 + i\phi(x, y) + \cdots \quad \text{(for } \phi \ll 2\pi\text{)} \tag{10}$$

It follows that the transform, which is displayed in the back focal plane of L_1, (Fig. 12), is given by

$$f(x, y) \longrightarrow F(u, v) = \delta(0) + iT\left[\phi(x, y)\right] \tag{11}$$

Now let a quarter-wave filter, $H(u, v)$, be placed in the focal plane of L_1, so that the quarter-wave $(\pi/2)$ section covers only the central (dc) region

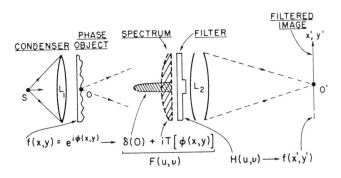

FIG. 12. Spatial filtering parameters.

of $F(u, v)$. Moreover, let the amplitude transmission of the quarter-wave region be t. We may write the filter equation

$$H(u, v) = \begin{cases} it & \text{over the region } \delta(0) \\ \\ 1 & \text{over the remaining spectrum} \end{cases} \tag{12}$$

where t is the amplitude transmission factor $(0 \le t \le 1)$.

The filtered spectrum is then

$$\begin{aligned} G(u, v) &= F(u, v)H(u, v) \\ &= it\,\delta(0) + iT\left[\phi(x, y)\right] \end{aligned} \tag{13}$$

By the second diffraction the image $f(x', y')$ is

$$\begin{aligned} f(x', y') &= T\left[G(u, v)\right] \\ &= it + i\phi(x, y) \end{aligned}$$

that is,

$$f(x', y') = i[t + \phi(x, y)] \qquad (14)$$

The intensity in the image is

$$I(x', y') = f(x', y') \cdot f(x', y')^* = t^2 + \phi^2 + 2\phi t \qquad (15)$$

In the case where $t = 1$ we have

$$I(x', y') \cong 1 + 2\phi \qquad (16)$$

and we see that the *image intensity variations* are proportional to the *object phase variations*. In general, when the amplitude transmission of the $\pi/2$ plate is t, so that the image intensity in the dc (background) portions of the image is t^2, the contrast in the phase portions of the image is

$$c \cong \frac{t^2 + \phi^2 + 2\phi t - (t^2 + \phi^2 - 2\phi t)}{t^2} = 4\frac{\phi}{t} \qquad (17)$$

With $t \leq 1$ it is seen that a very high contrast is obtained even for extremely small values of ϕ. (For instance, with $t = 10^{-3}$, $\phi = 10^{-3}$ radians, the contrast is $c \cong 4$. For $\phi = 10^{-3}$, $t = 10^{-2}$, the contrast is 0.4, both very good contrasts.

Equation (17) is to be compared with the contrast which would be obtained without filtering. One would have, from (10), the image

$$I(x', y')_{\text{unfiltered}} \simeq 1 + \phi^2(x, y) \qquad (18)$$

and the contrast of the phase portion would be

$$c_{\text{unfiltered}} \cong \frac{\phi^2}{1} = \phi^2 \qquad (19)$$

giving, for $\phi = 10^{-3}$, $c = 10^{-6}$.

The phase contrast method of selective filtering is a basic example of optical filtering.

7. Optical Filtering with Holographically Matched Spatial Filters

An example of optical correlation filtering is shown in Fig. 13, and a schematic representation of the arrangements used is shown in Figs. 14 and 15. Spatial filtering in optics was first described by Maréchal and Croce in 1953[1] and by O'Neill in 1954.[2] More recently, it has appeared that the principal problem in spatial filtering, that of realizing the complex filters, could be solved by recording the filters in the form of holograms, thus making use of the principles of wavefront-reconstruction imaging (holography) first described by Gabor in 1948.[6]

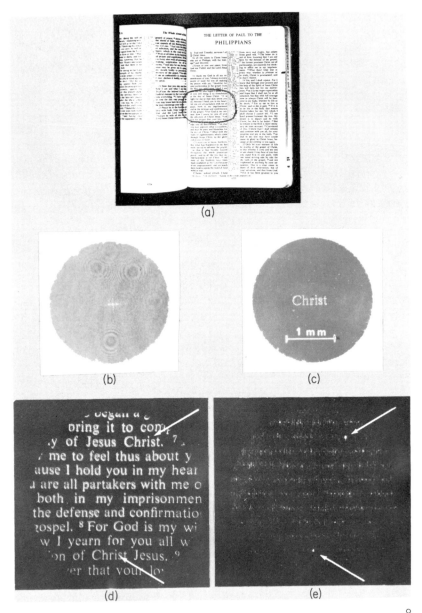

FIG. 13. Correlation filtering (see Section 7), entirely carried out in 6328-Å laser light, on Polaroid P/N film (G. W. Stroke and students, The University of Michigan, 1964). (a) Text being filtered. Input plane, $f(x, y)$. (b) Holographic filter for word "Christ." Filter plane, $H(u', v')$. (c) "Reconstruction" from filter of word "Christ." Output plane, $h(x, y)$. (d) Zero-order image. Output plane. (e) Filtered image (right-hand side band). Output plane $[f \circledast h]_{x+b, y+c}$.

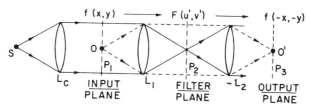

FIG. 14. Correlation filtering parameters (see Section 7).

Some historical background and a qualitative introduction to spatial filtering was given in Section 3, where it was noted that a correlation filtering arrangement in very wide use today is that of Figs. 14 and 15, based, in addition to the work by Maréchal, Croce, and O'Neill, on the work by Cutrona et al., as well as on an important paper by Vanderlugt.[7] We have also noted that true optical filtering, in the sense of adding or subtracting desired portions in an image, may require the use of the image-synthesis principles, first described by Gabor and Stroke et al.[5] (see Section 9), or indeed the use of optical compensation methods (see Chapter VI, Section 4), such as those first described by Stroke, Restrick, Funkhouser, and Brumm,[11] also in 1965.

We now describe the basic principles of optical correlation filtering with matched filters.[†]

The system shown in Figs. 14 and 15 is recognized as one of the coherent-light imaging systems. The object O in the P_1 plane is imaged one-to-one in the P_3 plane. If a suitable spatial filter is used in the P_2 plane, it is possible to selectively filter out of the image O' any desired frequency, in a way quite comparable in principle to the filtering methods used in electrical communication systems.

It should be clear that the entire process between S and O' must be carried out in coherent light, and, therefore, that the *spatial filter* $H(u', v')$ will in general be a complex filter.

As an example of a filtering operation, (see Fig. 13) one may wish to determine the location of the word CHRIST in a page of the "Letter of

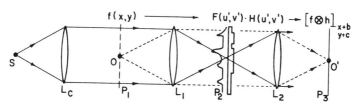

FIG. 15. Correlation filtering parameters (see Section 7).

[†]For general background and details see, for example, Ref. 7.

Paul to the Philippians." Because the correlation-filtering process involves, basically, a correlation between the unknown, or desired word (here CHRIST) and the rest of the page, the output image in the optical correlation filtering system will have maxima in the places where the correlation between the unknown word and the page has a maximum. The location of the word CHRIST will be marked by a *bright dot of light*, corresponding to the convolution [see Eq. (64), Chapter VII.]

$$f(-x, -y) \circledast h(-x, -y) \tag{20}$$

indicated in Fig. 15 where

$$h(x, y) = T[H(u', v')] \tag{21}$$

$H(u', v')$ is the complex filter for the word CHRIST, and (u', v') are the coordinates defined in (III.6a), page 30.

In addition to providing the maxima, wherever the correlation of the word CHRIST with the corresponding word CHRIST in the page is a maximum, the correlation filtering process also provides a graphical answer, in the photographic output, to the otherwise rather involved problem concerning the degree of correlation of the "unknown" word CHRIST with *other* words in the text. For example (see Fig. 13), we see that there are some correlation maxima of CHRIST with IMPRISONMENT, as one would expect, as well as of CHRIST with IS, but there is also noticeable correlation with ABOUT, IN MY, CONFIRMATION, IS MY, etc.

The photographs in Fig. 13 illustrate the three principal steps in a coherent-light filtering operation. Figure 13a shows the section of the transparency in which the location of the word CHRIST is to be determined. Figure 13b shows the "filter" obtained by recording the diffrac-

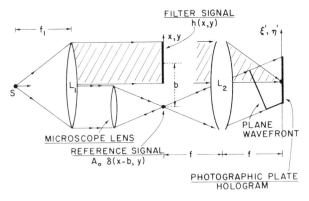

FIG. 16. Holographic Fourier-transform recording of complex filter (see Section 7).

tion pattern of the word CHRIST in the Fourier-transforming arrangement illustrated in Fig. 16, together with a coherent background. The similarity between the complex filter and a Fourier-transform hologram is readily recognized. A Fourier-transform reconstruction of the word CHRIST, obtained in the focal plane of a lens by illuminating the filter in collimated light, is shown in Fig. 13c.

Finally, the filtered image is shown in Fig. 13e, together with a zero-order image (Fig. 13d), both in the output plane.

The entire work was carried out on P./N Polaroid film. Among the many possible applications of correlation filtering, such as that described above, the possibilities of automatic reading and library automation have attracted particular attention.

7.1. Realization of the Complex Filter and of Filtering Operation

Let the desired filter be

$$H(u', v') = |H| e^{i\theta} \tag{22}$$

Let $h(x, y)$ be a transparency of the word CHRIST (which is to be filtered). We may define

$$h(x, y) = \text{CHRIST} = \text{filter function} \tag{23}$$

Let

$$h(x, y) \longrightarrow H(u', v') = |H| e^{i\theta} \tag{24}$$

by Fourier transformation, in the focal plane of L_2. Let the reference signal be

$$|A_0| e^{i\phi} \tag{25}$$

produced as shown in the Fig. 16, where we recall that the coordinates are

$$\xi' = \frac{f}{k} u' \qquad \eta' = \frac{f}{k} v' \tag{26}$$

according to (3.6a). Let b be the offset, along x, between the filter signal function $h(x, y)$ and the reference point R (see Fig. 16). One has, in the focal plane of L_2, by Fourier transformation, the complex amplitude

$$A(u', v') = H(u', v') + P(u', v') \tag{27}$$

according to the equation

$$h(x, y) + |A_0| \delta(x - b, y) \longrightarrow H(u', v') + P(u', v') \tag{28}$$

We may write

$$H(u', v') = |H| e^{i\theta} \tag{29}$$

and, because of the b offset, the function $P(u', v')$ is [see (VII.9a)]

$$P(u', v') = |A_0| e^{i\phi} = |A_0| \exp[i\,(bu')] \qquad (30)$$

where the phase ϕ is

$$\phi = bu' \qquad (31)$$

The phase ϕ is seen to be a *linear* function of u', and the function $P(u', v')$ is therefore seen to represent a plane wave, forming an angle

$$\psi = \tan^{-1} \frac{b}{f} \cong \frac{b}{f} \qquad (32)$$

with the ξ' axis (see Fig. 16).

The *intensity* $I(u', v')$ recorded on the filter is

$$I(u', v') = A(u', v')\, A(u', v')^* =$$

$$[H(u', v') + P(u', v')]\,[H(u', v') + P(u', v')]^* \qquad (33)$$

that is,

$$I(u', v') = [|H|e^{i\theta} + A_0 e^{i\phi}]\,[|H|e^{-i\theta} + A_0 e^{-i\phi}] \qquad (34)$$

or

$$I(u', v') = [|H|^2 + A_0{}^2] + |H|e^{i\theta} A_0 e^{-i\phi} + |H|e^{-i\theta} A_0 e^{+i\phi} \qquad (35)$$

However, the *desired* filter is not $I(u', v')$ but rather $H(u'v') = |H|e^{i\theta}$, according to (22).

We ask now: How can the desired filter $|H|e^{i\theta}$ be separated from the recorded filter $I(u', v')$? The answer is: The separation is carried out automatically by diffraction in the optical filtering system shown in Fig. 15. This can be shown as follows.

We first assume, for simplicity, that the complex amplitude transmission through the filter is proportional with a factor 1 to the intensity $I(u', v')$ recorded on the filter. [In reality, this assumption is correct for the case where the reference beam intensity $A_0{}^2$ is some 10 times the value of $|H|^2$, and where the exposure and development of the photographic plate are made with a gamma $\gamma = 2$ (see, for example, Chapter VI). For other values of gamma, the complex amplitude of the field transmitted through the filter is merely multiplied by $\gamma/2$, and the complex amplitude transmitted through the filter is still proportional to $I(u', v')$.] [See Eq. (4), Chapter VI.]

With reference to Fig. 15, we find, with (35), that the complex amplitude of the field transmitted through the filter, in the filtering arrangement, is

$$F(u', v')\, I(u', v') = F(u', v')\,[|H|^2 + A_0{}^2] + F(u', v')\, H(u', v')\, A_0 e^{-i\phi}$$

$$+ F(u', v')\, H(u', v')^* A_0 e^{i\phi} \qquad (36)$$

The field (36) transmitted through the filter is now the subject of a Fourier transformation by the lens L_2. We recognize three separate terms in (36), and we find that the corresponding images are obtained in the $\xi(u', v')$ plane, by Fourier transformation, with the use of the following equations (demonstrated in other sections) and now allowing also for a "c" offset in the y-direction:

$$F \cdot He^{i\phi} \longrightarrow T[F] \otimes T[He^{i\phi}] \tag{VII.41}$$

$$T[He^{i\phi}] = T\{H \exp[i(bu' + cv')]\} = h(x - b, y - c) \tag{VII.10a}$$

$$T[He^{-i\phi}] = T\{H \exp[-i(bu' + cv')]\} = h(x + b, y + c) \tag{VII.10a}$$

where we have now generalized the offset of the reference point R to $x = b$ and $y = c$ [see Fig. 16 and (31)].

We therefore find that the Fourier transformation of (36) gives the following image components [see (VII.41), (VII.62), with (VII.12), (VII.13)]:

$$F(u', v')[|H|^2 + A_0^2] \longrightarrow f(u', v') \otimes T[|H|^2 + A_0^2]$$

(a dc term, centered at $x = 0$, $y = 0$)

$$F(u', v') H(u', v') A_0 e^{-i\phi} \longrightarrow A_0[f \otimes h]_{x+b, y+c}$$

(centered at $x = -b$, $y = -c$) (37)

$$F(u', v') H(u', v')^* A_0 e^{+i\phi} \longrightarrow A_0[f \star h]_{x-b, y-c}$$

(centered at $x = +b$, $y = +c$)

where the low-frequency terms, centered at $x = 0$ and $y = 0$ are described as "dc term" and where we recall that $\otimes$ indicates a convolution and $\star$ a correlation, according to (VII. 22) and to (VII. 52), respectively. The two images, one involving the convolution and the other the correlation of the image $f(x, y)$ to be filtered, with the filter function $h(x, y)$ are seen to appear off axis, centered at points $x = -b$, $y = -c$, and $x = b$, $y = c$, respectively, and *separated* from the dc term (assuming that b and c have been chosen large enough)! It is this separation of the filtered images which was described as automatic or inherent in the filtering process. The separation of the filtered images from the dc is obtained by the use of the offset $(x = b, y = c)$ between the reference point R and the filter signal $h(x, y)$ in the holographic recording of the filter (Fig. 16).

The desired filtered image is, in this case [see (20)], the side-band image $f \otimes h$, obtained by convolution, rather than the other side-band image $f \star h$, obtained by correlation. Except for filter functions h having a twofold rotation symmetry (see Section 4 of Chapter VI and Ref. 11), only the convolution side-band image will be filtered as desired. It may be of interest to note (see Section 4 of Chapter VI), that there are other cases of optical filtering and synthesis, where it is the correlation side-band image, rather than the convolution side-band image, which is desired.

We may also interpret the holographic recording of the filter (i.e., the recording of the hologram of the filter function h) as a recording of the hologram on an interference-fringe grating carrier. We may indeed note that the *hologram*, which would result from the interference of the plane wave $P(u', v')$ with a plane wave parallel to the film plane [see Fig. 16 and (30) and (31)], would be an *interferogram with straight-line fringes*, with a fringe spacing

$$\xi_0' \cong \frac{\lambda}{f} b \qquad (38)$$

Such a fringe system can, of course, be obtained by introducing next to the point reference $R(x = b, y = c)$ another point reference $R(x = 0, y = 0)$, placed at the origin of the xy system.[†]

We may also qualitatively interpret the phase recording in the hologram as some local *shift* of the carrier fringes, according to the local phase in the diffraction pattern of the function of which the filter (i.e., hologram) is being recorded.

We shall further stress the close relationship between spatial filtering and holography in other sections, notably in Chapter VI, as well as in Section 9, where we deal with the problem of optical image synthesis by holographic Fourier transformation.

8. Optical Computing, Correlating, and Signal Processing

Among the image-processing systems which have found practical application, it may be of interest to single out the optical spectrum analyzers and the optical cross-correlators, in the form described by Cutrona et al.[3]

8.1   Spectrum Analyzers[‡]

Basically, an optical spectrum analyzer is comparable to an optical grating spectrometer using monochromatic radiation and an imperfectly spaced diffraction grating.[9, 10]

In an optical spectrum analyzer, the grating imperfections contain the significant, desired information, recorded on photographic film, for instance, with the help of a cathode-ray oscilloscope. Seismic vibration recordings are an example of signals recorded and analyzed in this manner. The analysis aims at determining the power spectrum of the recorded signal, forming the imperfect grating. The principle of spectrum

[†]Point references, for use in interferometers, appear to have been first introduced by Nomarski.[12,13]

[‡]For additional background and details see, for example, Ref. 9.

analyzers is based on the fact that the amplitude distribution of the light in the diffraction patterns formed by an optical grating is equal to the Fourier transform of the (complex) amplitude transmitted through the grating (perfect or imperfect!), when the grating is illuminated by a plane monochromatic wave. In other words, the intensity distribution in the diffraction patterns formed by the grating is nothing but the *spectrum* of the imperfections in the wavefronts diffracted by the grating. When the imperfections are deliberately introduced into the grating, for instance, in the form of an intensity-grating recorded on a photographic transparency, then the spectrum (formed by optical Fourier transformation) will immediately provide the spectral distribution of the signal-intensity variations as a function of the chosen coordinate on the film. (A classical example of this principle may also be found in sound recording; see, for example, Fischer and Lichte[8]).

A complete theoretical treatment of the relation between grating imperfections and the spectral diffraction patterns was given by Stroke in 1960,[9] in a form which is immediately applicable to the spectral analysis as carried out in spectrum analyzers. A recent treatment of this problem, with particular application to the diffractive processing of geophysical data is given by Jackson.[10]

A schematic diagram of an optical spectrum analyzer is given in Fig. 17.

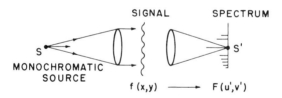

FIG. 17.   Spectral-analysis parameters.

We have, with reference to that figure [see (3.18)],

$$F(u', v') = T[f(x, y)] \qquad (39)$$

where $f(x, y)$ is the complex amplitude of the light transmitted through the film recording of which we wish to obtain the spectrum (i.e., the spectral intensity distribution). More explicitly, we have in the image plane (ξ', η') the complex spectrum [see (3.1a)]

$$F\left(k\,\frac{\xi'}{f},\ k\,\frac{\eta'}{f}\right) = \int\int f(x, y)\,\exp\left[\frac{2\pi i}{\lambda}\left(\frac{\xi'}{f}\,x + \frac{\eta'}{f}\,y\right)\right]\,dx\,dy \qquad (40)$$

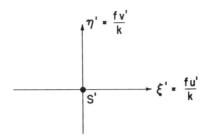

FIG. 18.   Spectral-analysis image coordinates.

where $\xi' = fu'/k$, $\eta' = fv'/k$, and $k = 2\pi/\lambda$ (Fig. 18). The spectral intensity distribution recorded in the focal plane (ξ', η') is determined by

$$I\left(k\frac{\xi'}{f}, k\frac{\eta'}{f}\right) = F \cdot F^* \tag{41}$$

An important feature of optical spectrum analyzers is that they can be designed to perform the spectral analysis *simultaneously* for dozens, hundreds, and in principle thousands of channels, and thus to provide a very simple means for correlating the signals represented in the different channels. A multichannel system is illustrated in Figs. 19 and 20.

Simultaneous analysis is based on the use of *cylindrical optics*. A two-plane view of a multichannel analyzer is shown in Fig. 20. For ex-

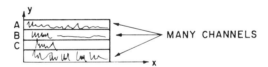

FIG. 19.   Multichannel spectral analysis, after Cutrona et al.[3]

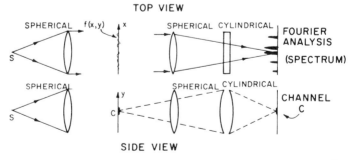

FIG. 20.   Multichannel spectral analysis, after Cutrona et al.[3]

ample, 120-channel analysis has been carried out in some oil-industry applications. Each of the channels may represent the signal from one seismograph, in this case.

8.2. Optical Cross-Correlators

Cross-correlation problems arise in such diverse fields as statistics, pharmaceutical-product development, and target detection and identification.

In an optical cross correlator,(see Fig. 21) using coherent light, both the known and the unknown signals are recorded on film. Both films are generally immersed in liquid gates, and indeed one of the films, say the reference film $g(x)$ is moved through the liquid gate, at a very uniform

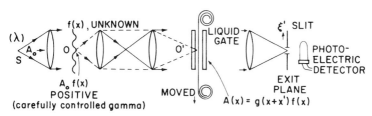

FIG. 21. Optical cross correlator, after Cutrona et al.[3]

rate, to carry out the cross correlation. The liquid gates contain a liquid of the same refractive index as the film substrate, so as to cancel out phase-variation effects, when necessary. One obtains in the exit plane the function $|A(k\xi'/f)|^2$, where

$$A\left(k\frac{\xi'}{f}\right) = T[A(x)] = \int_{-\infty}^{+\infty} f(x)\, g(x + x')\, \exp\left(\frac{2\pi i}{\lambda f}\, \xi' x\right) dx \qquad (42)$$

is called the *ambiguity function*.

Note that if the exponential term in (42) is made equal to unity, by limiting ξ' to the coordinate $\xi' = 0$, the photoelectric output becomes $|A(\xi' = 0),|^2$, where

$$A(\xi' = 0) = \int_{-\infty}^{+\infty} f(x)\, g(x + x')\, dx \qquad (43)$$

is recognized as the desired cross correlation. The desired reduction of the exponential term to unity is very simply accomplished by placing a slit (pinhole) on axis. Indeed, in this case $\xi' = 0$, and $\exp[0] = 1$. For multichannel cross correlations, cylindrical optics and a slit in the exit plane are used, together with multiple photodetectors, to obtain simultaneous analysis.

9. Interferometry and Optical Image Synthesis (Complex-Amplitude Addition) by Successive Addition of Holographic Intensities in a Single Hologram[5]

Heretofore in optics, it had generally been admitted that an interferogram of a scattered wavefront had to be recorded by comparing the scattered wavefront to a wavefront *simultaneously* reflected from a "reference" mirror, for example in a beam-splitting interferometer, such as that illustrated in Fig. 6a, Chapter VI. More generally, two different wavefronts can be added in beam-splitting interferometers, such as those introduced by Michelson and others. These interferometers permit one to add the complex amplitudes of the two wavefronts, provided that the two wavefronts are made to interfere simultaneously.

It was first pointed out by Gabor and Stroke *et al.*[5] that complex amplitudes of not just two, but indeed of several wavefronts could be added, by making each of the wavefronts first interfere, *in succession* (in the latent photographic image), with a "same" "coherent background," thus adding in succession the *intensities* of the individual "interferograms" (i.e. holograms), so as to retrievably store in the resultant hologram the complex addition of the several wavefronts.

As an example, Gabor and Stroke *et al.*[5] considered an "image synthesis" method, in which the complex amplitudes of two images (i.e. of the two corresponding wavefronts) were holographically subtracted according to this principle. We now give a description of the method, according to Ref. 5.

Figure 22a shows a schematic diagram of the experimental arrangement used for image synthesis, and the photographs in Fig. 22b and 22c show details of the apparatus as used in obtaining the experimental results of Fig. 23a and 23b.

The result shown is a first example of *subtractive addition* (i.e., optical erasing) of images obtained by adding, to the diffraction pattern of the function f_1, the diffraction pattern of the function f_2 with a phase shift of π (180°). (In all of the following description, *diffraction pattern* is taken to stand for the complex amplitude of the $\overline{E}$ field vector in the diffraction pattern.) In general, of course, functions can be added with any desired phase shift and, in particular, without any phase shift at all.

In the following description, $f_1, f_2, \ldots, f_n$ stand for the complex amplitude transmittance of a transparency (photographic film, etc.) or reflectance of an object (which can be three-dimensional, of course), when the object (or transparency) are illuminated in monochromatic light. Also

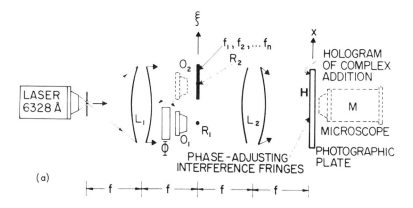

(a)

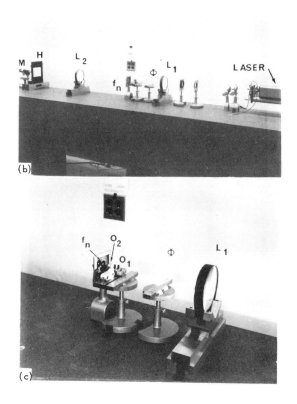

(b)

(c)

FIG. 22. Complex image synthesis by successive holographic addition of *intensities* in a hologram, after Gabor and Stroke *et al.*[5] (a) Schematic diagram. (b) Apparatus used in complex image synthesis by holographic Fourier transformation. (c) Details of apparatus.

$$f_1 \qquad f_2$$

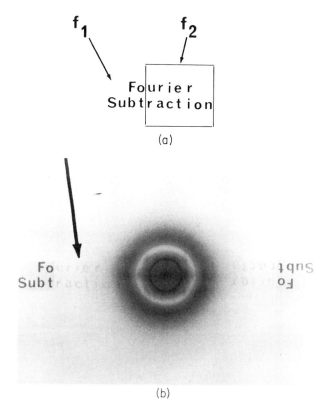

(a)

(b)

FIG. 23. (a) Functions f_1 = "Fourier subtraction" and f_2 = ".. urier....
raction," of which the *intensities* in the diffraction patterns were *successively*
added, in the holographic image-synthesis arrangement of Fig. 22, after Gabor
and Stroke *et al.*[5] A phase shift of $\Phi = 180°$ was introduced between the two exposures of the latent image, for the purpose of subtracting the complex amplitudes
in the diffraction patterns of f_1 and f_2. (b) Fourier-transform reconstruction (see
Fig. 19, Chapter VI) of the subtractive addition of the functions f_1 and f_2 (see
Fig. 23a), of which the hologram was recorded in the image-synthesis arrangement of Fig. 22a.

let F_n = Fourier transform of f_n. A one-dimensional notation is used for
simplicity.

The hologram exposure I is obtained by adding, in n successive exposures I_n, the diffraction patterns of the image-transmittance functions f_n
to the plane-wave coherent background produced by the reference delta
function $A \delta(0)$. The hologram exposure is (see Fig. 22a)

$$I = \sum_n t_n I_n \qquad (44)$$

where

$$I_n = F_n F_n{}^* + A F_n \exp(-2\pi i a x)$$
$$+ A F_n{}^* \exp(+2\pi i a x) \tag{45}$$

a is the offset between f_n and the point reference R_1, along ξ, in Fig. 22a, where t_n is the exposure time for each function.

In general, each exposure can be taken with an arbitrary phase shift Φ_n associated with each exposure. The phase shift can be conveniently associated with the reference delta function, giving

$$A_n \exp(i\Phi_n)\delta(0) \tag{46}$$

In the general case, the hologram exposure becomes, for each function,

$$I_n = A_n{}^2 + A_n F_n \exp(-i\Phi_n)\exp(-2\pi i a x)$$
$$+ A_n F_n{}^* \exp(+i\Phi_n)\exp(+2\pi i a x)$$
$$+ F_n F_n{}^* \tag{47}$$

and the total hologram exposure is still

$$I(x) = \sum_n t_n I_n \tag{48}$$

After processing, and assuming operation in the linear range of the H-D curve, the transmittance $H(x)$ of the hologram is (see Section 2 of Chapter VI)

$$H(x) = [I(x)]^{-\gamma/2} \tag{49}$$

If, as is now customary in holography, the reference-beam intensity is chosen sufficiently great, compared to the function intensity in the exposure, a binomial expansion of expansion of Eq. (49) gives

$$H(x) \propto 1 - \frac{\gamma}{2} X + \frac{(\gamma/2)\,[(\gamma/2)+1]}{2} X^2 + \cdots \tag{50}$$

where

$$X = \left(\frac{1}{\displaystyle\sum_n t_n \dot{A}_n{}^2}\right)\left\{\left[\sum_n t_n A_n F_n \exp(-i\Phi_n)\right]\exp(-2\pi i a x)\right.$$

$$+ \left[\sum_n t_n A_n F_n{}^* \exp(+i\Phi_n)\right]\exp(+2\pi i a x)$$

$$\left. + \left(\sum_n t_n F_n F_n{}^*\right)\right\} \tag{51}$$

and

$$X^2 = \left(\frac{1}{\sum_n t_n A_n^2}\right)\left\{\left[\sum_n t_n A_n F_n \exp(-i\Phi_n)\right]^2 \exp(-4\pi iax)\right.$$

$$+ \left[\sum_n t_n A_n F_n{}^* \exp(+i\Phi_n)\right]^2 \exp(+4\pi iax)$$

$$+ \left(\sum_n t_n F_n F_n{}^*\right)^2$$

$$+ 2\left[\sum_n t_n A_n E_n \exp(-i\Phi_n)\right]\left[\sum_n t_n F_n{}^* \exp(+i\Phi_n)\right]$$

$$+ 2\left(\sum_n t_n F_n F_n{}^*\right)\left[\sum_n t_n A_n F_n \exp(-i\Phi_n)\right] \exp(-2\pi iax)$$

$$+ 2\left(\sum_n t_n F_n F_n{}^*\right)\left[\sum_n t_n A_n F_n \exp(+i\Phi_n)\right] \exp(+2\pi iax) \qquad (52)$$

The hologram transmittance $H(x)$ may now be written

$$H(x) = H_0(x) + H_1(x)\exp(-2\pi iax) + H_1{}^*(x)\exp(+2\pi iax)$$

$$+ H_2(x)\exp(-4\pi iax) + H_2(x)^*\exp(+4\pi iax) \qquad (53)$$

where $H_0(x)$ is the 0th-order term, $H_1(x)$, the first side-bands, etc. For synthesis, the images of interest are those obtained by Fourier transformation of the first side-band terms [i.e., of $H_1(x)$ and of $H_1(x)^*$] in (53). The Fourier transformation is obtained by projecting a plane, spatially coherent, monochromatic wave through the hologram, and by recording the images obtained in the focal plane of a lens.

The $H_1(x)$ term of interest is

$$H_1(x) = \left(\sum_n t_n A_n^2\right)^{-\gamma/2}\left\{-(\gamma/2)\frac{[\Sigma_n t_n A_n F_n \exp(-i\Phi_n)]}{\Sigma_n t_n A_n^2}\right.$$

$$\left. + \frac{(\gamma/2)[(\gamma/2)+1](\Sigma_n t_n F_n F_n{}^*)[\Sigma_n t_n A_n F_n \exp(-i\Phi_n)]}{(\Sigma_n t_n A_n^2)^2}\right\} \qquad (54)$$

What should be noted in (54) is that $H_1(x)$ is proportional to

$$\sum_n t_n A_n \exp(-i\Phi_n)F_n \qquad (55)$$

In other words, the first side-band term of the hologram, $H_1(x)$, is indeed proportional, as desired for synthesis (complex addition), to the weighted complex sum of the Fourier transforms F_n of the functions f_n.

The above equations, and in particular the $H_1(x)$ [Eq. (54)] can be simplified (both theoretically and experimentally!) if all of the exposure times t_n are chosen to be equal, i.e., $t_n = t$, and if all reference-beam amplitudes A_n are also chosen to be equal, i.e., $A_n = A$.

In this case, the hologram transmittance in the first side-band is

$$H_1(x) = (ntA^2)^{-\gamma/2} \left\{ -(\gamma/2A) \left(\sum_n F_n \exp(-i\Phi_n) \right) \right.$$

$$\left. + [(\gamma/2) + 1](\gamma/2n^2A^3) \left(\sum_n F_n F_n{}^* \right)\left(\sum_n F_n \exp(-i\Phi_n) + \cdots \right) \right\}$$

$$(56)$$

Success of the complex addition $[\sum_n F_n]$ depends on the high signal-to-noise ratio, characteristic of the method.

With the first term in (56) being taken as the signal, S, where

$$S = (ntA^2)^{-\gamma/2} (-\gamma/2A_n)\left(\sum_n F_n \right) \exp(-i\Phi_n) \qquad (57)$$

and with the second term in (56) being taken as the noise, N, where

$$N = (ntA^2)^{-\gamma/2} [(\gamma/2) + 1](\gamma/2n^2A^3) \left(\sum_n F_n F_n{}^* \right)\left(\sum_n F_n \exp(-i\Phi_n) \right)$$

$$(58)$$

one has, as the signal-to-noise ratio, the equation

$$\frac{S}{N} = \frac{n^2 A^2}{[(\gamma/2 + 1]\left(\sum_n F_n F_n{}^* \right)} \qquad (59)$$

It can be readily verified that the excellent signal-to-noise ratio obtained in the experiments of Fig. 23a and 23b bears out the prediction of (59).

Some additional remarks are in order with regard to the image-synthesizing arrangement, shown in Fig. 22. An essential element is the phase plate Φ, used (in this case) to obtain the desired subtraction by adding to f_1 the function f_2 with a phase shift of π (180°).

The Φ adjustment is straightforward, and can be achieved, for instance, by means of the two point interference fringes formed in the x plane (see Fig. 22a) with the help of an auxiliary point reference R_2, introduced, temporarily, by an auxiliary microscope objective O_2. Phase plates with arbitrary phase shifts may also be produced by bleaching of suitably exposed optical-quality photographic plates.[14]

An example of interferometry by successive addition of intensities in a single hologram was also recently described by Stroke and Labeyrie[15]

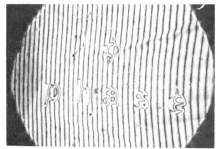

FIG. 24. Interferogram obtained by method of successive addition of holographic intensities in single hologram, after Stroke and Labeyrie. [15]

(Fig. 24). The interferogram shown was reconstructed by Fourier transformation (see Chapter VI) from a "lensless Fourier-transform hologram"[16] recorded by having a coherent background successively interfere (1) with the wave transmitted through the photographically produced phase object[14] and (2) with the wave incident on the phase object.

References

1. A. Maréchal and P. Croce, *Compt. Rend.* **237,** 607 (1953).
2. E. L. O'Neill, Optical Research Laboratory, Boston Univ., 1956.
3. L. J. Cutrona, E. N. Leith, C. J. Palermo, and L. J. Porcello, *IRE Trans. Inform. Theory* **6,** 386 (1960).
4a. E. Abbe's theory of image formation, as presented in his lectures at Jena. (See, e.g., S. Czapski, "Grundzüge der Theorie der Optischen Instruments nach Abbe," 2nd ed., Chap. II, pp. 27–64. Barth, Leipzig, 1904.)
4b. R. W. Wood, "Physical Optics," 3rd ed. Macmillan, New York, 1934.
4c. For example, A. Maréchal and M. Françon, "Diffraction." Revue d'Optique, Paris, 1960.
5. D. Gabor, G. W. Stroke, R. Restrick, A. Funkhouser, and D. Brumm, *Phys. Letters* **18,** 116 (1965).
6. D. Gabor, *Nature* **161,** 777 (1948).
7. A. Vander Lugt, *IEEE Trans. Inform. Theory* **10,** 139 (1964).
8. F. Fischer and H. Lichte, "Tonfilm Aufnahme und Wiedergabe nach dem Klangfilm-Verfahren." Hirzel, Leipzig, 1931.
9. G. W. Stroke, *Rev. Optique* **39,** 291–398 (1960).
10. P. L. Jackson, *Appl. Opt.* **4,** 419 (1965).
11. G. W. Stroke, R. Restrick, A. Funkhouser and D. Brumm, *Phys. Letters* **18,** 274 (1965).
12. G. Nomarski, in "Catalogue de la 53ᵉ Exposition de Physique," page 69. Paris, 1956; see also *Optik* **9–10,** 537 (1960).
13. J. Dyson, *J. Opt. Soc. Am.* **47,** 386 (1957).
14. D. Gabor, G. W. Stroke, D. Brumm, A. Funkhouser and A. Labeyrie, *Nature* **208.** 1159 (1965).
15. G.W. Stroke and A. Labeyrie, *Appl. Phys. Letters* **8,** 42-44 (1966).
16. G.W. Stroke, *Appl. Phys. Letters* **6,** 201 (1965).
17. R.E. Brooks, L.O. Heflinger, and R.F. Wuerker, *Appl. Phys. Letters* **7,** 248-249 (1965).
18. J.M. Burch, *Production Engineer* **44,** 431-432 (1965).
19. R.J. Collier, E.T. Doherty, and K.S. Pennington, *Appl. Phys. Letters* **7,** pp. 223-225 (1965).
20. R. Powell and K. Stetson, *J. Opt. Soc. Am.* **55,** 1570 (1965).
21. K.A. Stetson and R.L. Powell, *J. Opt. Soc. Am.,* **56,** 1161-1166 (1966).

VI. THEORETICAL AND EXPERIMENTAL FOUNDATIONS OF OPTICAL HOLOGRAPHY (WAVEFRONT-RECONSTRUCTION IMAGING)[†]

Recent developments have generated a new interest in extensions of the wavefront-reconstruction imaging method first described by Gabor in 1948.[1-5]

Some qualitative aspects of the two-step holographic imaging principle were introduced in Chapter I for introductory reasons, and several other applications of holography were incorporated in Chapter V because of the important role that holography has come to play in optical image-filtering and optical image-synthesis systems.

In this chapter we deal more particularly with the image-forming principles of holography, and present several new advances (theoretical and experimental) which we have recently made in view of realizing image-forming systems, and of attaining the highest possible resolution in the domains of the electromagnetic spectrum (notably at x-ray wavelengths), where no image-forming systems other than those based on holography now seem to hold a promise of success. In a general way, we may say that the principles of holography which we are introducing here are basic to all systems where holographic imaging and holographic image processing methods are involved. For example, it now appears possible to actually generate a three-dimensional image from a set of computed image coordinates, by generating the hologram, capable of displaying the desired image, from the set of computed image coordinates!

A great deal of work has been devoted to wavefront reconstruction since 1948. Much progress has been made in developing methods for superposing the coherent background on the field scattered by the object, and for obtaining well-separated reconstructed wavefronts and images. By and large, the optimism and foresight expressed by Gabor in his 33-page paper in the 1949 *Proceedings of the Royal Society* have appeared to be justified in the 15-odd years which have followed his work. It now appears that some of the principles required for new extensions of wavefront-reconstruction methods may require additional clarification.

In this chapter we consider the theoretical and experimental principles of wavefront-reconstruction imaging, as they now appear, in the light of

[†]This chapter is based in some part on material presented on Nov. 9, 1964, in Boston at the Symposium on Optical Information Processing. See also the chapter Theoretical and Experimental Foundations of Optical Holography (Wavefront-Reconstruction Imaging) by G. W. Stroke in "Optical Information Processing" (J. T. Tippett, L. C. Clapp, D. Berkowitz, and C. J. Koester, eds.), M.I.T. Press, Cambridge, Massachusetts, 1965.

new theoretical and experimental evidence. Four recent advances, two experimental and two theoretical, may be singled out particularly.

(1) The reconstruction of wavefronts scattered from three-dimensional macroscopic scenes illuminated with 6328-Å laser light.

(2) The attainment by lensless wavefront reconstruction of greatly magnified ($\simeq$ 150 X) microphotographs of biological samples, illuminated in 6328-Å laser light.

(3) The new theoretical evidence, which we indicate, that considerably greater resolution can be obtained with x-ray holograms then had in the past been considered possible. Real rather than empty magnifications on the order of 1 million and more appear attainable, and should permit one to obtain the highly resolved x-ray pictures which have been sought.

(4) The simple interpretation of the spatial and temporal coherence requirements, which led to the three-dimensional laser holograms, and which is necessary for the extensions of the method to such problems as x-ray microscopy.

It might be in order to recognize Gabor's unique role in introducing a new method of image formation in optics.[1-5] In analogy with photography, where lenses are used to form images, the author has suggested *holography* as the description for a process where holograms are used as aids of image formation.

1. Background and Experimental Foundations

The term *wavefront reconstruction* refers to a process in which the amplitude and phase of a scattered electromagnetic wavefront is recorded (usually photographically) together with a suitable coherent background in such a way that it is possible to produce at a latter time a reproduction of the electromagnetic field distribution of the original wavefront. The coherent background is necessary for the recording of the negative and complex values of the electric field distribution on the hologram. The various wavefront recording methods differ by the manner in which the coherent background is provided, although the general idea of introducing a coherent background may be shown to be directly related to the methods introduced by Zernike in 1934 with specific application to phase-contrast microscopy.[7-10]

Much similarity can be found between the manner in which the phase in the scattered field is recorded in a hologram (Figs. 1–3), on one hand, and the manner in which the phase is being recorded in an ordinary two-beam interferogram, on the other (Fig. 6). This analogy is almost complete in the method which we use for illustration.

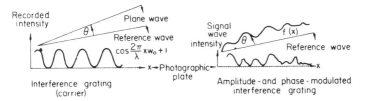

FIG. 1. Amplitude and phase-modulated interference grating.

In fact, it can be readily shown[11] (see Figs. 4 and 6 of Chapter I.) that an interference grating is formed on the photographic plate, both in the case of a two-beam interferogram (Fig. 6), and in the case of a hologram (Fig. 3), where the background or reference wave is made to fall at a suitable angle on the plate with respect to the scattered wave. In the case of a plane scattered wave (such as that reflected by a mirror) and a plane background wave (Fig. 2), the hologram or interferogram is simply formed of a grating with sinusoidally varying, spatial straight-line interference fringes.

When this grating, forming the hologram, is illuminated by a plane wave (Fig. 4), it will produce by diffraction a set of plane diffracted waves, which are readily seen to be the reconstructions of the original plane scattered wave.

In the case of a scattered wave containing both amplitude and phase variations, the fringes in the hologram will still form a grating in the general sense (Fig. 1). The fringes in the grating will be suitably modulated in position and in intensity, according to the distribution of the electromagnetic field in the scattered wave near the photographic plate. When the modulated interference-fringe grating is now illuminated by a plane wave, it will reproduce (see Section 2) in two *distinct* sets of diffracted waves, precisely the phase and amplitude modulations which were

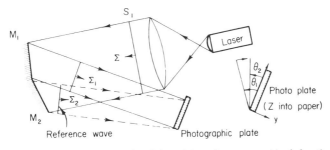

FIG. 2. Recording of hologram (modulated interference grating) for the case of a plane-wave generating object (M_1).

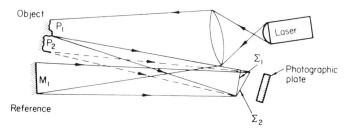

FIG. 3.   Recording of hologram of three-dimensional object, after Stroke.[11]

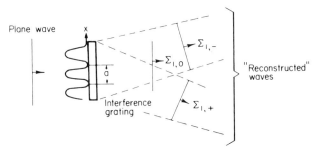

FIG. 4.   Reconstruction of plane waves from hologram of Fig. 2.

present in the original scattered wave at the plate. An observer "look-ing" at one of the diffracted waves would "see" the same object which he would have seen by looking at the waves scattered by the original ob-ject. The other wave has the property of actually forming a real image of the object without the aid of any auxiliary lenses.

Interference and diffraction principles of gratings are not only basic but sufficient to explain the physical aspects of wavefront reconstruc-tion.[11,12] It may frequently appear convenient to visualize the gen-eralized hologram as modulated interference gratings, or diffractograms.

The Fresnel-zone plate interpretation of some aspects of holography was already given by Gabor,[1,2] and later amplified by several authors (e.g., Rogers[13] and El-Sum[14]).

For the purposes of clarity we shall briefly review those elements in the theory which are required for the further development of the theory we present.

2. Theoretical Foundations

Strictly speaking, rigorous electromagnetic theory of scattering, dif-fraction, and polarization is required for an exact treatment of holography. Under the conditions indicated in Section 2 of Chapter II (see also Ref. 12) the physical optics approximations used in this chapter are generally found to be sufficient.

Let Σ_2 be a wavefront, such as the one scattered by an illuminated object (Fig. 3). Then the complex amplitude of an electromagnetic wavefront may be decomposed into two parts: its magnitude $A(x)$ and its phase $\phi(x)$, each of which is essential to the structure of the wavefront. To be able to reconstruct this wavefront at a later time, care must be taken to lose neither the magnitude nor the phase of the scattered amplitude during the recording process. The magnitude, or rather some power of it, can be recorded by simply photographing the wavefront; however, the phase is invariably lost in such a process, since photographic emulsions are sensitive only to the absolute value of the scattered amplitude.

We may note here, parenthetically, that the difficulty in reconstructing images from ordinary x-ray diffraction patterns results precisely from this irretrievable loss of phase information. Fortunately, holographic recording of the diffracted (i.e., scattered) field does provide a means for a retrievable recording of the phase as well as of the magnitude of the scattered field, both of which are necessary for a reconstruction of the image. Recording of the phase in a scattered field may perhaps appear somewhat less surprising, at first sight, if we recall the many well-known interferometric methods used to record the phase in optical wavefronts.

Perhaps the best-known interferometric methods for recording the phase in a wavefront are the methods of *two-beam interferometry*. For instance, it is possible to photograph the phase distribution in a wavefront diffracted by a ruled optical grating (Fig. 5) by means of the interferometer shown in (Fig. 6). These figures illustrate clearly the place that both grating diffraction and interferometric recording of diffracted wavefronts hold in holography. The spatial displacements of the interference fringes in the interferogram are linearly related to the phase distribution in the diffracted wavefront, where the interval from one fringe to the next corresponds to a phase difference of 2π. Using this interferogram it is possible to reconstruct the diffraction pattern, either empirically or by Fourier-transform computation (Fig. 7). Many other two-beam interferometer systems are known to permit similar recordings of the phase in a wavefront. It may suffice to recall the Lloyd mirror, the Fizeau interferometer, the Michelson-Twyman-Green interferometer, and so on. Heterodyning methods, using lasers, also have many points in common with two-beam interferometry.[11]

An important method of recording the amplitude and the phase distribution in scattered wavefronts was introduced by Zernike in 1934 in connection with microscopy.[7-10] Zernike's method of phase-contrast mi-

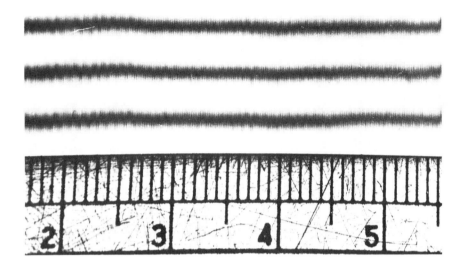

FIG. 5. Interferogram of wavefront diffracted by optical grating, after
Stroke.[15] The interferogram is obtained by photographing the complex amplitude
distribution in the wavefront diffracted by the grating, in the interferometer shown
in Fig. 6. To photograph the wavefront a wedge is formed between the wavefront
diffracted by the grating and the wavefront reflected from the reference mirror,
very much as in the recording of a hologram (see Figs. 2 and 3).

croscopy is based on bringing a suitably phased and attenuated background to interfere with the wavefront scattered by the object, the interference taking place before the recording. The coherent background
originates in the object itself and is, in fact, nothing but the undiffracted
portion of the field scattered by the object (Fig. 8). The coherent background is superposed on the widely scattered field, produced by small-
sized object regions in the object. The superposition as it occurs in the
Fourier space before the second imaging by the lens L_2 is shown in
Fig. 8. The entire process then amounts to a one-step imaging process.
The role of the complex filter in the Fourier space is to suitably shift
the phase and to attenuate the generally strong coherent background
(concentrated near the axis) with respect to the field scattered by the
small object regions under study. Principles similar to those illustrated
in Fig. 8 are basic to the methods of spatial filtering (originated by
Maréchal in 1953,[21] and developed by several authors[22,23]), which we
have described in Chapter V. In a general sense, the two-step hologram
imaging process introduced by Gabor was already noted by him to have
some significant basic similarities with Zernike's phase-contrast mi-

croscopy.[2] As in Zernike's method, the coherent background is introduced by means of scattering from the object itself. The important difference between the two methods results from Gabor's prediction, and successful demonstration that the hologram obtained by interference between the background and the scattered light can in fact be first photographed and subsequently used to form an image by a second diffraction, rather than proceeding directly from the filtered diffraction pattern to the image, as in the method by Zernike.

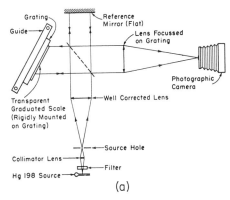

(a)

(b)

FIG. 6 (a) Interferometric arrangement used to photograph complex amplitude distribution in wavefronts diffracted by optical gratings, after Stroke.[15] Note the similarity between this interferometer and the arrangements used to record holograms (Leith and Upatnieks[16,17] and Stroke[11]). (b) Interferometer used to photograph complex amplitude distribution in wavefronts diffracted by optical gratings, after Stroke.[15]

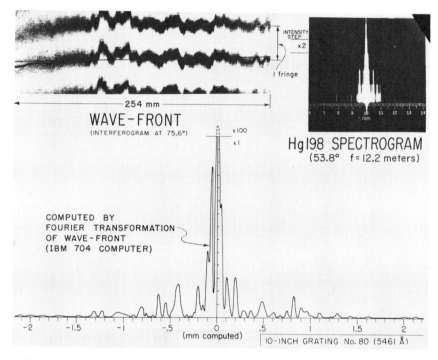

FIG. 7. Fourier-transform computation of point-source image (diffraction pattern) corresponding to diffracted-wavefront interferogram of a 300 groove/mm
grating, after Stroke.[18,19] The spectrogram of the green line (5461 Å) of Hg-198
is shown. Note the similarity of relation between diffraction pattern (spectrum)
and wavefront with optical data processing of data recorded on film, for instance,
in diffractive processing of geophysical data (P. Jackson[20]).

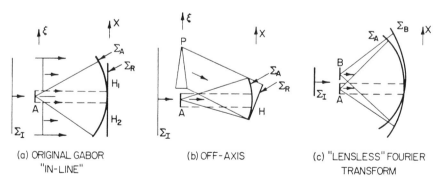

FIG. 7A. The three basic arrangements used for the recording of transmission holograms. Note that the off-axis hologram is, in effect, one-half
of an in-line hologram.

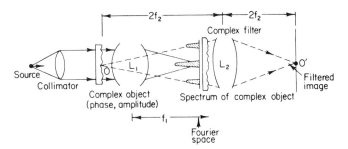

FIG. 8. Arrangement used in phase-contrast imaging, after Zernike,[7-10] and in spatial filtering, after Maréchal and Croce.[21]

As we show in Section 2.2, *two* images are formed by reconstruction from the holograms, in all types of holographic imaging. In contrast to a general belief (e.g., Ref. 85), Stroke (with students)[86] has recently demonstrated that perfectly separated images may indeed be obtained in the original "in-line" Gabor recording scheme, notably in high-resolution imaging and with diffused illumination. Experimental convenience, rather than any basic considerations, calls for an "off-axis" (offset) holography arrangement in the general case. The in-line introduction of the coherent background in Gabor's original arrangement had at first been considered to be a basic limitation in some applications of holography, in that it was assumed to result in an unseparable superposition of the two "twin" (real and virtual) images. Some recent descriptions[85] of the off-axis arrangements (e.g., Fig. 9) have seemed to imply that their introduction was an essential improvement, required *sine qua non* for complete phase recording in holography and separation of the reconstructed images. However, Stroke *et al.*[86] showed that an asymmetrical off-axis hologram (such as that of Fig. 9) is simply one half of a corresponding symmetrical in-line Gabor hologram, recorded by making the reference beam incident normally onto the photographic plate after passing by the two sides of the object. The offset method, as first suggested by Lohmann,[24] and described by Cutrona *et al.*[25] requires the introduction of a plane (or spherical) reference beam, produced by bending a portion of the incident beam with the help of a prism,[16] or, as first suggested by Stroke,[11] by reflecting the reference beam from a mirror. At x-ray wavelengths, where no suitable prisms or mirrors are available to produce a suitable offset reference beam (or suitably small offset point-reference source), the original methods appeared to present insurmountable difficulties, until it was first suggested by Stroke and Falconer,[26] in late 1964, that *Fourier-transform holograms*, rather than the previously used Fresnel-transform holograms, were needed for attaining high resolutions in wavefront-reconstruction imaging (see Section 3). However, in the past (see Section 7 of Chapter V), Fourier-transform holograms had only appeared attainable in the focal plane of focusing lens or mirror systems, by Fourier transformation of the scattered object wavefront, and by interference with an offset plane wavefront. The need for a focusing system to obtain the desired Fourier-transform holograms, necessary for high-resolution holography, again appeared to present an insurmountable obstacle, until Stroke[28] showed in early 1965 that it was possible to obtain *lensless Fourier-transform holograms* without the intervention of any focusing lens or mirror system, between the object and the hologram (see Section 3)! Gains in hologram recordability in excess of 10^3 result from the use of lensless Fourier-transform holograms, when compared to Fresnel-transform holograms, at the limit of resolution, in 1-$\overset{\circ}{\mathrm{A}}$ x-ray applications (see Section 3). However, even this considerable gain in recordability (and the corresponding intrinsic gain in resolution) again appeared unattainable, because of the need for having a point-reference source of the same dimensions as the desired resolution (1 $\overset{\circ}{\mathrm{A}}$ at x-rays), until Stroke *et al.*[30] finally showed, in 1965, that the "spreading" (or smearing) of every point image, which would result from the use of an extended (rather than a point) source in the recording of the hologram, could be compensated, and the resolution retrieved, by reconstructing the image from the hologram in a correlative compensating arrangement, using a suitably structured *resolution-retrieving reconstruction source* (see Section 3).[30]

We shall now describe the principles of hologram recording and image reconstruction, which are common to all holographic imaging systems.

2.1. The Recording Process

The magnitude and the phase of a scattered wavefront can be recorded photographically by superposing a coherent reference beam or background wave on the field striking the photographic plate. Perhaps the simplest technique for carrying out this superposition is the one illustrated in Figs. 3 and 9, wherein a plane wave illuminates a region containing scattering object and a plane mirror[11] or simple triangular prism,[16] respectively. The object, of course, diffracts the incident radiation to generate a field with some magnitude $A(x)$ and some phase $\phi(x)$ at the photographic plate, while the prism simply turns the incident plane wave through a small angle θ to contribute a field with a uniform magnitude A_0 and a linear phase variation αx, where α is a constant relating the angle θ and the wavelength λ according to the equation

$$\alpha\lambda = 2\pi\theta \tag{1}$$

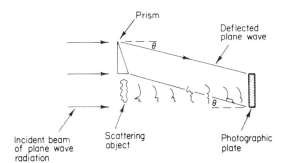

FIG. 9. Schematic arrangement to illustrate recording of hologram (see also Figs. 2, and 3.

valid for small angles θ (otherwise, use $\sin \theta$ in place of θ). Thus the total amplitude striking the plate is

$$A_0 e^{-i\alpha x} + A(x) e^{i\phi(x)} \tag{2}$$

Hence, the intensity, i.e., the quantity to which the emulsion is sensitive, is

$$I(x) = A_0^2 + A(x)^2 + 2A_0A(x) \cos [\alpha x + \phi(x)] \tag{3}$$

It will be noted that the phase $\phi(x)$ of the scattered wavefront has not been lost in computing the intensity, as it would be if the reference beam were not present.

The emulsion, of course, records some power of the intensity; that is, the *amplitude* transmittance $T(x)$ of the resulting photographic plate, provided that one works in the linear range of the H-D curve, is proportional to

$$
\begin{aligned}
T(x) &\propto [I(x)]^{-\gamma/2} \\
&= [A_0^2 + A(x)^2 + 2A_0A(x) \cos [\alpha x + \phi(x)]]^{-\gamma/2} \\
&\simeq A_0^{-\gamma-2} [A_0^2 - \tfrac{1}{2}\gamma A(x)^2 - \gamma A_0 A(x) \cos (\alpha x + \phi(x))] \\
&\propto 2A_0^2 - \gamma A(x)^2 - 2\gamma A_0 A(x) \cos [\alpha x + \phi(x)] \\
&= 2A_0^2 - \gamma A(x)^2 - \gamma A_0 A(x) e^{i\phi(x)+i\alpha x} - \gamma A_0 A(x) e^{-i\phi(x)-i\alpha x} \tag{4}
\end{aligned}
$$

where γ is the slope of the H-D curve. It has been assumed that the intensity of the reference beam greatly exceeds that of the radiation scattered by the object, so that the approximation made in dropping the higher-orders terms of the binomial expansion is justified. The photograph described by (4) is called a *hologram* after Gabor. [1,2]

There are two aspects of (4) that should be pointed out. The first involves the role of γ: Contrary to the requirements of many similar pro-

cesses, neither the sign nor the exact magnitude of y is of any consequence in the recording process; that is, making a contact print of the
hologram, which is equivalent to changing the sign of y, serves only to
shift the phase of the nonconstant portion of the transmittance of an in-
consequential $180°$, while changing slightly the magnitude of y serves
only to enhance or to suppress the magnitude of this same portion of the
transmittance. The second facet involves the relationship between $A(x)$
and $\phi(x)$: It will be noted that the magnitude $A(x)$ and the phase $\phi(x)$ of
the scattered wave appear in the natural way, i.e., as $A(x)\exp[i\phi(x)]$, in
the third term of (4), and with the sign of the phase reversed in the fourth
term. Cathey[87] and Pennington[88] recently noted that the thickness variations in the emulsion (proportional under suitable circumstances[89] to the
hologram exposure) may be used to produce *phase holograms* (upon bleaching of the emulsion), so that the complex amplitude transmission of the
hologram becomes

$$T(x) = e^{iI(x)} \cong 1 + iI(x) \tag{4a}$$

for small-enough values of the exposure.

2.2. The Reconstruction Process

With the hologram of (4) the reconstruction process is simple, involving no lens systems, schlieren disks, half-plane filters, or the like. In
fact, to reconstruct the original wavefront it is only necessary to illuminate the hologram with a plane wave of radiation, as illustrated in Fig.
10. As the plane wave passes through the photographic plate, it is multiplied by the transmittance $T(x)$, thereby producing four distinct components of radiation corresponding to four terms of (4). The first term,
being a constant, attenuates the parallel beam uniformly, but otherwise
does not alter it. The second term also attenuates the beam, but not uni-

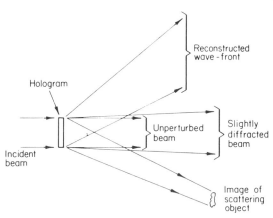

FIG. 10. Wavefront reconstruction and image formation from a hologram in the
case of plane-wave illumination.

formly, so that the plane wave suffers some diffraction as it passes
through the photograph.

The patterns produced by the third and fourth terms are more complicated. To understand how they affect the incident plane wave, it is
necessary to recall that a common triangular prism shifts the phase of an
incident ray by an amount proportional to its thickness at the point of incidence (Fig. 11), a positive phase shift deflecting the ray upward and a
negative one deflecting the ray downward. Thus the third term of (4) may
be interpreted as the product of the amplitude of the scattered wavefront
and a positive prismatic phase shift; similarly, the fourth term of (4) may
be viewed as a composite of the complex conjugate of the amplitude of
the original wavefront and a negative prismatic phase shift.

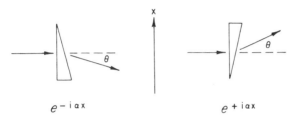

$$e^{-i\alpha x} \qquad\qquad e^{+i\alpha x}$$

FIG. 11. Phase shifts and deflection angles corresponding to prism terms in
Eq. (4). Note $\alpha\lambda = 2\pi\theta$.

By virtue of these prismatic phase shifts the third and fourth terms of
(4) deflect the incident beam upward and downward, respectively, through
an angle θ, as defined by (1). Furthermore, in the case of the third term,
the upward deflected beam is also multiplied by the scattered amplitude
$A(x) \exp[i\phi(x)]$, and hence reconstructs a copy of this wavefront. On
the other hand, the fourth term multiplies the downward beam by the complex conjugate of the scattered amplitude, and hence constructs a copy
of the scattered wavefront except that it travels backward in time and
consequently constructs a three-dimensional image of the scattering object. (The physical principles underlying this process are explained in
Section 2.3.)

2.3. Physics of the Method

The physical principles of the process described above can be illustrated by tracing the history of a vanishing small object through the
recording and reconstruction process. This method was originally introduced by Gabor[2] and later clarified by Rogers[13, 31–33] and El-Sum.[14]

Our approach is similar to theirs in that we suppose that the object
used in the recording process is an opaque plate with a very small hole
in it. When this aperture is illuminated with a plane wave it will act as
a simple spherical radiator according to Huygens' principle. Thus the

amplitude striking the photographic plate will be of the form

$$A_0 e^{-i\alpha x} + A \exp[i(\pi/\lambda f) x^2] \tag{5}$$

where A is some constant, λ is the wavelength of the radiation, and f is defined in Fig. 12. Hence according to (4) the transmittance of the hologram corresponding to this elementary source will be of the form

$$\begin{aligned} T(x) \propto \ & 2A_0{}^2 - \gamma A^2 \\ & - \gamma A_0 A \exp[i\,(\pi/\lambda f)\,x^2 + i\,\alpha\,x] \\ & - \gamma A_0 A \exp[-i(\pi/\lambda f)\,x^2 - i\alpha x] \end{aligned} \tag{6}$$

The relative simplicity of this formula together with that of the diffracting object permits one to understand the mechanism of the recon-

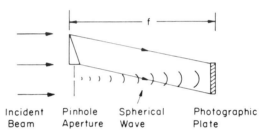

Incident   Pinhole   Spherical   Photographic
Beam   Aperture   Wave   Plate

FIG. 12. Hologram of point object (pinhole aperture). Arrangement used also in discussion of magnification of Eq. (6) and in the discussion regarding Figs. 13 and 14.

struction process in the following way. When the hologram described by (6) is placed in a parallel beam, three distinct components of radiation are generated, as shown in Fig. 13. The first component arises from the first two terms of (6), which, being constants, uniformly attenuate the incident waves producing another parallel beam to the right of the hologram. The third and fourth terms produce two additional components by deflecting the incident waves upward and downward, respectively, by virtue of the linear phase shift in their exponents.

To understand how the quadratic phase shifts in the terms $\{-\gamma A_0 A \exp[i\,(\pi/\lambda f)\,x^2 + i\alpha x]\}$ and $\{-\gamma A_0 A \exp[-i\,(\pi/\lambda f)\,x^2 - i\alpha x]\}$ act on the incident radiation, one simply recalls that a thin spherical lens shifts the phase of an impinging ray by an amount proportional to the square of the distance between the optic axis and the point of incidence, a positive phase shift describing the action of a negative lens (Fig. 14). Thus the third term of (6) acts on the incident radiation not only as an upward deflecting prism, but also as a negative lens; i.e., an incident plane wave

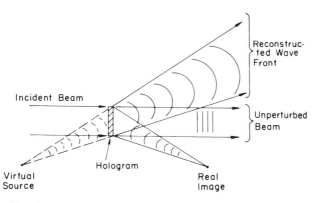

FIG. 13. Wavefront reconstruction and image formation from a hologram of a point scatterer (see Fig. 12).

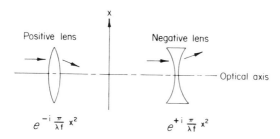

FIG. 14. Lenslike effects and phaseshifts associated with terms in Eq. (6).

will be deflected upward and given a convex spherical wavefront, and this spherical wavefront will in fact be identical to the one which exposed the hologram. Similarly, the fourth term of (6) acts on the incident radiation not only as a downward deflecting prism, but also as a positive lens on, i.e., an incident beam will be deflected downward and be given a concave spherical wavefront, and this spherical wavefront will in the normal way come to focus at a distance f from the hologram. Thus in illuminating the hologram with a *plane wave*, one not only reconstructs the scattered wavefront but also obtains "focused" images of the object, which in this case was a point source. The image formation with *point-source* illumination of holograms, and the associated very considerable magnification characterics inherent in holography, are discussed in Section 2.4.

2.4. Magnification

A surprisingly large magnification is attainable with wavefront-reconstruction systems, especially if one uses a longer wavelength radiation in the reconstruction process than in the recording process. To obtain a formula for the degree of magnification we suppose that the object is again an opaque plate, but now with two identical and vanishingly small holes in it, which are separated by a distance 2δ. Then, since each of the holes will act as a simple spherical radiator according to the Huygens' principle, the amplitude of the wavefront at the photographic plate will be

$$A_0 \exp(-i\alpha x) + A \exp[i(\pi/\lambda f)(x - \delta)^2] + A \exp[i(\pi/\lambda f)(x + \delta)^2] \quad (7)$$

Hence, according to (4), the corresponding hologram will have a transmittance of the form

$$T(x) \propto 2A_0^2 - 2\gamma A^2 \left[1 + \cos\left(\frac{4\pi}{\lambda f}\delta x\right)\right]$$

$$-\gamma A_0 A \left\{\exp\left[i\frac{\pi}{\lambda f}(x - \delta)^2\right] + \exp\left[i\frac{\pi}{\lambda f}(x + \delta)^2\right]\right\}\exp(i\alpha x)$$

$$-\gamma A_0 A \left\{\exp\left[-i\frac{\pi}{\lambda f}(x - \delta)^2\right] + \exp\left[-i\frac{\pi}{\lambda f}(x + \delta)^2\right]\right\}\exp(-i\alpha x)$$

$$(8)$$

At this point we depart[6] from the usual method of reconstruction and use instead the system shown in Fig. 15.[†] Here a *point source* of wavelength λ' is used to illuminate the hologram, rather than a plane wave of

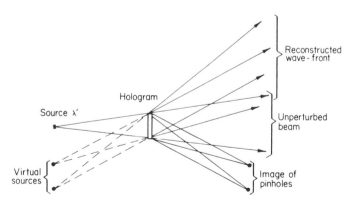

FIG. 15. Wavefront reconstruction and image formation with point-source illumination of hologram, in the case of an object formed by two point scatterers (see also Fig. 16).

[†] For a more extensive treatment of Fresnel holography based on the general principles given here see, for example, Armstrong[34] and H. H. Stroke.[35]

wavelength λ. The fourth term of (8) still focuses the radiation from the point source according to the lens plus prism interpretation introduced in Section 2.3. However, since we have changed the wavelength used in the reconstruction, the focal length of the lens will no longer be f but f', where

$$\lambda' f' = \lambda f \tag{9}$$

Of course (see Fig. 16), the object distance p will be related to the image distance q according to the classical formula, namely,

$$\frac{1}{f'} = \frac{1}{p} + \frac{1}{q} = \frac{\lambda'}{\lambda f} \tag{10}$$

The linear magnification M attained with this method of reconstruction can be seen from Fig. 16 to be

$$M = \frac{2\Delta}{2\delta} \tag{11}$$

To reduce this formula to one involving known parameters we observe from (Fig. 16) that

$$\frac{2\Delta}{2\delta} = \frac{p+q}{p} \tag{12}$$

because of the similar triangles involved. (The prism deflects the two rays through an equal angle and thus does not affect the similar-triangles argument.) Hence, we immediately obtain the following formula for the linear magnification:

$$M = \frac{\lambda'}{\lambda} \frac{q}{f} \tag{13}$$

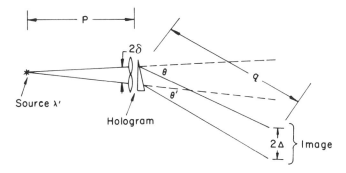

FIG. 16. Magnification property inherent in holograms. In comparing the geometry shown with Fig. 15, it should be noted that the hologram of Fig. 15 has been replaced by its equivalent lens-prism arrangement. (The central and upper beam components have been omitted for clarity.)

In this equation, f is the distance of the original object from the holo-gram (Fig. 12), when photographed in the wavelength λ, and q is the dis-tance of the hologram from the final image plane (Fig. 16), when the wavelength used in the reconstruction is λ'. The magnification M could exceed 10^6 in applications to x-ray microscopy.[†]

2.5. Resolution

It is well known that magnification alone is "empty" unless it is ac-companied by a corresponding degree of resolution. As several writers, notably Baez[36,37] and El-Sum,[14] have pointed out, a resolution capabil-ity of conventional Fresnel-transform holography[1-5,13,14,16,17,31-33,36,43] by (1) to that of the photographic plate used to make the hologram, and by (2) the diameter of the source used in the recording process. Since holographic recording systems are fundamentally interferometers, the source diameter and emulsion limitations enter into holography as they do into interferometry in general (see Ref. 15 and Fig. 6).

Inasmuch as a hologram is basically an interferogram, its recordability will be determined by two conditions, other than the monochromaticity of the source: (1) the angles between the reference wave and the scattered-field wave, and (2) the dimensions and structure of the source aperture used to generate the plane or spherical coherent-background reference wave. The important gains to be achieved by improvements of these con-

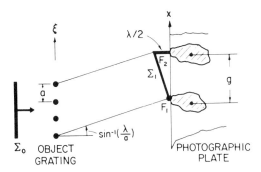

FOR RECORDING OF FRINGES (F) NEED:

$$\frac{\lambda}{a} \cdot g \lesssim \frac{\lambda}{2}$$

i.e. $\boxed{a \geq 2g}$

FIG. 17. Resolution limitation, caused by photographic emulsion graininess, in conventional Fresnel-transform holography.

[†]In Fourier-transform holography (Section 3) an object point P (coordinate ξ) produces in the hologram a "grating" of spacing constant $a_P = \lambda f/\xi$. In the re-construction (Fourier-transforming lens with focal length f'), the "grating" pro-duces an image point P' at a distance $\xi' = \lambda' f'/a_P = (\lambda' f'/\lambda f)\,\xi$ from the axis, corresponding to the magnification $(\lambda' f'/\lambda f)$.

ditions are dealt with in Section 3. Here we deal more particularly with the classical resolution limitations, which are characteristic of Fresnel-transform holography.

The classical resolution limitations [36,37,40] in projection holography can be shown to result from the nonrecordability of the interference fringes, with available photographic emulsions and source sizes. For instance (see Fig. 17), an emulsion with a $\frac{1}{2}$ micron resolution will only resolve about 10,000 Å at a 1-Å x-ray wavelength (if a minimum of one interference fringe is recorded in the $\frac{1}{2}$ micron distance, assuming plane-wave illumination).

We may obtain the values for the classical resolution limitations in Fresnel-transform holography in a somewhat more formal way. Let $I(x)$ be the intensity which is recorded by the photographic plate, with a plane reference wavefront, when recording the hologram of a single point:

$$I(x) = A_0{}^2 + A^2 + 2A_0A \cos \left(\alpha x + \frac{\pi}{\lambda f} x^2 \right) \tag{14}$$

Now the frequency ν of the term of interest, i.e., the third term, is a function of x, since by definition

$$\nu(x) = \frac{1}{2\pi} \frac{d}{dx} \left(\alpha x + \frac{\pi x^2}{\lambda f} \right)$$

$$= \frac{\alpha}{2\pi} + \frac{x}{\lambda f} \tag{15}$$

Thus if the emulsion used in the recording process has a resolution of N lines per unit length, the only frequencies which will register on the plate will be those which satisfy

$$|\nu(x)| = \left| \frac{\alpha}{2\pi} + \frac{x}{\lambda f} \right| \leq N \tag{16}$$

In other words, an oscillating pattern will be recorded only if x falls in the range defined by

$$-N_0 - \frac{\alpha}{2\pi} \leq \frac{x}{\lambda f} \leq N_0 - \frac{\alpha}{2\pi} \tag{17}$$

where the length $2x_{max}$ of this range is clearly equal to the x dimension of the hologram. Accordingly, we have

$$2x_{max} = 2N\lambda f \tag{18}$$

Physically, the finiteness of this range means that the positive lens which brings the incident plane wave to focus in Fig. 13 has a finite width $2x_{max}$. Moreover, as is well known from classical diffraction theory [Eq. (V.8)], the Rayleigh limit of resolution, corresponding to a lens of width $2x_{max}$, when illuminated by a plane wave, is

$$\varepsilon = \frac{\lambda}{2x_{max}} f \tag{19}$$

Thus in light of (18) we see that the classical Fresnel-transform holography limit of resolution is

$$\varepsilon \cong \frac{1}{2N} \tag{20}$$

in agreement with the qualitative estimate obtained with the help of Fig. 17. Because of the approximations involved, (20) remains valid, also for the case when an offset spherical wavefront is used as a reference beam, and where the source diameter is of the same order as the resolution limit $(1/N)$ of the emulsion (which is the case with presently available pinhole sources and photographic emulsions). A more rigorous discussion of the effect of the sources (both that used in the recording, and the one used in the reconstruction), as well as means for compensating for the resolution loss, which would result from the use of sources with an extended dimension (as compared to the desired resolution) is given in Section 4.

The ultimate resolution of the conventional Fresnel-transform projection wavefront-reconstruction technique is seen to be approximately one-half that of the recording media. Since the best emulsions, e.g., Kodak spectroscopic plate 649 F, have a resolution of about $\frac{1}{2}$ micron, the conventional Fresnel-transform projection wavefront-reconstruction technique is limited to resolutions to the order of 1 micron independently of the wavelength used in the recording process. However, as we show in the next section our recent theoretical investigations[26-30] have revealed that considerably higher real resolutions may be attained by using the method of lensless Fourier-transform holography[28] first described by us in 1965, together with the resolution-retrieving compensation of source effects, by a correlative reconstruction, which we also first described[30] in 1965.

3. High-Resolution Fourier-Transform Holography

3.1. Introduction

For attaining high resolutions in holography, it is necessary to fulfill not only the wide-angle diffraction conditions characteristic of optical systems in general [see Abbe and Rayleigh resolution conditions, Eqs. (V.7) and (V.8)], but it is also necessary that the waves, diffracted (i.e., scattered) by the object into the directions forming the highest angles with respect to the direction of illumination (and the object), be recordable by interference with a coherent-background reference wave (see Section II.5). According to the Abbe resolution criterion spatial resolutions in the object of the order of a grating constant a equal to the wavelength (i.e., $a = \lambda$) require a recording over an angular domain up to $i' =$

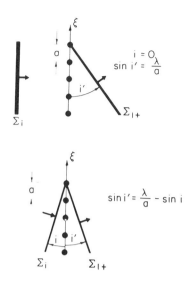

FIG. 18. Grating-diffraction parameters.

$\sin^{-1}(\lambda/a) = \pi/2$ (see Fig. 18). The size $2x_{max}$ of the hologram plate, needed to attain a given resolution in the object domain, is given by the same consideration [see (18)].

A very important advance in the principles of holography, particularly important for attainment of high resolutions, has resulted, when it was first shown (Stroke and Falconer[26] and Stroke[28]) that a considerable gain in the ultimate resolution attainable in holographic imaging would result from recording the holograms in a "Fourier-transforming" arrangement, rather than in the conventional Fresnel-transform projection hologram recording arrangement. In Fourier-transform holography, the real and the virtual images, reconstructed from the holograms, are formed at $+\infty$ and $-\infty$, respectively; the images are simply reconstructed by projecting a plane monochromatic wave through the Fourier-transform hologram, and by recording the images (obtained by a second Fourier transformation) in the focal plane of a lens (see Fig. 19).

The following analysis is based on the assumption that point sources are used, both to produce the coherent background in the recording of the hologram, and for illumination of the hologram in the reconstruction. This assumption is justified, in particular, because the effects produced by the use of sources having a finite extent can be compensated, according to the principle described by Stroke et al.[30] (see Section 4).

In the past, Fourier transformation, and indeed Fourier-transform recording of holograms, required the intervention of a Fourier-transforming

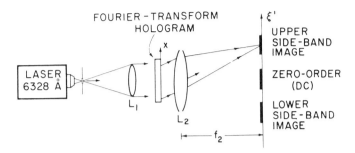

FIG. 19. Fourier-transform reconstruction of images from Fourier-transform hologram, after Stroke. [11]

lens (or other focusing system) *in the recording*; the Fourier transformation was obtained between the complex amplitude (say of the $\bar{E}$ field) in the pupil of the lens, on the one hand, and the complex amplitude distribution of the field in the focal plane of a lens or other focusing system (see Fig. 20a) on the other. In x-ray applications, the need for the intervention of any focusing lens or mirror, in the recording of the holograms, clearly vitiates any advantage of lensless photography (holography).

A second very important advance in the principles of holography resulted when it was shown (Stroke[28]) that it was possible to record the high-resolution Fourier-transform holograms without the intervention of

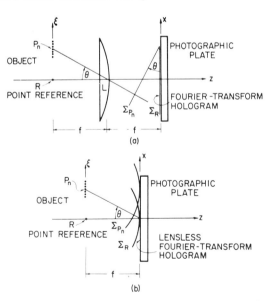

FIG. 20. (a) Fourier-transform hologram recording. (b) Lensless Fourier-transform hologram recording, after Stroke. [28]

any focusing lenses or mirrors, in what Stroke describes as a lensless-Fourier-transform-hologram recording arrangement (Fig. 20b). Higher resolutions, attainable by means of these arrangements, have now been obtained by Stroke and his associates in a number of applications, in particular with three-dimensional objects (Fig. 21).

The high-resolution advantage of Fourier-transform holography holds both when lenses are used (as in certain image filtering and synthesis applications), and of course in the lensless form. The advantages become particularly significant in the recording of the high-angle waves which characterize "high" spatial resolutions. Examples of Fourier-transform holographic imaging, carried out entirely on Polaroid P/N film, are shown in Fig. 22

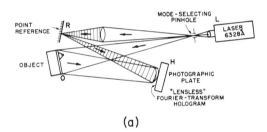

(a)

(b)

FIG. 21. (a) Recording to lensless Fourier-transform hologram of three-dimensional object, after Stroke *et al.* [29] (b) Fourier-transform reconstruction (see Fig. 19) from lensless Fourier-transform hologram, recorded according to Fig. 21a.

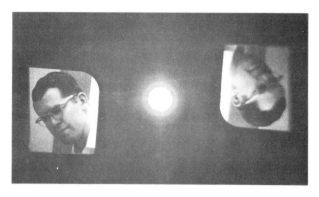

FIG. 22. Fourier-transform reconstruction from Fourier-transform hologram, recorded, on Polaroid P/N film, according to Fig. 20a.

3.2. Physical Principles of High-Resolution Lensless Fourier-Transform Holography

We have already noted that a hologram is basically an interferogram, formed by interference between the field scattered by the object, on the one hand, and a reference or background wave, on the other. Generally, the background (reference) wave is taken as an extended wave; it may be either plane or spherical, depending on the arrangement.

For the convenience of the analysis, the scattering object may be considered either as a generalized two- or three-dimensional grating, or as having been formed by an array of re-radiating point dipoles. In holography it is assumed that the various object points scatter coherently, in the sense of a temporal, steady-state coherence; each object point is considered as a steady-state oscillator, radiating at the same frequency as the frequency of the coherent-background reference wave. The complex amplitude of the field in the interferogram (i.e., hologram) is obtained by interference of the reference wave with the field scattered by the object. The resultant field can be computed according to the superposition principle as (1) the resultant sum of the scattered field and the reference field, as projected onto the hologram, or (2) the sum of the fields resulting from the interference of the background field with each of the object waves. We shall take the second point of view in the development which follows.

In the now conventional form of Fresnel-transform projection holography, the hologram is recorded when the spherical waves, scattered by the various object points, are made to interfere with either a plane reference wave, or a spherical reference wave, of a curvature different from the mean curvature of the object waves, so that a Fresnel-zone fringe system is recorded for each object point. The Fresnel-zone carrier fringes had generally been considered essential for a faithful recording of the *phase information* in the field scattered by the object.

An important advance in the principles of holography was achieved when it first appeared (see Stroke and Falconer[26] and Stroke[28]) that the phase information in the scattered field could also be recorded, essentially *without Fresnel-zone carrier fringes*, by making the various spherical waves, scattered from the various object points, interfere with a *spherical reference wave of the same curvature*, and centered in the near vicinity of the object (Fourier-transform hologram).

That it should be possible to record *phase* information in a hologram without the Fresnel-zone "carrier fringes" becomes immediately clear if we recall the analogy of holograms with interferograms. We further recall that we may distinguish two general groups of interferograms, used for the purpose of recording the phase in reflected, diffracted, or scattered wavefronts: (1) interferograms with *localized fringes* (such as Newton's rings or Fresnel-zone interferograms, and straight-line fringe interferograms, formed by interference between two plane waves), and (2) interferograms with *nonlocalized fringes* (fringes "at infinity") formed by interference between two plane waves, or indeed *two spherical waves of the same curvature*, having the same direction of propagation. In the case of two waves having the same curvature and the same direction of propagation, the phase information is uniquely translated in the interferogram simply by the local intensity across the wavefront, without the intervention of any interference fringes.

Even though we have recently noted[86] that the reference wave may indeed be centered on the object, the separation of the two side-band images (one at $+\infty$ and one at $-\infty$) is obtained, in most recent forms of holography, by using the now-conventional offset separation between the mean scattered field and the reference wave. It might be useful to note that the need for an offset separation between the center of the object and the point reference (Fig. 20b) arises as a natural requirement for the formation of a lensless Fourier-transform hologram; accordingly, separation of the two symmetric images is automatically achieved, with Fourier-transform holograms, in the Fourier-transform reconstruction.

3.3. Mathematical Analysis of Lensless Fourier-Transform Holography

Let $T(\xi)$ be the complex amplitude in the electric field scattered by the object, as expressed at the object. Let

$$A_r = A_0 \exp\left[i\,(k/2f)x^2\right] \tag{21}$$

be the complex amplitude produced by the spherical reference wave, cen-

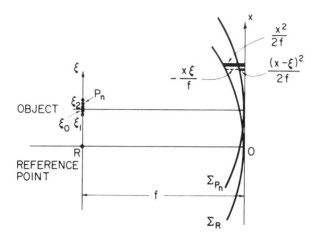

FIG. 23. Hologram-recording parameters.

tered at $\xi = 0$, as measured in the plane of the photographic plage (Fig. 23). Each spherical wave, originating from each of the object points, will produce, in the plane of the photographic plate, a field

$$T(\xi) \exp \left[i\,(k/2f)(x - \xi)^2\right]d\xi \tag{22}$$

The total field, produced by the object in the plane of the photographic plate is then clearly

$$A_s(x) = \int_{\xi_1}^{\xi_2} T(\xi) \exp \left[i\,(k/2f)(x - \xi)^2\right]d\xi \tag{23}$$

For the case of a fairly small object, and the recording of the hologram on a large plate, as required for high resolutions, one has (neglecting ξ^2),[†]

$$(x - \xi)^2 \cong x^2 - 2x\xi \tag{24}$$

and therefore (23) gives, with (24), the field

$$A_s(x) \cong \int_{\xi_1}^{\xi_2} T(\xi) \exp \left[i\,(k/2f)x^2\right] \exp \left[-i\,(k/f)x\xi\right] d\xi \tag{25}$$

The intensity recorded on the hologram is given by

$$I(x) = (A_r + A_s)(A_r + A_s)^* \tag{26}$$

that is,

$$I(x) = |A_r|^2 + |A_s|^2 + A_r^*A_s + A_rA_s^* \tag{27}$$

[†]Inclusion of the $\exp\left[i\,(k/2f)\xi^2\right]$ factor merely makes the reconstructed object appear as recorded through a thin negative "field" lens (see Fig. 14), for which the compensation was given by Stroke and Falconer.[26]

It follows from (21), (25), and (27) that the intensity recorded on the hologram is

$$I(x) = |A_0|^2 + |A_s|^2 + A_0 \int_{\xi_1}^{\xi_2} T(\xi) \exp\left[-i(k/f)x\xi\right] d\xi$$

$$+ A_0 \int_{\xi_1}^{\xi_2} T(\xi)^* \exp\left[+i(k/f)x\xi\right] d\xi \quad (28)$$

It is the intensities in the two integrals in the hologram recording of (28) which are recognized, respectively, as the Fourier transform of the complex amplitude $T(\xi)$ in the object, and of its conjugate $T(\xi)^*$.

We have shown [see (4)] that the complex amplitude $H(x)$ of the field transmitted through the photographic plate, when the hologram is illuminated by a plane wave, is given by the equation

$$H(x) \cong [I(x)]^{-\gamma/2} \quad (29)$$

when the hologram is recorded in the linear range of the H-D curve, where

$$\tan^{-1} \text{ (slope of H-D curve)} = \gamma \quad (30)$$

If, as usual, the reference-beam intensity is sufficiently large compared to the scattered field in the plane of the hologram, we obtain the hologram transmission $H(x)$ by a series expansion from (28). With (29) we obtain

$$H(x) \cong A_0^{-\gamma-2} \left[A_0^2 - \frac{\gamma}{2}|A_s|^2 - \frac{\gamma}{2} A_0 \int_{\xi_1}^{\xi_2} T(\xi) \exp\left[-i(k/f)x\xi\right] d\xi \right.$$

$$\left. - \frac{\gamma}{2} A_0 \int_{\xi_1}^{\xi_2} T(\xi)^* \exp\left[+i(k/f)x\xi\right] d\xi \right] \quad (31)$$

as the equation for the hologram transmission, under the stated conditions, when the hologram is illuminated by a plane wave (see Fig. 24).

The first two terms inside the brackets of (31) are constants, while the two terms with the integrals are again recognized as the Fourier transforms of the object and of its complex conjugate, respectively.

The reconstruction of the images $T(\xi)$ and $T(\xi)^*$ is clearly obtained by taking the Fourier transform of $H(x)$, for instance, by placing the hologram in a collimated beam (illumination by a plane wave Σ_i, Fig. 24) and photographing the images in the focal plane of a lens (L in Fig. 24). Two separated images are obtained off-axis, one centered at $+\xi_0'$ and one at $-\xi_0'$, because the original object was off-axis by the amount ξ_0, so as to obtain the Fourier-transform relation of (25) and (28). Indeed,

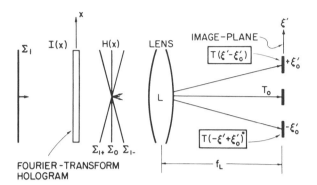

FIG. 24. Fourier-transform reconstruction from Fourier-transform hologram $H(x)$.

we recall that a shift ξ_0 in the object plane may be expressed as

$$T(\xi - \xi_0) \tag{32}$$

in the equation for the object. Accordingly, if we use the symbolism

$$T(\xi) \to t(x) \to T(-\xi) = T(\xi) \rightleftharpoons t(x) \tag{33}$$

with the arrow indicating "by Fourier transformation," then [see VII. 9a)]

$$T(\xi - \xi_0) \rightleftharpoons t(x) \exp(2\pi i \xi_0 x) \tag{34}$$

Similarly, since

$$T(\xi)^* \rightleftharpoons t(-x)^* \tag{35}$$

according to (7.1b), we also have

$$T(\xi - \xi_0)^* \rightleftharpoons t(-x)^* \exp(2\pi i \xi_0 x) \tag{36}$$

Finally, because of the minus and plus signs in the $\exp(-ikx\xi/f)$ and $\exp(+ikx\xi/f)$ factors in the Fourier transforms of (31), it follows from (34) and (36) that the images obtained by Fourier transformation of (31) consist two symmetrical images, with the complex amplitude

and

$$T(\xi' - \xi_0')$$
$$T(-\xi' + \xi_0')^* \tag{37}$$

respectively.

3.4. Comparison of Fourier-Transform and Fresnel-Transform Holography for High-Resolution Imaging

Fourier-transform holography, in particular its lensless form, becomes of particular interest, as noted, in the recording of holograms with high-angle diffracted waves, characteristic of high *spatial* resolutions in the object.

For a two-dimensional grating (object) the relation between the spacing constant a, and the angles of incidence i and of diffraction (i') is given by the grating equation [see (1.1)]

$$\sin i + \sin i' = \frac{m\lambda}{a} \tag{38}$$

For spatial resolutions on the order of λ, and for normal illumination $(i = 0)$, the angle of diffraction

$$i' = \sin^{-1}\left(\frac{\lambda}{a}\right) \tag{39}$$

becomes $\pi/2 = 90°$.

For wide-angle diffraction, the path difference Λ, in a Fresnel-transform hologram, cannot be (any more) obtained from the approximate expression, used in (24), with a plane reference wave. One has (see Fig. 25) rather,

$$\Lambda = f(1 - \cos i') \tag{40}$$

For the case of a lensless Fourier-transform hologram, on the other hand (see Fig. 26),

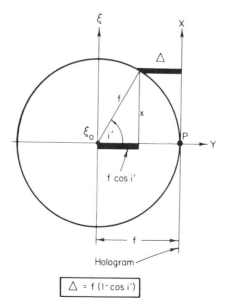

FIG. 25. Path difference in recording of Fresnel-transform hologram.

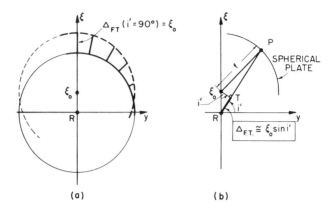

(a)   (b)

FIG. 26. Path difference in recording of lensless Fourier-transform hologram
on spherical plate, after Stroke.[28]

$$\Delta_{FT} = \xi_0 \sin i' \qquad (41)$$

where we note that (41) holds for all angles, when the Fourier-transform
hologram is recorded on a *spherical photographic plate* (see Fig. 26), and
for small angles, as well, when the Fourier-transform hologram is recorded on a flat plate (see Fig. 27).

In keeping with the analysis given in Eqs. (14)-(20), we may note
(following G. Toraldo di Francia[†]) that the quantity of interest for as-
sessing the resolution "gain" is not the path difference Δ, but rather
its derivative, which specifies the spacing of the "carrier" fringes.
Indeed the spacing p of the fringes is given by

$$p \, \frac{d\Delta}{dx} = \lambda \qquad (42)$$

In the case of the Fresnel-transform hologram, if we take the reference wavefront parallel to the hologram plate, the path difference,
apart from a constant, is

$$\Delta = \sqrt{f^2 + x^2} \qquad (40a)$$

and the fringe spacing is readily found to be

$$p_{\text{Fresnel}} = \lambda \sqrt{f^2 + x^2} \Big/ x = \lambda \Big/ \sin i' \qquad (42a)$$

For the case of the lensless Fourier-transform hologram, the path
difference is

$$\Lambda_{FT} = \sqrt{f^2 + x^2} - \sqrt{f^2 + (x - \xi_0)^2} \cong \xi_0 x \Big/ \sqrt{f^2 + x^2} \qquad (41a)$$

[†]Private communication of Professor G. Toraldo di Francia to G.W. Stroke
(March 28, 1966).

and the fringe spacing is found to be

$$p_{Fourier} = f\lambda/\xi_0\cos^3 i' \tag{42b}$$

Hence we may conclude that the "gain" in resolution capability of the "lensless Fourier-transform hologram," in comparison to the Fresnel transform holograms is

$$G = \frac{p_{Fourier}}{p_{Fresnel}} = \frac{f}{\xi_0}\frac{\sin i'}{\cos^3 i'} \tag{43}$$

The "gain" is seen to be very considerable indeed. It is seen to become increasingly great at high angles and with an increasing neighborhood of the reference point R to the center of the object (at ξ_0). This may perhaps be most readily seen by writing Eq. (43) in the form

$$G = g\frac{f}{\xi_0} \tag{43a}$$

which singles out the significant nondimensional "gain factor" $g = \sin i'/\cos^3 i'$. Thus for $i' = 10°$, the gain factor is already 0.18 which, with an f/ξ_0 ratio of 100, would result in a gain G exceeding 18. In x-ray microscopy, where an f/ξ_0 ratio of 1000 may be more representative, the gain would be 180, even for this small angle. Most significantly, for wide-angle high-resolution microscopy, where i' tends to 90°, the gain tends to *infinity* !

4. Resolution-Retrieving Compensation of Source Effects in Holography

When sources other than point sources' are used in holography, either for the recording or for the reconstruction, the resolution of the image-forming process is reduced, because of a spreading or smearing of the image of a point object, in a manner quite similar to that described by (III.17). The resolution loss enters, according to (III.17) (and as we further show in what follows), in the form of a convolution between the object and the image of the source, so that we may say that the image of a point object is spread out to the width of the source image. In the past it had generally been assumed that the use of an extended, rather than a point, source in the recording of a hologram would result in an irretrievable loss of resolution. The resolution loss which would result from an uncompensated source effect would be very considerable, indeed, in x-ray holography, where the smallest available x-ray sources may not be less than some 100 Å in diameter for image formation at wavelengths where a 1-Å resolution may be sought.

A very important advance was made in holography, and indeed in image formation in coherent light, when Stroke et al.[30] showed that the loss of resolution which would result from the recording of a hologram with an extended source could, paradoxically be retrieved, in the reconstruction, by illuminating the hologram also with an extended source, provide that the correlation function of the two suitably structured sources had a narrow central peak, of a width comparable to the resolution limit sought in the two-step process. This is the case for structured sources having a broad spatial-frequency representation, for example, a Fresnel-zone plate, or other such structure, possibly obtained by interference.

4.1. Concise Analysis

Because of the importance of this conclusion, we first give the analysis in a very concise form, as it was given in the original paper.[30]

We consider as a model the now-usual one-dimensional Fourier-transform holography arrangements (see Section 3), which would normally use a point source $T_s(\xi) = \delta(0)$ in the plane of the object $T_0(\xi - a)$ (see Fig. 28). However, in place of the point reference we now use a spatially coherent extended source, of complex amplitude given by the equation

$$T_s(\xi - a) \tag{44}$$

and an object

$$T_0(\xi) \tag{45}$$

where a is now taken as the usual offset between the centers of gravity of T_s and T_0. According to the same procedure used to obtain (28), we find that the hologram intensity is obtained under these conditions in the form

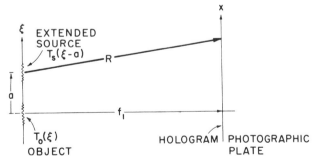

FIG. 27. Holography with *extended* sources used in place of conventional point-reference sources, after Stroke et al.[30] Note the dualism between source and object in holography with extended sources, as well as the *coding* in the recording, and the *decoding* in the reconstruction (see Fig. 28 and 29).

$$I(x) = [t_0 t_0{}^* + t_s t_s{}^*] + t_0 t_s{}^* \exp(2\pi i a x)$$
$$+ t_0{}^* t_s \exp(-2\pi i a x) \tag{46}$$

where t_0 and t_s are the Fourier transforms of T_0 and T_s, respectively [see (33)].

Under the usual Fourier-transform reconstruction conditions (see Section 3), illumination of the hologram *with a point source* will produce an image, in which the two side-band images of interest are, respectively, the Fourier transforms of the complex amplitude transmittance corresponding to the second and third terms in (46). By making use of (VII.57) and (VII.9), we obtain, by Fourier transformation of amplitude transmittance in (46), the equations

$$(T_0 \star T_s{}^*)_{\xi - a} \tag{47}$$

and

$$(T_0{}^* \star T_s)_{\xi + a} \tag{48}$$

for the amplitudes in the upper-side-band and lower-side-band images, respectively, where $\star$ indicates a correlation [see (52)], and where the image coordinate is taken as ξ, under unit magnification. The images are seen to be "spread" out, by correlation with the extended source T_s.

However, if we now illuminate the hologram, not with a point source, but with an extended source $T_{s'}$ (which may or may not equal to T_s), and if we let t_s, be the Fourier transform of $T_{s'}$, then the complex amplitude transmission through the hologram in the upper side-band of (46) becomes

$$t_{s'}[t_0 t_s{}^* \exp(2\pi i a x)] = [t_0 \exp(2\pi i a x)](t_{s'} t_s{}^*) \tag{49}$$

The corresponding upper-side-band image, obtained by Fourier transformation of (49), is equal to [see (VII.34a), (VII.57), and (VII.9)]

$$[T_0 \otimes (T_{s'} \star T_s{}^*)]_{\xi - a} \tag{50}$$

where $\otimes$ indicates a convolution [see (VII.22)]. We note that if

$$(T_{s'} \star T_s{}^*) = \delta$$

a delta function, then
$$\tag{51}$$

$$(T_0 \otimes \delta)_{\xi - a} = (T_0)_{\xi - a}$$

Equation (51) states that the reconstructed upper-side-band image will be identical to the object if the correlation of the reconstructing source with the complex conjugate of the recording source is equal to a delta function. Equation (51) gives the condition under which the use of extended sources in holography will not result in any loss of resolution.

For instance, if $T_{s'}$, is equal to T_s, then

$$(T_s \bigstar T_s{}^*) \tag{52}$$

is the autocorrelation function of the source. $(T_s \bigstar T_s{}^*)$ will be a function with a very narrow central peak, for example, if T_s can be represented by a very broad spectrum of spatial frequencies. We may note that such a broad spatial frequency spectrum representation would *require* an extended source field, so as to represent the various spatial frequencies with sufficient accuracy; such an extended source would therefore also result in a desirable great luminosity in the holographic system. We have already mentioned a Fresnel-zone plate as one example of such an extended structured source.

Examples of resolution-retrieving compensation using the above principles are shown in Figs. 28 and 29. We see that even in the case of such an arbitrarily selected source as the $\int d\xi$ used in Fig. 28, a very clear compensation of the spreading was obtained in the upper-side-band image. We also note, in Fig. 28, that the lower-side-band image has remained uncompensated, as can be predicted from the transformation of the second term in (46).

Indeed, when the hologram is illuminated by a source $T_{s'}$, as before, we find that the second term in (46) will result, by Fourier transformation, in an image term

$$[T_0{}^* \bigstar (T_{s'} \otimes T_s)]_{\xi + a} \tag{53}$$

equal, for $T_{s'} = T_s$, to

$$[T_0{}^* \bigstar (T_s \otimes T_s)]_{\xi + a} \tag{54}$$

The convolution $(T_s \otimes T_s)$ will, in general, not be a delta function, except for sources having a twofold rotation symmetry about the optical axis (see Fig. 7.2). The importance of the distinction between the generally desirable correlation of the source functions, $(T_s \bigstar T_s)$, and the not always desirable convolution, $(T_s \otimes T_s)$, appears in the compensated reconstruction of Fourier in the upper-side-band image (Fig. 28), and the uncompensated reconstruction in the lower-side-band image (Fig. 28).

We may finally note that there are cases where an advantage might be taken of this difference between the image compensated by source cor-

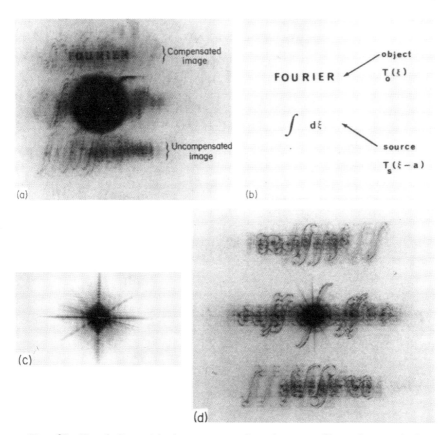

FIG. 28. Resolution-retrieving compensation of source effects, by correlative reconstruction, in Fourier-transform holography, after Stroke et al.[30] (a) Compensated image, reconstructed from coded hologram (c), by illuminating the hologram with the same arbitrary *extended* source $\int d\xi$ used in the recording [see (b)]. Note that the resolution in the compensated image is retrieved by a correlation of the source functions, while the lower-side-band image remains uncompensated, because of the convolution of the source functions (see also Fig. VII.2). (b) Object = FOURIER and source = $\int d\xi$, used in the recording of the coded hologram (c). The arbitrary source $\int d\xi$ was used in place of the conventional point source for the recording of the Fourier-transform hologram of part (c) to demonstrate that the resolution (corresponding to a point-reference-source hologram) could be retrieved, a posteriori, by correlation compensation in the reconstruction [see part (a)]. Note analogies with apodisation [e.g., P. Jacquinot and B. Roizen-Dossier in "Progress in Optics" (E. Wolf, ed.), Vol. III, pp. 29–186, North-Holland Publ., Amsterdam, 1964; in particular, see pp. 136–146. See also Stroke, Vol. II, pp. 1–72]. (c) Coded Fourier-transform hologram, recorded with $\int d\xi$ in place of point source. (d) Completely uncompensated reconstruction, obtained by illuminating the coded hologram of part (c) by a plane wave, according to Fig. 19.

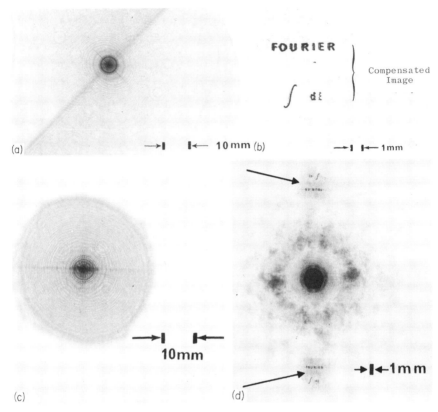

FIG. 29. Resolution-retrieving a posteriori compensation of resolution loss
which would result from the use of an extended source in place of the point-
reference source in holography, after Stroke et al.[30] (a) Fourier-transform hologram (with focal length $f = 600$mm) obtained with a 19-mm-diameter Fresnel-zone
half-plate source (the portion on one side of a diameter of a Fresnel-zone plate,
itself previously obtained by interference) used in place of the conventional
point source in the recording. Note that the two halves of the hologram (separated by a diagonal in the figure) were each illuminated by one-half of a spherical wave, originating, respectively, from the real and virtual foci of the zone
plate. Alternate equivalent nonstructured extended sources (e.g., two half-field
lenses), used in the place of the structured Fresnel half-zone plate may also be
suitable, as are full Fresnel-zone plates [see parts (c) and (d)]. (b) Compensated
image, reconstructed by Fourier transformation, by illuminating the hologram of
part (a) with the *same* extended source as used in the recording. (c) Coded
Fourier-transform hologram, obtained with the *full* Fresnel-zone plate (in this
case a 10-mm-diameter portion) used in place of the conventional point source
[compare with part (a)]. (d) Compensated reconstruction of side-band images
from hologram of part (a) by illuminating the hologram with the same full Fresnel-
zone plate as used in the recording [compare with part (b)], after Stroke et al.[30]

relation and the image remaining uncompensated, because of source convolution. This might in particular be the case in *suppressing* the twin image, in the original arrangement of the Gabor microscope, by use of a suitably structured source, having *no* two fold rotation symmetry!

Before drawing some other conclusions, we may now give a somewhat more formal proof of the results, which we have just obtained in a concise form.

4.2. More-Formal Analysis[†]

Consider Fig. 27. Let the transmittance of the object be

$$T_0(\xi) \tag{45}$$

and the transmittance of the source

$$T_s(\xi - a) \tag{44}$$

as before. The electric-field amplitude at the hologram plate, due to the object, is given by

$$E_0(x) = \int T_0(\xi) \exp(ik_1 R) \, d\xi \tag{55}$$

where $k_1 = 2\pi/\lambda_1$ and where R is indicated in Fig. 27. By a binomial expansion, we have (see Fig. 27)

$$R = [f_1{}^2 + (x - \xi)^2]^{1/2} = f_1 \left[1 + \frac{(x - \xi)^2}{f_1{}^2}\right]^{1/2}$$

$$= f_1 \left[1 + \frac{(x - \xi)^2}{2f_1{}^2} + \cdots\right]$$

$$\cong f_1 + \frac{x^2}{2f_1} - \frac{x\xi}{f_1} \qquad \text{if } f_1 \gg x \gg \xi \tag{56}$$

By letting t_0 and t_s, be the Fourier transforms of T_0 and T_s, respectively, we obtain for the fields produced by the object and the source in the hologram plane the equations

$$E_0(x) = \exp\{ik_1[f_1 + (x^2/2f_1)]\} \, t_0(x) \tag{57}$$

and

$$E_s(x) = \exp\{ik_1[f_1 + (x^2/2f_1)]\} \, t_s(x) \exp[-ik_1(xa/f_1)] \tag{58}$$

[†]In collaboration with R. Restrick.[30]

The intensity $I(x)$ recorded in the hologram is given by

$$I(x) = [E_0(x) + E_s(x)] [E_0(x) + E_s(x)]^*$$
$$= t_0 t_0^* + t_s t_s^* + t_0 t_s^* \exp [i(k_1 xa/f_1)]$$
$$+ t_0^* t_s \exp [i(k_1 xa/f_1)] \tag{59}$$

The complex amplitude $h(x)$ of the field transmitted through the hologram will be some function $h(I)$ of the intensity I, as given in (59). For the case when the hologram is illuminated by a plane wave (corresponding to the use of a point source in the reconstruction) we have, by expanding $h(I)$ in a power series about the point of average intensity I_0, the equation

$$h(I) = h(I_0) + h'(I_0)(I - I_0) + \frac{h''(I_0)}{2!}(I - I_0)^2 + \dots \tag{60}$$

By introducing (59) in (60) we obtain

$$h(I) = h_0 + h_1 \exp [-i(k_1 xa/f_1)] + h_{-1} \exp [i(k_1 xa/f_1)]$$
$$+ h_2 \exp [-2i(k_1 xa/f_1)] + \dots + h_n \exp [-in(k_1 xa/f_1)] \tag{61}$$

The complex coefficient h_1 of the first-order side-band is given by the equation

$$h_1 = h'(I_0) t_s t_0^* + \frac{h''(I_0)}{2!} 2 t_s t_0^* (t_0 t_0^* + t_s t_s^* - I_0) + \dots \tag{62}$$

We shall now show the conditions under which the first term in (62) will dominate. Let

$$(I - I_0) = A + \{B^* \exp [i(k_1 xa/f_1)] + B \exp [-i(k_1 xa/f_1)]\} \tag{63}$$

where

$$A = t_0 t_0^* + t_s t_s^* - I_0 \tag{64}$$

and

$$B = t_s t_0^*$$

By using a binomial expansion twice, we obtain from (63),

$$(I - I_0)^n = \sum_{\alpha=0}^{n} \sum_{\beta=0}^{\alpha} \frac{n!}{(n-\alpha)!(\alpha-\beta)!\beta!} A^{n-\alpha} (B^*)^{\alpha-\beta} B^{\beta}$$
$$\times \exp [(-ik_1 xa/f_1)(2\beta - \alpha)] \tag{65}$$

However, for the first side-band term,

$$2\beta - \alpha = 1 \qquad (66)$$

or

$$\alpha = \text{odd} \qquad \text{and} \qquad \beta = \frac{1 + \alpha}{2} \qquad (67)$$

In this case,

$$[(I - I_0)^n]_{1\text{st side-band}} = \sum_{\alpha\,\text{odd}} n! \Big/ (n - \alpha)! \left(\frac{\alpha - 1}{2}\right)! \left(\frac{\alpha + 1}{2}\right)!$$
$$\times A^{n-\alpha} (B*)^{\frac{\alpha-1}{2}} B^{\frac{\alpha+1}{2}} \qquad (68)$$

Let us now consider the Fourier transformation of a typical term in (68). We have (using here FT for T)

$$FT\,[A\,(A^{n-\alpha-1}(B*)^{\frac{\alpha-1}{2}} B^{\frac{\alpha+1}{2}}$$
$$= FT(A) \otimes FT[A^{n-\alpha-1}\,(B*)^{\frac{\alpha-1}{2}} B^{\frac{\alpha+1}{2}} \qquad (69)$$

But, according to (64),

$$FT(A) = FT[t_0 t_0{}^* + t_s t_s{}^* - I_0] \qquad (70)$$

We now let

$$FT(t_0 t_0{}^*) \cong \delta \text{ function} \qquad (71)$$

and

$$FT(t_s t_s{}^*) \cong \delta \text{ function} \qquad (72)$$

where

$$FT(t_0 t_0{}^*) = T_0 \star T_0{}^* \qquad (73)$$

and

$$FT(t_s t_s{}^*) = T_s \star T_s{}^* \qquad (74)$$

We also note that

$$FT(I_0) = \delta \text{ function} \qquad (I_0 = \text{constant}) \qquad (75)$$

It follows that

$$FT(A) \cong \delta \text{ function} \qquad (76)$$

By n applications of the preceding reasoning, together with the fact that

$$\delta \otimes W = W \qquad (77)$$

for an arbitrary W, we obtain for (69) the expression

$$FT[(B*B)^{(\alpha-1)/2} B] \qquad (78)$$

which reduces, by a similar method to that used above, to

$$FT[B] = FT[t_s t_0{}^*] = T_s \divideontimes T_0{}^* \tag{79}$$

In comparing (79) with the first term in (62), we recognize that two expressions have the same form. We may therefore conclude that the higher terms in (62) may be neglected, as long as (71), namely,

$$FT(t_0 t_0{}^*) = T_0 \divideontimes T_0{}^* \cong \delta \text{ function} \tag{71a}$$

is a good approximation.

Under the assumption of (71), (62) becomes

$$h_1 \cong h'(I_0) \, t_s t_0{}^* \tag{80}$$

We may complete the analysis by taking the Fourier transform of (80), to obtain the reconstructed image. For the sake of generality, we shall assume that the reconstruction is carried out in another wavelength, namely λ_2, and with another focal length, namely f_2. We have $k_2 = 2\pi/\lambda_2$.

Let $H_1(\xi)$ be the Fourier transform of $h_1(x)$. We have

$$H_1(\xi') = \int h_1(x) \exp[-i(k_2 x\xi'/f_2)] \, dx \tag{81}$$

where ξ' is the image coordinate in the focal plane of the Fourier-transforming lens, of which the focal length is f_2. Equations (80) and (81) give by suitable integration and use of dummy variables the equation

$$H_1(\xi') = \int h'(I_0) \, t_s t_0{}^* \exp\left(-i\frac{k_1 \, xa}{f_1}\right) \exp\left(-i\frac{k_2 \, xa}{f_2}\right) dx$$

$$= h'(I_0) \iiint T_s(\nu) \exp\left(-i\frac{k_1 \, x\nu}{f_1}\right) T_0{}^*(\eta) \exp\left(i\frac{k_1 \, x\eta}{f_1}\right)$$

$$\times \exp\left(-i\frac{k_1 \, xa}{f_1}\right) \exp\left(-i\frac{k_2 \, x\xi'}{f_2}\right) d\nu \, d\eta \, dx$$

$$= h'(I_0) \int T_s(\nu) \int T_0{}^*(\eta) \int \exp\left\{-i\left[\frac{k_1\nu}{f_1} + \frac{k_2 a}{f_1} - \frac{k_1\eta}{f_1} + \frac{k_2\xi'}{f_2}\right]x\right\}$$

$$\times dx \, d\eta \, d\nu$$

$$= h'(I_0) \int T_s(\nu) \int T_0{}^*(\eta) \, \delta\left[\frac{k_1\nu}{f_1} + \frac{k_1 a}{f_1} - \frac{k_1\eta}{f_1} + \frac{k_2\xi'}{f_2}\right] d\eta \, d\nu$$

$$= h'(I_0) \int T_s(\nu) \, T_0{}^*\left(\nu + a + \frac{k_2 f_1}{k_1 f_2}\,\xi'\right) d\nu \tag{82}$$

Equation (82) states that the indicated side-band image is equal to the correlation function of T_s with $T_0{}^*$, shifted by a units from the optical axis, and magnified by the factor

$$\frac{f_2 k_1}{f_1 k_2} = \frac{f_2}{f_1} \frac{\lambda_2}{\lambda_1} \tag{83}$$

in agreement with (48), and indeed with the magnification equation (13). We may therefore conclude that the simplifying assumptions made in the concise analysis given above were indeed suitable, and that the concise analysis does give the correct results, under the assumption of (71) and (72).

4.3. Concluding Remarks

We may note that the source-compensation principle may have applications in interferometry, as well as in holography. We might also note that the compensation scheme could itself be used for optical filtering, either in place, or indeed in association with the methods described in Chapter 5.

Finally, we may note the existence of analogies between the correlative source-compensation method in *coherent* light, which we have described here, and the matched filtering methods, used in *noncoherent* light, for instance, in spectroscopy (as in the grille spectrometer, first described by Girard). In this connection we may note that it had not at all seemed obvious, a priori, that we would succeed in what, in the final analysis, may be considered matched filtering in *coherent* light.

5. Coherence Requirements in Holography

Strictly speaking, there is only one coherence requirement necessary for the recording of a hologram in spatially coherent light: *Every object point must be capable of interference with every source point.*[†]

The coherence condition has several implications. First of all, the object T_0 and the source T_s must be made to scatter light of the same frequency. Next, the object and the source must be stationary with respect to each other, as seen from the hologram recording plate. (Scattering from any nontrivial object will always tend to produce a spatially "diffused" field in the hologram plane; accordingly, the object may be illuminated with diffuse light, without additional violation of the above condition.) Finally, when source compensation is used (see Section 4), there is no essential difference between the source and the object; accordingly, both the source and the object may be diffusely illumi-

---

[†]In spatially incoherent (noncoherent) light, the contrary must hold (see Section 9.2).

nated (provided, of course, that the source T_s', used in the reconstruction, has a cross correlation $T_s \star T_s'$, with the recording source T_s, equal to a delta function, as indicated in (Section 4).

We may also note that illumination of the object with a diffuser moved during the exposure, and the success of the experiments performed with a moving diffuser, first reported by Stroke and Falconer[27] (see Fig. 38) may be readily interpreted in terms of the theory of intensity superposition by successive exposures of the same hologram, before development, which we have given, following Gabor and Stroke et al.,[44] in Section 9 of Chapter V.

When sources other than lasers are being used to record a hologram, and when the spectral line width may be sufficiently wide to restrict the spatial domain over which interference can be obtained with the given source, it is of some interest to define a *coherence length L*, which can be taken as a measure of the dimensions (or depth) of the object, of which a hologram can be recorded by interference with a plane (or spherical) wave, reflected by a mirror placed near the object, for example.[†]

Consider two waves, one at the frequency ν and the other at the frequency $(\nu - \Delta\nu)$. Let c be the free-space velocity of light and λ the wavelength at the frequency ν. We have the equation

$$c = \nu\lambda \qquad (84)$$

We also note that the circular frequency ω is given by

$$\omega = 2\pi\nu \qquad (85)$$

and that the phase ϕ in the wave equation is

$$\phi = \omega t = 2\pi\nu \qquad (86)$$

Let the two waves of frequency ν and $(\nu - \Delta\nu)$ be in phase at some time $t = 0$. After a propagation time t, to which corresponds a propagation length L, given by

$$L = ct \qquad (87)$$

the two waves will be out of phase by an amount $\Delta\phi$ equal to

$$\Delta\phi = 2\pi[\nu t - (\nu - \Delta\nu)t] = 2\pi(\Delta\nu)t \qquad (88)$$

By differentiating (86) and (84), we obtain from (88),

$$\Delta\phi = -2\pi\frac{c\Delta\lambda}{\lambda^2}t \qquad (89)$$

where $\Delta\lambda$ is the wavelength difference corresponding to the frequency difference $\Delta\nu$.

[†]The following derivation was prepared in collaboration with Professor H. H. Stroke.

We may now chose to define, arbitrarily, as coherence length $L_{\pi/2}$, the length L for which the phase difference between the two waves is equal to $\pi/2$. By comparing (89) and (87) we obtain for the coherence length the value

$$L_{\pi/2} = \frac{1}{4} \frac{\lambda}{\Delta\lambda} \, \lambda \tag{90}$$

Equation (90) may be used to evaluate the dimensions over which coherence can be obtained under particular geometric conditions, with a source of which the spectral line width can be represented according to the preceding equations.

6. Summary and Results

The theoretical and experimental foundations of holography are implicit in the discussions of the previous sections. We recall here for clarity the principles involved, notably:

(1) Interferometry.
(2) Diffraction gratings.
(3) Coherence requirements.

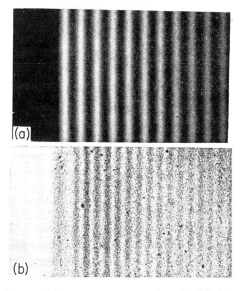

FIG. 30. Interference fringes at x-ray wavelengths (b), illustrating likelihood of success of x-ray holographic microscopy. Interference fringes produced with Lloyd's mirror. (a) Taken with visible light $\lambda = 4358$ Å, after F. A. Jenkins and H. E. White, "Fundamentals of Optics," McGraw-Hill, New York, 1957. (b) Taken with X-rays, $\lambda = 8.33$ Å, after Kellström[46].

(a)

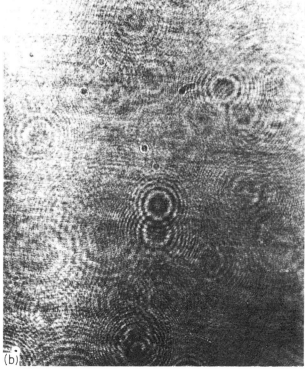

(b)

In view of extensions, such as to the x-ray domain, it is important to note that the interferometric criteria permit an a priori evaluation of the likelihood of success of a holography method, provided that suitable experimental evidence is available.

In Fig. 30 we show, following Kellstrom,[46] interference fringes produced with $\lambda = 8.33$-$\overset{\circ}{A}$ x-rays, by means of a Lloyd's mirror system. Such evidence can be taken as an indication of the orders of magnitude of attainable coherence, and therefore of the attainable holograms, according to the principles discussed in the preceding section. Reconstructions of a three-dimensional scene illuminated according to the principles which we have described[11,17,26,47,] (see preceding sections) are also shown in Fig. 31a and 31b. Holograms corresponding to three-dimensional scenes are also shown in Fig. 32 (holograms and reconstruction in 6328-$\overset{\circ}{A}$ laser light).

A photomicrograph of a crystal-like grating (magnification $\simeq$ 6X) obtained by holography, entirely without lenses, according to the principles discussed in Section 2.4, is shown in Fig. 33, and the corresponding hologram, also obtained without any lenses, is shown in Fig. 34 (photograph in Fig. 35). A schematic diagram of the arrangement used in obtaining the hologram is shown in Fig. 36, and a photograph of the apparatus used in Fig. 37. The lensless reconstruction of the real image of Fig. 38 was obtained in the arrangement sketched in Fig. 39, and a photograph of the actual apparatus used is shown in Fig. 40. In extending holography to x-ray microscopy applications, it may be necessary to illuminate the object by means of a mirror or scatterer, moved so as to direct the reference beam (or portions of it) into suitable directions, for the purpose of providing the coherent background over the required parts of the scattered field, when a single extended background wave may not be achievable. A photographic reproduction of a lensless reconstruction obtained from a hologram photographed with a scatterer *moved* during the exposure is shown in Fig. 38, together with a reproduction of the actual hologram and object, all to scale. A sketch of the apparatus used for obtaining the hologram of Fig. 38 is shown in Fig. 39, and a photo-

---

FIG. 31. (a) Images of a three-dimensional scene reconstructed from a hologram according to Leith and Upatnieks[17] and Stroke.[11] All four images were photographed by means of a camera looking through the hologram, and so focused and inclined to reveal the three-dimensional nature of the reconstruction. The top two images are photographs with a small camera aperture, at two angles, to show parallax. The lower two images are photographs with a large camera aperture, at the same angle with respect to the hologram (b) to show the three-dimensional field depth. (Hologram and reconstructions in 6328-$\overset{\circ}{A}$ laser light.)

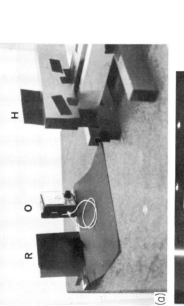

(a)

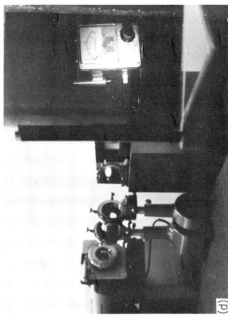

(b)

(c)

(d)

FIG. 32. Holographic recording and reconstruction of a three-dimensional object according to the principles first suggested by Stroke.[11] (a) Recording: R = reference mirror, O = object, H = hologram (on Kodak 649F photoplate). (b) Hologram. (c) Lensless reconstruction of real image. (d) Reconstruction of virtual image, as photographed through hologram (holographic work in 6328-Å laser light) (Electro-Optical Sciences Laboratory, University of Michigan, Ann Arbor.)

graph of the apparatus in Fig. 40. Among the principal experimental requirements for the attainment of good holograms are:

(1) Interferometric (mechanical and thermal) stability between the object, reference mirror (or lens), and the photographic plate. Waxing down
of the various elements on a solid support, and waiting for the object,
mirrors, and photographic plate to come to thermal equilibrium has been
found to be as important in holography as it is in interferometry in
general.

(2) An intensity in the reference beam about 5 to 10 times greater than
the intensity in the scattered field, at the photographic plate, so as to
permit us to maintain only the first side-band terms in (4) [see also (62)
and Section 4].

We may finally note that any one of a number of Fourier-transforming
recording arrangements should permit one to realize a high-resolution
Fourier-transform hologram in addition to the one indicated in Section 3.
Regardless of the manner in which the high-resolution hologram of (28) is
recorded, the reconstruction is then clearly obtained in the focal plane of
a Fourier-transforming lens. The high magnification of (13) is, of course,
maintained in the use of high-resolution holograms [with the geometric
part of the magnification being simply equal to the ratio of the focal
length f_2 used in the reconstruction to the focal length f_1 used in the recording; see (83)].

7. Electron Microscopy and X-Ray-Hologram Microscopy

Some remarks may be in order concerning the possible applications of
hologram microscopy, in the light of the remarkable results obtained with
modern electron microscopes.

Modern electron microscopes appear to readily provide resolving powers on the order of angstroms (1 Å $= 10^{-10}$ meter). Even the most opti-
mistic present estimate makes it unlikely that these values would be exceeded by an x-ray hologram microscope, or for that matter by any x-ray
microscope, using the wavelengths of the x-rays contemplated. On the
other hand, it is not at all unlikely that even these remarkable resolutions will be exceeded by future electron microscopes, and that sample-
heating problems will be satisfactorily eliminated.

There are, however, several aspects in which an x-ray hologram microscope, if developed, would fill a role which cannot at the moment be

filled by electron microscopes. Considerably greater penetrations without heating of the samples can be obtained with x-rays than with even very high energy electron beams. This would be of a particular interest in areas such as metallurgy, and especially biophysics, in particular perhaps with live tissues and so on. It is also of interest to note that an x-ray microscope would not necessarily require a vacuum, which is a necessity with electron microscopes. Considerably greater resolutions might be attainable if holography were to be extended to gamma rays.

8. Conclusion

The determination of the phase of a scattered wave has long been a significant and formidable problem for the x-ray crystallographer. Without some knowledge of both the magnitude and the phase of the scattered waves, it does not readily appear possible to obtain a complete, well-resolved image of the crystal specimen. The first attempts to solve the phase problem appear to have been made in 1939 by Bragg[51,52] and by Buerger,[53-55] who noted an early suggestion by Boersch.[56]

Buerger and Bragg demonstrated that an image of a crystal could be obtained by placing suitably manufactured optical phase plates over the various diffraction points in the reciprocal lattice, and then optically Fourier-transforming the composite arrangement in what amounts to an optical image synthesizer. The technique has been very successful, but it clearly requires some a priori knowledge of the phases of the scattered waves, such as that which is available with centrosymmetrical crystals, for example.[51-55] More recently, Kendrew[57,58] and co-workers have successfully synthesized images of crystals by electronic-computer Fourier transformation of x-ray diffraction patterns, in particular in those cases where the "heavy-atom" or isomorphous replacement technique was applicable, for instance, in protein crystals such as the myoglobin molecule.[59] A general applicability to x-ray microscopy of either the Kendrew or of the Buerger and Bragg methods appears to have been restricted, in their present forms, by the difficulty of ascertaining the phases of the various diffraction spots in the reciprocal lattice, in the general case.

A new approach toward the solution to the phase problem in x-ray microscopy has recently been proposed by Stroke and Falconer,[26] and was further extended by Stroke[28] and by Stroke et al.[30] The new ap-

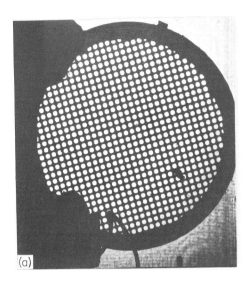

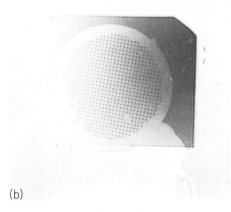

(b)

FIG. 33. Lensless microscopy. (a) Lensless reconstruction of image of a crystal-like grating. The 6× enlarged image was reconstructed from the hologram (b) by means of the arrangement and apparatus shown in Figs. 36 and 37. Note the remarkably fine resolution of details (e.g., the 0.1-mm-diameter wire supporting the grating). The 6× enlargement was used here to show the reconstruction in its entirety, but enlargements of well over 100× have been obtained without any difficulty.[26] (b) Hologram obtained by projection microscopy arrangement shown in Figs. 34 and 35, and used in reconstructing the image in (a). The entire re- cording and reconstruction process shown was carried out without the aid of lenses in 6328-Å laser light. An additional magnification of 6328× would result if the recording were carried out in 1-Å x-ray light, leading to magnifications in excess of 1 million in x-ray microscopy. (Electro-Optical Sciences Laboratory, University of Michigan, Ann Arbor.)

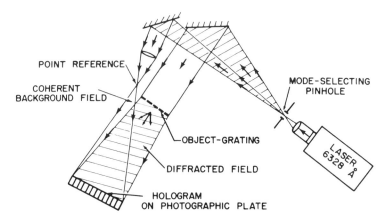

FIG. 34. Lensless microscopy. Projection holography arrangement used for lensless microscopic recording of the hologram of Fig. 33 (see photograph of apparatus in Fig. 35).

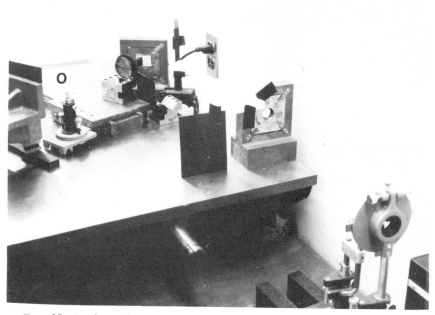

FIG. 35. Lenless microscopy. Projection holography arrangement, after Stroke[11] (see Fig. 34). The three-dimensional character of holographic microscopy and the associated great depth of field are of particular interest, not only for x-ray and electron-microscopy applications, but for visible and ultraviolet applications as well. The object grating (O) in the photograph has a diameter of about 14 mm.

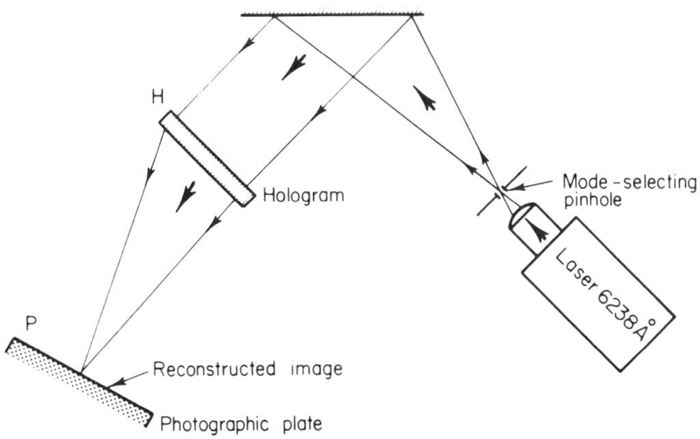

FIG. 36. Holographic microscopy. Apparatus used in the reconstruction of
images in projection microscopy. P = plate holder, H = hologram, L = laser
(photograph of apparatus in Fig. 37).

proach may be described as lensless Fourier-transform holography.
It is based on the general principles of wavefront-reconstruction imaging
first proposed by Gabor in 1948.[1-5] Gabor proposed that a coherent
background be superposed onto the diffraction pattern, so as to provide
a reference wavefront for the recording of both the amplitude *and* the
phase of the scattered waves. Success in Gabor's method of holographic

FIG. 37. Holographic microscopy (projection reconstruction apparatus, as
used in the Electro-Optical Sciences Laboratory; schematic diagram in
Fig. 36).

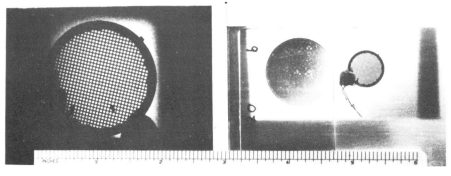

FIG. 38. Holographic microscopy using multidirectional illumination and moving scatterers, after Stroke and Falconer.[27] Shown to scale are, left to right: image (magnified about 3×), hologram, and object. The hologram was recorded in the arrangement shown in Fig. 39 (photograph in Fig. 40), while a diffusing scatterer was actually moved during the exposure. [In connection with the complex-image-synthesis work, described by Gabor and Stroke et al.[30] (see Figs. 22 and 23 of Chapter V), one may note that complex amplitudes may be superposed in the latent image in the hologram by a *successive superposition of intensities* in the recording under suitable conditions!] The remarkable quality of the reconstructed image obtained under these (at first sight) unusual conditions serves to illustrate some important coherence and superposition properties in holography. In some cases the use of a moving mirror, or scatterer, may be required to attain high-resolution holograms.

microscopy in the visible domain was immediately demonstrated by Gabor himself[1-5] and has been verified by many others since that time. However, it quickly appeared that high resolutions at x-ray wavelengths would be unattainable, in the application of conventional Fresnel-transform holography, because of the difficulties associated with film resolution and source dimension discussed in Sections 3 and 4. Indeed, resolutions of only 5000 to 10,000 Å, rather than 1 Å, appeared attainable by means of conventional Fresnel-transform holography. In noting the basic similarity between the Buerger and Kendrew image-synthesizing methods, on the one hand, and microholographic image reconstruction, on the other,[†] Stroke and Falconer argued[26] that high resolutions in microholography should be attainable by a suitable modification of the early prin-

---

[†]A formal proof of this analogy, first noted by W. L. Bragg,[90] has now been obtained in the collaboration[91] of Professor M. G. Rossmann and his co-workers with Stroke and his co-workers, when they showed a complete formal analogy between (1) Fourier-transform holography with extended-source compensation[30] (Section 4), on one hand, and, on the other, (2) the Raman-Ramachandran α-synthesis (in the heavy-atom method)[92,93] to deconvolute the Patterson function when the structure is partially known.

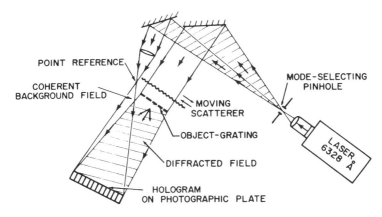

FIG. 39. Recording of hologram with multidirectional illumination and moving scatterer, after Stroke and Falconer.[27] The use of a scatterer in holographic imaging makes it possible not only to obtain multidirectional illumination, but is also a means for beam-splitting (for instance with the aid of partially diffusing scatter-plates, such as those already mentioned for interferometry by Newton in his *Optics*; see also Burch[48] and Dyson[49]).

ciples of holography. In particular, they showed that Fourier-transform holograms should permit one to overcome the source-emulsion problem characteristic of conventional projection holography. Shortly afterward, Stroke[28] showed how to obtain Fourier-transform holograms by *lensless* Fourier transformation, thus preserving the original advantages of lensless photography first suggested by Gabor. More recently still, Stroke *et al.*[30] also showed that the resolution loss which would result from the use of extended sources, in the recording of holograms, can be compensated, paradoxically, by use of another (suitably structured) extended source in the reconstruction of the image from the hologram. Scattered-light illumination and structured source apertures, based on interferometric resolution-luminosity-coherence considerations, are thus likely to have a primary role in the complete solution of the source-aperture problem[11,27,28,30,47] in microholography, as are lensless Fourier-transform recordings of holograms.[28,29]

9. Holography with Noncoherent Light

The principles of Fourier-transform holography with coherent light (introduced, in particular, by the author in 1963[11], and presented in the preceding sections of this chapter) have permitted very considerable mathematical and experimental simplicity to be introduced into the field of wavefront-reconstruction imaging, originated by Gabor in 1948. Spa-

FIG. 40. Apparatus used in holographic microscopy with diffused illumination and moving scatterer, after Stroke and Falconer[27] (schematic diagram in Fig. 39). The medium-power laser (He-Ne, cw), used in obtaining the hologram of Fig. 38, appears on the right. Only cw lasers were used in the work described in this text, but pulsed lasers may help, in some cases, to surmount vibration difficulties and temperature-stability requirements, which tend to appear in interferometric arrangements used for the recording of holograms of large-scale objects. (Holographic photography of high-speed phenomena with conventional and Q-switched ruby lasers was first described by Brooks et al.[50]) However, the holographic work described in this text (carried out in the Electro-Optical Sciences Laboratory, University of Michigan, with the assistance of D. Brumm, D. G. Falconer, A. Funkhouser, and R. Restrick) has demonstrated that completely satisfactory stability in the interferometers used for holographic recording is readily achieved.

tial coherence in the objects (i.e., the capability of the various object points to interfere with each other, see Section 2.8 of Chapter IV, or with a reference field, see Section 5, this chapter) has generally appeared to be essential to the recording of holograms. A possibility of an exception to this assumption was first indicated by Stroke and Funkhouser[68] in 1965, and is related to their method (discussed below, in Section 9.1) in which a spectroscopic Fourier-transform hologram is recorded with *spectrally noncoherent* light in a two-beam Michelson-Twyman-Green interferometer and is used to produce spectra directly (without computing) by a second Fourier transformation in the focal plane of a lens.

Many schemes for noncoherent light holography, other than those given in Section 9.2, had been suggested in the literature[69-71] and elsewhere, and the general desirability of achieving holography in noncoherent light had been recognized, but no results appear to have been reported until 1965, when Stroke and Restrick[72] first showed theoretically, and demonstrated experimentally, that image-forming Fourier-transform holograms could indeed be recorded with extended *spatially noncoherent* monochromatic objects, and that high-quality images could be reconstructed from them by a second Fourier transformation (e.g., in the focal plane of a lens).

Details and an analysis of the Stroke-Restrick method of holography with noncoherent light are given in Section 9.2. The essence of the method consists in producing, in the hologram plane, one sinusoidal interference-fringe grating per object point, with the correct spatial frequency and orientation, in such a way that the various gratings are made to add in *intensity*, thus recording the spatial Fourier transform of the intensity distribution in the object. The desired intensity summation in the hologram is made possible precisely because of the noncoherence of the various object points with each other. Paradoxically, therefore, noncoherence between the various object points is a requirement for holography with noncoherent light.

Because of its simplicity and special importance in holography, we first discuss the method of holographic Fourier-transform spectroscopy with spectrally noncoherent light.

9.1. Holographic Fourier-Transform Spectroscopy with Spectrally Noncoherent Light[68,73]

The remarkable properties and the important advantages of several forms of Fourier-transform spectroscopy (see pages 35–57) have now been established,[74-79] following initial work by Jacquinot,[77] Fellgett,[78] Strong,[79] and others. Among the principal advantages are a simultaneous recording of all spectral elements (recording time independent of the spectral width) and high luminosity. The method requires very accurate (ruling-engine quality) moving-mirror motion (or scanning) and computation (by Fourier transformation) of the spectrum from the photoelectrically recorded interferogram. The recent advances in holographic (wavefront-front-reconstruction) imaging, and in its Fourier-transform formulation, have made it reasonable to investigate possible simplifications which might result from extensions of holography to such applications as spectroscopy and astronomy.

The theoretical principles and experimental verifications of a holographic method of Fourier-transform spectroscopy were first presented in Ref. 68. The method permits one to obtain the spectrum without any computation and, indeed, in an interferometric system having completely stationary elements and medium. Various methods for transcribing onto film the photoelectric recordings of the Fourier transforms (or even Fresnel transforms) of spectra have been suggested for the purpose of optically reconstructing the spectrum. In our method, no transcription of a photoelectrically recorded spectrum nor any interferometer scanning are required.

As a model, we may discuss the recording of the spectral Fourier-transform hologram in a two-beam interferometer, in which the wavefronts form a small angle θ with each other (See Fig. 41). The interference-fringe system photographically recorded in a plane "parallel" to the bisector of the wavefronts is proportional (in the holographic sense) to a noncoherent superposition of the monochromatic fringe systems corresponding to each wavelength λ. As we show below, the equation of the fringe system in the plane of the photographic plate is readily found to be equal to

$$I(x) = \int_0^\infty I(\sigma)[1 + \cos 2\pi\sigma\theta x]\, d\sigma \qquad (91)$$

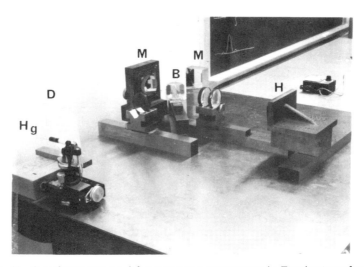

FIG. 41. Interferometer used for recording spectroscopic Fourier-transform holograms with a spectrally noncoherent source (after Stroke and Funkhouser[68]). Hg = cold mercury arc, D = diffusing glass, M,M = interferometer mirrors, B = beam splitter, H = hologram photographic plate.

where $\sigma = 1/\lambda$ (cm) and $I(\sigma)$ is the spectral intensity distribution (the spectrum) of the source. In Eq. (91), the cosine Fourier transform of $I(\sigma)$ is recognized as the equation on the spectral Fourier-transform hologram of $I(\sigma)$ (i.e., a "noncoherent" superposition of sinusoidal intensity gratings, one per each λ). Consequently, illumination of the hologram by a spatially coherent monochromatic plane wave (as in Fig. 24) will produce in the focal plane of a lens the spectrum $I(\sigma)$ symmetrically displayed on the two sides of the optical axis (one pair of spectral lines for each grating recorded, plus zero-frequency terms).

The compensated Michelson-Twyman-Green interferometer used in our experiments[68,73] to record the hologram is shown in Fig. 41, a hologram of the spectrum of a cold mercury arc is shown in Fig. 42, and a reconstruction of the spectrum obtained by illuminating the hologram in 6328 Å laser light (see Fig. 24) is shown in Fig. 43. A very wide, diffusely il-

FIG. 42. Spectroscopic Fourier-transform hologram recorded in interferometer of Fig. 41.

luminated source "aperture" was used in the recording, with an interferometer wedge of about 30 white-light (mercury) fringes per millimeter, near zero-path difference, and with Kodak 649F high-resolution plates. (Aligning experiments were carried out on Polaroid P/N film. Initial experiments with a flash-light source have shown that a continuous spectrum does produce a recordable interferometric hologram and a reproducible spectrum).

Although Eq. (91) can be obtained almost by inspection, it may be in order to give some of the steps of a more detailed analysis.

Let $E(x, y, z)$ be the electric field in a plane wave of wavelength λ, traveling at an angle $(\theta/2)$ with respect to the z-axis in the interfero-

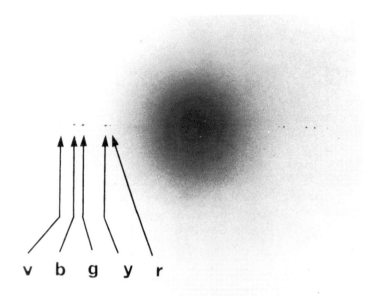

FIG. 43. Fourier-transform reconstruction of mercury spectrum, obtained in the focal plane of a lens (see Fig. 2) by projecting a plane wave of 6328 Å laser light through the spectroscopic Fourier-transform hologram of Fig. 42. Note the characteristic symmetrical display of the spectrum into a left and right "sideband," which permits wavelength determination by a simple measurement of distance ratios. Identified, in the left side band, are the mercury lines[84]: v (4047 Å, 4078 Å), b (4348 Å-4359 Å), g (5461 Å), y (5770 Å, 5790 Å), r (6152 Å, 6234 Å).

meter, such that (in an x-plane, normal to the z-axis)

$$E(x, y) = E_0 \exp [2\pi i \sigma x \sin (\theta/2) + \omega t] \tag{92}$$

where $\sigma = 1/\lambda$, $\omega = 2\pi f$, $c = f\lambda = $ velocity of light, and $t = $ time.

Let a second plane wave of wavelength λ, traveling at the angle $-\theta/2$ with respect to the z-axis, be such that in the x-plane we have

$$E(x, y) = E_1 \exp [2\pi i \sigma x \sin (-\theta/2) + \omega t + \phi] \tag{93}$$

where ϕ is a phase difference. When the two waves of Eqs. (92) and (93) are made to interfere on a photographic plate in the x-plane, the resultant intensity recorded on the plate is

$$[I(x, y)]_\lambda = \tfrac{1}{2} E_0^2 + \tfrac{1}{2} E_1^2 + E_1 E_0 \cos [2\pi \sigma x \sin (\theta/2) - \phi] \tag{94}$$

We may now arrange the interferometer so as to record the interference pattern at zero-path difference, for which $\phi = 0$. We may further simplify the analysis by assuming $E_0{}^2 = E_1{}^2 = I(\sigma)$, where $I(\sigma)$ is the intensity of the source at the wave number σ, and also let θ be a small angle. Under these simplifying assumptions, we readily obtain from Eq. (94) the equation

$$[I(x, y)]_\lambda = I(\sigma) [1 + \cos 2\pi\sigma\theta x] \tag{95}$$

which is the intensity of the interference fringe pattern (i.e. hologram) recorded in a bandwith $\Delta\sigma$, centered at the wave number σ.

Because of the noncoherence between spectral elements (characteristic of sources generally studied by Fourier transform spectroscopy, see, e.g., Ref. 74), the interferogram (i.e., hologram) formed when the interferometer is illuminated by a polychromatic source having an intensity distribution $I(\sigma)$ is

$$I(x) = \int_0^\infty I(\sigma) [1 + \cos 2\pi\sigma\theta x] \, d\sigma \tag{96}$$

which is indeed identical to Eq. (91).

We have previously shown (see, e.g., Sections 2.1 and 4) that the complex amplitude $H(x)$, transmitted through a hologram on which has been recorded an intensity $I(x)$, may be given by an equation of the form

$$H(x) \cong I(x)^{-\gamma/2} \tag{97}$$

where γ is the gamma of the photographic plate. Under the usual holographic conditions, when $I(x)$ may be written in the form

$$I(x) \cong 1 + \varepsilon \tag{98}$$

we obtain, by a series expansion, the equation

$$H(x) \cong 1 - \frac{\gamma}{2} \varepsilon \tag{99}$$

which shows that the complex amplitude in the terms of interest in $H(x)$ (for Fourier-transform holography reconstruction) is linearly related to the recorded intensity under these conditions, and the reconstructed image intensity is independent of the gamma of the hologram.

To fulfill the condition of Eq. (98) in the strictest sense might require, according to Eq. (94), the equation

$$\tfrac{1}{2}(E_0{}^2 + E_1{}^2) > E_0 E_1 \tag{100}$$

implying that the intensity in one of the beams must be somewhat different from the intensity of the other beam; this condition is readily achieved in practice (for instance, by different coatings of the mirrors in the two interferometer arms). We have shown, however (see Figs. 41–43), that spectra can be reconstructed with an interferometer in which both mirrors have the same reflectivity, because the effect of having the same intensity in the two beams, in holography, generally only results in the appearance of higher order "spectra," somewhat like the higher order spectra in quadratic detection in electrical engineering, or indeed in optical gratings.[12] Further details of this analysis and detailed conditions are given in Ref. 80, together with additional experimental results, showing in particular high-resolution spectra which we have obtained by this method of holographic Fourier-transform spectroscopy.

Here, we may simply recall (see Section 4) that the relation between $H(x)$ and $I(x)$ may more generally be given in the form of a Taylor series expansion about the intensity I_0.

We may also wish to note here a dualism[68] between the recording of spectral frequencies in holographic spectroscopy and the recording of spatial frequencies in (single-spectrum frequency) holographic imaging, in particular in the case of spatially *noncoherent sources*.

A discussion of the limitations, characteristic of holography, associated with the recordability of the interference-fringe components, with special application to spectroscopy, and a comparison of the "luminosity-resolution" parameters with other methods of spectroscopy, with the purpose of singling out possible areas of application, will also be given in more detail in Ref. 80.

We may, however, give a very simple indication of the resolutions which are attainable by our method of holographic Fourier-transform spectroscopy.[73] Let the hologram be recorded within an aperture W along x. Let a_λ be the period of the "interference grating" corresponding to the wavelength λ in the hologram. According to the grating equation[12]

$$\sin i + \sin i' = m \frac{\lambda}{a} \qquad (101)$$

Illumination of the hologram with a plane wave Σ_i at the angle $i = 0$ will produce a first order image at

$$(\xi_0')_1 \cong f \frac{\lambda_L}{a_\lambda} = f \lambda_L v_\lambda \qquad (102)$$

(see Fig. 24), where λ_L is the wavelength of the laser light used in the reconstruction, and where we let $v_\lambda = 1/a_\lambda$ be the spatial frequency of the "interference grating" corresponding to the wavelength λ in the hologram. By differentiation of Eq. (102) it readily follows that

$$\frac{d(\xi_0')_1}{dv_\lambda} = f\lambda_L \tag{103}$$

i.e.

$$\Delta(\xi_0')_1 = f\lambda_L \, \Delta v_\lambda \tag{104}$$

We now further recall [see Eq. (94)] that the grating frequency v_λ is inversely proportional to λ. Let C be the constant of proportionality. We have

$$v_\lambda = \frac{C}{\lambda} \tag{105}$$

and therefore, with Eq. (104),

$$\Delta(\xi_0')_1 = -Cf\lambda_L \, \frac{\Delta\lambda}{\lambda^2} \tag{106}$$

According to the Rayleigh criterion (see Section 4 of Chapter 5 and Ref. 12), the spectral limit of resolution is obtained when (Fig. 24)

$$\Delta(\xi_0')_1 = f \, \frac{\lambda_L}{W} \tag{107}$$

where, we recall, W is the *width* of the hologram. In equating Eqs. (106) and (107), and by recalling Eq. (105), we finally have, for the *resolving power* of the holographic Fourier-transform spectroscopy method at the wavelength λ (to which corresponds a hologram fringe frequency v_λ), the equation

$$\text{R.P.} \equiv \frac{\lambda}{\Delta\lambda} = W v_\lambda = N \tag{108}$$

where N is *the total number of fringes recorded on the hologram* in the wavelength λ. The resolving power given by Eq. (108) is recognized to be identical to that of a grating of width W.[12] For example, let $v_\lambda = 2000$ fringes/mm (readily recordable on Kodak 649F spectroscopic plates). The resolving power is then

$$\text{R.P.} = 2000\,W \qquad (\text{per mm of } W) \tag{109}$$

A hologram as small as $W = 20$ mm (such as that of Fig. 42) may therefore yield a resolving power of 40,000 with this fringe frequency.

We may now note that the attainment of high resolution may at first sight appear to present some problems when an extended source aperture is used in the holographic interferometer. The use of extended source apertures (such as that shown in Figs. 41–43 which we used in our work) is of particular interest when very high luminosity is also important, beyond even that attainable in the basic form (small source aperture interferometric and Fourier-transform photoelectric spectroscopy) of which the high luminosity was first stressed by Jacquinot.[74,75,81,82]

Among the several advantages of holographic Fourier-transform spectroscopy methods, such as the one described, appears the possibility of recording spectra on a lensless, slitless, and completely static interferometer assembly of two mirrors (or their equivalent) simply pointed in the general direction of a diffuse source, useful for instance in astronomy or in the study of plasmas.

9.2. Image-Forming Holography with Spatially Noncoherent Light[72]

Stroke and Restrick[72] first noted (in analogy with the method of holographic Fourier-transform spectroscopy) that image-forming Fourier-transform holograms could be recorded with extended, *spatially noncoherent* monochromatic objects, and showed that high-quality images could be reconstructed from them by a second Fourier transformation (for example, by illuminating the hologram with a plane monochromatic wave (see Fig. 24), and by recording the images in the focal plane of a lens). The dualism between the recording of the spectral frequencies in the holographic Fourier-transform holograms (Section 9.1) and the recording of spatial frequencies in single-frequency holographic imaging had already been stressed previously by Stroke and Funkhouser[68] and by Stroke,[73] with particular reference to spatially noncoherent sources.

Before proceeding with the analysis, we recall[72] that an image-forming Fourier-transform hologram will be obtained, with a spatially noncoherent object, in an arrangement where one succeeds in producing, in the hologram, one sinusoidal interference-fringe grating per object point, with the correct spatial frequency and orientation (in the hologram), in such a way that the various gratings add in *intensity* and, therefore, result in the recording of the spatial Fourier-transform of the intensity distribution in the object. We also recall that the desired Fourier-transform summation in the hologram is made possible precisely because of the noncoherence of the various object points with each other.

For conciseness, we now give the analysis in the form first presented by Stroke and Restrick.[72]

As a model, we discuss a beam-splitting arrangement in which the projection of the object onto a plane (vector-coordinate $\bar{\xi}$) produces two equally intense images $I(\bar{\xi})$ and $I(-\bar{\xi})$ related to each other by a twofold rotation (see Section 4)[30] about a z-axis normal to the plane $\bar{\xi}$ (see, e.g., Fig. 44). Because of the spatial noncoherence of the various object points $I(\bar{\xi}_1)$, $I(\bar{\xi}_2)$, ..., $I(\bar{\xi}_r)$, each object point interferes only with its respective "mirror" image $I(-\bar{\xi}_1)$, $I(-\bar{\xi}_2)$, ..., $I(-\bar{\xi}_r)$, to produce, on a hologram plane $\bar{x}$ normal to the z-axis at a distance $z = f$ from the $\bar{\xi}$ plane, a fringe system of a frequency and orientation characteristic of that object point only. The resultant intensity $I(\bar{x})$ recorded on the hologram is given by the equation

$$I(\bar{x}) = \int I(\bar{\xi}) \left[1 + \cos 2\pi \frac{\bar{x}}{\lambda} \bullet \frac{2\bar{\xi}}{f} \right] d\bar{\xi} \tag{110}$$

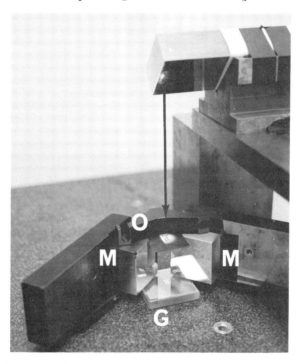

FIG. 44. Beam-splitting arrangement used for the recording of "lensless" Fourier-transform holograms in *noncoherent* light (see text), as seen from the hologram plane region (after Stroke and Restrick[72]). The object 0 (equal to the letter "R") is illuminated from above (indicated by vertical black line) and is made to produce two images by the mirrors M,M, upon "beam-splitting" diffraction by the grating G. Note the two-fold rotation with respect to each other of the two images seen on the mirrors M,M. (The image "R" appearing on the grating surface, directly below the object, is due to surface scattering, and is not seen by the hologram.) The grating dimensions (55 × 55 mm) give the scale of the arrangement. The Fourier-transform hologram was recorded on a Kodak 649F plate at a distance $f = 1$ meter from M,M, without the intervention of any additional optical elements. (One of the "wavefront-folding" arrangements recently independently published for purposes of stellar interferometry by L. Mertz[94] might also be suitable as an alternate beam-splitting arrangement for our method of noncoherent-light holography.)

(where ● indicates a vector dot product, and λ is the wavelength). The hologram intensity $I(\bar{x})$ given by Eq. (110) is recognized as formed of a constant ("dc") term, plus the cosine Fourier transform of the intensity distribution $I(\bar{\xi})$ in the object. Consequently, illumination of the hologram with a spatially coherent, monochromatic plane wave (Fig. 24) will produce, by Fourier transformation in the focal plane of a lens, two images $I(\bar{\xi})$ symmetrically displayed on the two sides of the optical axis (together with a "dc" image centered on the axis).

The beam-splitting arrangement used in our experiments is shown in Fig. 44, together with the object (letter "R") illuminated with spatially noncoherent light. A reconstruction of the image of the letter "R" is shown in Fig. 45. The actual beam splitter used in this case was an op-

FIG. 45. Fourier-transform reconstruction of the images of the letter "R" from the hologram recorded in *spatially noncoherent light* in the arrangement of Fig. 44.

tical diffraction grating[12] with a spacing of 1180 grooves/mm, and having a groove form shaped to produce two equally intense first orders, without polarization.[12,83] The spatial noncoherence with the monochromatic 6328 Å laser light used was achieved by imaging a rapidly moving diffuser onto the object plane. The success in having indeed achieved the desired noncoherence with the moving diffuser and the laser used was verified by also recording a hologram with the diffuser held stationary, so as to maintain spatially coherent light (see also Figs. 38–40). The reconstruction, in this case, can be readily shown (see Section 4) to consist of a convolution of $I(\bar{\xi})$ with itself, rather than of images of $I(\bar{\xi})$. A reconstruction of the "images" from the hologram recorded with spatially coherent light under these conditions is shown in Fig. 46.

Stroke and Restrick[72] noted the possible applicability of this method of "lensless" Fourier-transform holography with noncoherent light to image-forming x-ray microscopy, especially because of the possible use of grating-like and beam-splitting properties of crystals. We have previously stressed (Sections 3 and 4) the high-resolution advantages of Fourier-transform holograms compared to Fresnel-transform holograms. The

FIG. 46. Fourier-transform reconstruction of "images" (central "dc" term
and two side-band images) from hologram recorded in spatially *coherent* light
(obtained by keeping stationary the diffuser focused onto the object in Fig. 44)
to verify that spatial noncoherence was indeed obtained in the recording of the
hologram for Fig. 45 by moving the diffuser during the exposure (see text).

"plane-wave" summation property (or its spherical-wave equivalent) (see
Fig. 20) inherent in Fourier-transform holography, and its related absence of aberrations, is apparent, and analogous to similar advantages
of plane versus concave gratings. [12]

It is the author's hope that the general principles of coherent and noncoherent optics and holography, presented in this book, may help in
stimulating further inquiry and result in success in many exciting applications, some of which can now only be barely envisioned.

References

1. D. Gabor, *Nature* **161**, 777 (1948).

2. D. Gabor, *Proc. Roy. Soc. (London)* **A197**, 454 (1949).

3. D. Gabor, *Proc. Roy. Soc. (London)* **B64**, 449 (1951).

4. D. Gabor, *Research* **4**, 107 (1951).

5. D. Gabor, *in* Proceedings of the Congrés de Microscopie Electronique,
1952.

6. G. W. Stroke, *in* "Optical Information Processing" (J. T. Tippett, L. C.
Clapp, D. Berkowitz, and C. J. Koester, eds.). M.I.T. Press, Cambridge,
Massachusetts, 1965.

7. F. Zernike, *Physica* **1**, 43 (1934).

8. F. Zernike, *Z. Tech. Phys.* **16**, 454 (1935).

9. F. Zernike, *Physik. Z.* **36**, 848 (1935).

10. H. Wolter, *in* "Handbuch der Physik" (S. Flügge, ed.) Vol. 24, pp. 555–
645. Springer, Berlin, 1956.

11. G. W. Stroke, "An Introduction to Optics of Coherent and Non-Coherent
Electromagnetic Radiations," (Engineering Summer Conferences Text).
Univ. Michigan, Ann Arbor, Michigan, 1964 (2nd ed., 1965).

12. G.W. Stroke, "Diffraction Gratings," in "Handbuch der Physik" (S. Flügge, ed.),
Vol. 29, pp. 426-754.Springer, Berlin, 1967.

13. G. Rogers, *Nature* **166**, 237 (1950).

14. H.M.A. El-Sum, "Reconstructed Wavefront Microscopy," Ph.D. thesis,
Stanford Univ., Stanford, California, November 1952; available from
University Microfilms, Inc., Ann Arbor, Michigan (Dissertation Abstracts
4663, 1953).

15. G. W. Stroke, *J. Opt. Soc. Am.* **45**, 30 (1955).

16. E. Leith and J. Upatnieks, *J. Opt. Soc. Am.* **53**, 1377 (1963).

17. E. Leith and J. Upatnieks, *J. Opt. Soc. Am.* **54**, 1295 (1964).

18. G. W. Stroke, *Rev. Optique* **39**, 291 (1960).

19. G. W. Stroke, *J. Opt. Soc. Am.* **51**, 1321 (1961).

20. P. L. Jackson, *Appl. Opt.* **4**, 419 (1965).

21. A. Maréchal and P. Croce, *Compt. Rend.* **237**, 607 (1953).

22. A. Maréchal and P. Croce, *Compt. Rend.* **237**, 607 (1953).

23. E. L. O'Neill, "Selected Topics in Optics and Communication Theory," Optical Research Laboratory," *Boston Univ. Tech. Note* **133**, 1957.

24. A. Lohmann, *Opt. Acta* **3**, 97–99 (1956).

25. L.J. Cutrona, E.N. Leith, C.J. Palermo, and L.J. Porcello, *IRE Trans. Inform. Theory* **6** (3), 386-400 (1960).

26. G. W. Stroke and D. G. Falconer, *Phys. Letters* **13**, 306 (1964).

27. G. W. Stroke and D. G. Falconer, *Phys. Letters* **15**, 238 (1965).

28. G. W. Stroke, *Appl. Phys. Letters* **6**, 201 (1965).

29. G. W. Stroke, D. Brumm, and A. Funkhouser, *J. Opt. Soc. Am.* **55**, 1327 (1965).

30. G. W. Stroke, R. Restrick, A. Funkhouser, and D. Brumm, *Phys. Letters* **18**, 274 (1965); *Appl. Phys. Letters* **6**, 178 (1965).

31. G. Rogers, *Proc. Roy. Soc. (Edinburgh)* **A58**, 193 (1950-1951).

32. G. Rogers, *Proc. Roy. Soc. (Edinburgh)* **A63**, 193 (1952).

33. W. Bragg and G. Rogers, *Nature* **167**, 190 (1951).

34. J. A. Armstrong, *IBM J. Res. Develop.* **9**, 3 (1965).

35. H. H. Stroke, "Principles of Holography." Laser, Inc., Briarcliff Manor, New York, 1965.

36. A. V. Baez, *J. Opt. Soc. Am.* **42**, 756 (1952).

37. A. V. Baez and H. M. A. El-Sum, *in* "X-Ray Microscopy and Microradiography" (V.E. Cosslett, A. Engstrom, and H. H. Pattee, Jr., Eds.), pp. 347-366. Academic Press, New York, 1957.

38. M. Haine and J. Dyson, *Nature* **166**, 315 (1950).

39. M. Haine and T. Mulvey, *J. Opt. Soc. Am.*, **42**, 763 (1952).

40. P. Kirkpatrick and H. El-Sum, *J. Opt. Soc. Am.* **46**, 825 (1956).

41. U.S. Patent 3,083,615.

42. H. M. A. El-Sum, *in* "Optical Processing of Information" (D. K. Pollack, C. J. Koester, and J. T. Tippett, eds.), pp. 85-97. Spartan Books, Baltimore, Maryland, 1963.

43. E. Leith and J. Upatnieks, *J. Opt. Soc. Am.* **52**, 1123 (1962).

44. D. Gabor, G. W. Stroke, R. Restrick, A. Funkhouser, and D. Brumm, *Phys. Letters* **18**, 116 (1965).

45. G. W. Stroke, *Intern. Sci. Tech.* **41**, (1965).

46. G. Kellstrom, *Nov. Acta Reg. Soc. Sci. Uppsaliensis* **8**, 5 (1932).

47. G. W. Stroke, private communications to E. N. Leith and associates, 1963-1964.

48. J. M. Burch, *Nature* **171**, 889 (1953).

49. J. Dyson, *in* "Concepts of Classical Optics" (J. Strong, ed.), p. 383. Freeman, San Francisco, California, 1958.

50. R. E. Brooks, L. O. Heflinger, R. F. Wuerker and R. A. Briones, *Appl. Phys. Letters* **7**, 92 (1965).

51. W. L. Bragg, *Nature* **149**, 470 (1942).

52. W. Bragg, *Nature* **149**, 470 (1942).

53. M. J. Buerger, *Proc. Natl. Acad. Sci. U.S.* **27**, 117 (1941).

54. M. J. Buerger, *J. Appl. Phys.* **21**, 909 (1950).

55. M. J. Buerger, *Proc. Natl. Acad. Sci. U.S.* **36**, 330-335 (1950).

56. H. Boersch, *Z. Tech. Physik*, 337–338 (1938), especially footnote 3, p. 338.

57. J. C. Kendrew, G. Bodo, H. M. Dinitz, R. G. Parrish, W. Wyckoff, and D. C. Phillips, *Nature* 181, 662 (1958).

58. J. C. Kendrew, *in* "Biophysical Science—A Study Program" (J. L. Oncley, ed.), p. 94. Wiley, New York, 1959.

59. M. Perutz, *Sci. Am.* 211, 64 (1964).

60. G. W. Stroke and D. G. Falconer, *Phys. Letters* 15, 238 (1965).

61. G. W. Stroke, *Appl. Phys. Letters* 6, 201 (1965).

62. G. W. Stroke, D. Brumm, and A. Funkhouser, *J. Opt. Soc. Am.* 55, 1327 (1965).

63. D. Gabor, G. W. Stroke, R. Restrick, A. Funkhouser, and D. Brumm, *Phys. Letters* 18, 116 (1965).

64. D. Gabor, G. W. Stroke, R. Restrick, A. Funkhouser, and D. Brumm, *Phys. Letters* 18, 274 (1965).

65. G. W. Stroke, *Intern. Sci. Technol.* No. 41 (May 1965).

66. J. A. Armstrong, *IBM J. Res. Develop.* 9, No. 3, 171 (1965).

67. H. H. Stroke, "Principles of Holography." Laser, Inc., Briarcliff Manor, New York, 1965.

68. G. A. Stroke and A. Funkhouser, *Phys. Letters* 16, 272 (1965).

69. L. Mertz, advertisement, *J. Opt. Soc. Am.* 54, No. 10, iv (1964).

70. J. T. Winthrop and C. R. Worthington, *Phys. Letters* 15, 124 (1965).

71. G. Cochran, abstract, *J. Opt. Soc. Am.* 55, 615 (1965).

72. G. W. Stroke and R. C. Restrick, *Appl. Phys. Letters* 7 (1 November 1965).

73. G.W. Stroke, *in Proc. Zeeman Centennial Conference, Amsterdam, September 1965. Physica* 33, 253-267 (1967).

74. For example, P. Jacquinot, *Rep. Progr. Phys.* 23, 267–312 (1960).

75. Colloquium papers, in *J. Phys. Radium* 19, No. 3 (1958).

76. J. Connes, *Rev. Optique* 40, 45, 116, 171, 231 (1961).

77. P. Jacquinot and Ch. Dufour, *J. Rech. Centre Nat. Rech. Sci., Lab. Bellevue (Paris)* 6, 91 (1948).

78. P. Fellgett, Thesis, Cambridge University, 1951.

79. J. Strong, *J. Opt. Soc. Am.* 47, 354 (1957).

80. G.W. Stroke, *J. Phys. Radium* 28, Nos. 3-4, 196-203 (1967).

81. P. Jacquinot, *in* XVIIe Congress du G. A. M. S. Paris, 1954.

82. P. Jacquinot, *J. Opt. Soc. Am.* 44, 761 (1954).

83. G. W. Stroke, *Phys. Letters* 5, 45 (1963).

84. G. R. Harrison, R. C. Lord, and J. R. Loofbourow, "Practical Spectroscopy." Prentice-Hall, Englewood Cliffs, New Jersey, 1949.

85. E. Leith and J. Upatnieks, *Physics Today* 18, 26 (1965).

86. G. W. Stroke, D. Brumm, A. Funkhouser, A. Labeyrie, and R. Restrick, *Brit. J. Appl. Phys.* (submitted for publication).

87. W. T. Cathey, *J. Opt. Soc. Am.* 55, 457 (1965).

88. K. S. Pennington, *Microwaves*, p. 35 (October 1965). ⌊This paper also discusses Bragg-diffraction in thick holographic emulsions and presents a method of color holography⌋.

89. D. Gabor, G. W. Stroke, D. Brumm, A. Funkhouser, and A. Labeyrie, *Nature* 208, 1159 (1965).

90. W. L. Bragg, *Nature* 166, 399 (1950).

91. P. Tollin, P. Main, M.G. Rossmann, G.W. Stroke, and R. Restrick, *Nature* 209, 603-604 (1966).

92. G. N. Ramachandran and S. Raman, *Acta Cryst.* 12, 957 (1959).

93. G. N. Ramachandran, "Advanced Methods of Crystallography," p. 37 f. Academic Press, New York, 1964.

94. L. Mertz, "Transformations in Optics," Wiley, New York, 1965.

VII. FOURIER TRANSFORMS, CONVOLUTIONS, CORRELATIONS, SPECTRAL ANALYSIS, AND THE THEORY OF DISTRIBUTIONS†

Fourier transforms, and their various applications to convolutions, correlations, and distributions, are found to enter "naturally" into optics, notably into the theory of image formation, interferometry, spectroscopy, and, indeed, into holography. It is also found that even a basic treatment of Fourier-transform theory, such as that which we give below, provides a most powerful tool for dealing with the various problems in physical optics, diffraction theory, and interferometry. In many cases, a mere application of theorems, such as the Fourier-transform translation theorem or the convolution theorem and correlation theorem (which we give in the following sections) permits one to obtain the solution to problems which had in the past required the application of especially developed and frequently much more lengthy treatment.

1. Fourier Transforms

Given a function $f(x)$, where x is a real variable (with values ranging from $-\infty$ to $+\infty$), and such that $f(x)$ may be real or complex, but such that

$$\int_{-\infty}^{+\infty} |f(x)|\, dx \tag{1}$$

exists, i.e., that $f(x)$ belongs to a class L (one writes: if $f(x) \, \epsilon L$), then the Fourier transform $F(u)$ of the function $f(x)$ is by definition given by the equation

$$F(u) = \int_{-\infty}^{+\infty} f(x) \exp\,(2\pi i u x)\, dx \tag{2}$$

One frequently writes (2), for compactness, as

$$f(x) \longrightarrow F(u) \tag{3}$$

or

$$T[f(x)] = F(u) \tag{4}$$

where T stand for "Fourier transform of" and $\longrightarrow$ means "gives by Fourier transformation."

†Much of the mathematical material in Sections 1 and 2 of this chapter is based on the developments given by J. Arsac in his remarkable treatise.[1] Many other excellent treatises have been found equally useful—see Refs. 2-4. Section 3 on correlation functions was prepared by G. W. Stroke with R. Restrick, III.

One may also further define an inverse Fourier transform, obtained when a minus sign is taken in the exponent exp $(-2\pi iux)$.

For the function $F(u)$, and provided that $F(u)$ belongs to the class L, that is, if

$$F(u) \in L$$

we have

$$\int_{-\infty}^{+\infty} F(u) \exp (-2\pi iux) \, du = f(x) \tag{5}$$

that is, in compact notation

$$F(u) \longrightarrow f(x) \tag{6}$$

and

$$IT[F(u)] = f(x) \tag{7}$$

where IT stands for "inverse Fourier transform of." Henceforth only T will be used for indicating both FT and IT.

1.1. Some Properties of Fourier Transforms

Only some of the properties which are found to be very frequently useful in optics are given, many without proof.

1.2. Linearity

Let a_1 and a_2 be any complex constants. It readily follows from the definition in (2) that if

$$\left.\begin{array}{l} a_1 f_1(x) \\ a_2 f_2(x) \end{array}\right\} \in L \tag{8}$$

then

$$a_1 f_1(x) + a_2 f_2(x) \longrightarrow a_1 F_1(u) + a_2 F_2(u)$$

1.3. Translation Theorem

This most important property is given in the following equation, which can be readily demonstrated, as we do below. The translation property is: If

$$f(x) \longrightarrow F(u)$$

then

$$f(x - x') \longrightarrow F(u) \exp (2\pi iux') \tag{9}$$

when x' is a constant (translation, or shift), and $f(x - x')$ is equal to the function $f(x)$ with the origin shifted to $+ x'$. Indeed, one has

$$\int_{-\infty}^{+\infty} f(x - x') \exp (2\pi iux) \, dx = F(u)$$

With a change of variables (that is, with the implicit substitution of a "dummy" variable)

$$x - x' \quad \text{"becomes"} \quad y \equiv x \quad (y = \text{dummy variable})$$
$$dx \quad \text{"becomes"} \quad dx' = dx$$

and

$$\int_{-\infty}^{+\infty} f(x - x') \exp (2\pi iux) \, dx$$

"becomes"

$$\int_{-\infty}^{+\infty} f(y) \exp [2\pi iu (y + x')] \, dy$$

$$= \int_{-\infty}^{+\infty} f(y) \exp (2\pi iuy) \exp (2\pi iux') \, dy$$

$$= F(u) \exp (+ 2\pi iux') \qquad\qquad \text{Q.E.D.}$$

1.4. Reciprocal Translation Theorem

This property, symmetrical of that just demonstrated, readily follows, in the form: If

$$f(x) \longrightarrow F(u)$$

then

$$f(x) \exp (-2\pi iu'x) \longrightarrow F(u - u')$$

(10)

where u' is a constant.

Under the conditions where the Fourier transforms exist, and using the symbolism

$$f(x) \rightarrow F(u) \rightarrow f(-x) = f(x) \rightleftharpoons F(u)$$

(9) and (10) can be written:

If

$$f(x) \rightleftharpoons F(u)$$

then

$$f(x - x') \rightleftharpoons F(u) \exp (2\pi iux')$$

(9a)

and

$$f(x) \exp (-2\pi iu'x) \rightleftharpoons F(u - u')$$

(10a)

We may note that it is sometimes useful to use the notations (see Fig. 1)

$$[f]_{x-x'} \equiv f(x - x')$$

(11a)

$$[f]_{x+x'} \equiv f(x + x')$$

(11b)

$$[F]_{u-u'} \equiv F(u - u')$$

(11c)

$$[F]_{u+u'} \equiv F(u + u')$$

(11d)

where, for example, $[f]_{x+x'} \equiv f(x + x')$ is equal to the function $f(x)$ with its origin shifted from $x = 0$ to $x = -x'$. With (11a) to (11d) we may write

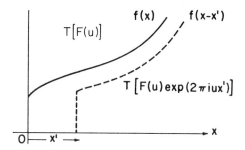

FIG. 1. Fourier-transform translation property, following Eq. (9).

the set of (9) and (10) in the form

$$[f]_{x-x'} \rightleftharpoons F \exp (2\pi i u x')$$
$$[f]_{x+x'} \rightleftharpoons F \exp (-2\pi i u x') \tag{9b}$$
$$f \exp (2\pi i u' x) \rightleftharpoons [F]_{u-u'}$$
$$f \exp (-2\pi i u' x) \rightleftharpoons [F]_{u+u'} \tag{9c}$$

We may finally note that the distinction between Fourier transformation, which involves $\exp (2\pi i u x)$, and inverse Fourier transformation, which involves $\exp (-2\pi i u x)$, may be of importance, when it is of importance to group two such sets of transformations, as (9b) and (9c), next to each other. An example is given in (58) to (64).

1.5. Table of Fourier-Transform Properties

The following table of "properties" is given without further proof, because of its importance in many optical applications. Most of the proofs can be readily obtained along the lines of the proofs given in the preceding section. x' and u' are constants.

$$f(x) \rightleftharpoons F(u)$$
$$f(x - x') \rightleftharpoons F(u) \exp (2\pi i u x')$$
$$f(x + x') \rightleftharpoons F(u) \exp (-2\pi i u x')$$
$$f(x - x') + f(x + x') \rightleftharpoons 2F(u) \cos (2\pi u x')$$
$$f(x - x') - f(x + x') \rightleftharpoons 2iF(u) \sin (2\pi u x')$$
$$2f(x) - f(x - x') - f(x + x') \rightleftharpoons 4F(u) \sin^2 \pi u x'$$
$$f(x) \exp (-2\pi i u' x) \rightleftharpoons F(u - u')$$
$$f(x) \exp (2\pi i u' x) \rightleftharpoons F(u + u')$$
$$f(x) \cos (2\pi u' x) \rightleftharpoons \tfrac{1}{2} [F(u - u') + F(u + u')]$$
$$f(x) \sin (2\pi u' x) \rightleftharpoons -\tfrac{1}{2} i [F(u + u') - F(u - u')]$$
$$f(x) \sin^2 \pi u' x \rightleftharpoons \tfrac{1}{4} [2F(u) - F(u + u') - F(u - u')]$$
$$\frac{df(x)}{dx} \rightleftharpoons -2\pi i u F(u) \qquad \frac{d^n f(x)}{dx^n} \rightleftharpoons (-2\pi i u)^n F(u) \tag{12}$$

1.6. Two-Dimensional and Multidimensional Fourier Transforms

Wherever x and u are used in the above equations, it is possible, with only the obvious restrictions [as in (1)] to use the variables x, y, z, etc., and u, v, w, etc., with corresponding multiple integrals and transforms.

1.7. Scale Change in Fourier Transforms

We shall prove that if

$$f(x) \longrightarrow F(u)$$

then   (13)

$$f(ax) \longrightarrow \frac{1}{|a|} F\left(\frac{u}{a}\right)$$

where a is a complex constant, with a magnitude $|a|$, different from zero.

Proof of

$$T = \int_{-\infty}^{+\infty} f(ax) \exp(2\pi iux)\, dx = \frac{1}{|a|} F\left(\frac{u}{a}\right) \tag{14}$$

Let $ax = y$ and $dx = (1/a)dy$. First let $a > 0$. $y \rightarrow \infty$, as $x \rightarrow \infty$ and

$$T = \frac{1}{a} \int_{-\infty}^{+\infty} f(y) \exp(2\pi iuy/a)\, dy = \frac{1}{a} F\left(\frac{u}{a}\right)$$

Now let $a < 0$. $y \rightarrow -\infty$, as $x \rightarrow \infty$, and

$$T = \frac{1}{a} \int_{+\infty}^{-\infty} f(y) \exp(2\pi iuy/a)\, dy = -\frac{1}{a} F\left(\frac{u}{a}\right)$$

For the general case, where $a > 0$ and $a < 0$, one has

$$f(ax) \longrightarrow \frac{1}{|a|} F\left(\frac{u}{a}\right) \qquad \qquad \text{Q.E.D.}$$

A special case of scale change arises for $a = -1$. One has

$$f(-x) = F(-u) \tag{15}$$

The important case of the transform of the complex conjugate of a function is dealt with in the next section.

1.8. Fourier Transform of the Complex Conjugate of a Function

This important case arises very frequently in optics. Let $f(x)^*$ describe the complex conjugate of the function $f(x)$. In other words, $f(x)^*$

is obtained by replacing all $i = \sqrt{-1}$ in the function $f(x)$ by $-i$.
One finds that

$$f(x)^* \longrightarrow F^*(-u) \qquad (16)$$

Indeed, we immediately have

$$\int_{-\infty}^{+\infty} f(x)^* \exp\,(2\pi iux)\,dx = \left\{\int_{-\infty}^{+\infty} f(x)\,\exp\,(-2\pi iux)\right\}^*$$

$$= F^*(-u) \qquad \text{Q.E.D.}$$

1.9. Fourier Cosine Transforms and Fourier Sine Transforms

The following expansion of the Fourier transform is frequently of use:

$$\int_{-\infty}^{+\infty} f(x)\,\exp\,(2\pi iux)\,dx = \int_{-\infty}^{+\infty} f(x)\,\cos\,(2\pi ux)\,dx$$

$$+ i \int_{-\infty}^{+\infty} f(x)\,\sin\,(2\pi ux)\,dx \qquad (17)$$

1.10. Symmetry Relations

The following important relations can be readily demonstrated, with the
the help of the previous equations.

If $f(x)$ is real

$$f^*(x) = f(x)$$

and

$$F^*(-u) = F(u) \qquad (18)$$

or

$$F^*(u) = F(-u)$$

If $f(x)$ is real and even

$$f(x) = f^*(x) = f^*(-x)$$

and $\qquad\qquad\qquad\qquad\qquad\qquad\qquad\qquad\qquad\qquad$ (19)

$$F(u) = F^*(-u) = F^*(u)$$

If $f(x)$ is real and odd

$$f(x) = -f(-x) = f^*(x)$$

and $\qquad\qquad\qquad\qquad\qquad\qquad\qquad\qquad\qquad\qquad$ (20)

$$F(u) = -F(-u) = F^*(-u)$$

2. Convolution Integrals

Convolution integrals are found to appear "naturally" in optics, as
they do in linear systems, in general. The following brief treatment is
therefore of considerable importance, throughout many parts of this text.

2.1. Definition of Convolution

By definition, the convolution integral $f(t)$ is obtained from two functions $f_1(t)$ and $f_2(t)$ according to the equation

$$f(t) = \int_{-\infty}^{+\infty} f_1(\tau) \, f_2(t - \tau) \, d\tau \qquad (21)$$

in which τ is taken as the independent (or running variable) and t represents successive values of a shift of $f_2(t)$ with respect to $f_1(t)$, for
which the integral of the product of the "overlapping" portions of $f_1(t)$
and of $f_2(t)$ is evaluated. In other words, the function $f(t)$ is obtained, in
analytical (or graphical) form, by successive evaluation of the integral
of the product of the two functions $f_1(t)$ and $f_2(t)$, as the two functions
are successively shifted with respect to each other. One value of the
integral is obtained for each position (or shift) t of $f_2(t)$ with respect to
$f_1(t)$, including a value $f(0)$, which is obtained for the value $t = 0$.

For compactness it is frequently convenient to write the convolution
integral of (21)

$$f(t) = f_1(t) \circledast f_2(t) \qquad (22)$$

2.2. Multidimensional Convolutions and Convolutions of Several
Functions

A two-dimensional convolution of two functions, with x and y as the
independent variables, can be written

$$f(x, y) = \int_{-\infty}^{+\infty} \int_{-\infty}^{+\infty} f_1(x_0, y_0) \, f_2(x - x_0, y - y_0) \, dx_0 \, dy_0 \qquad (23)$$

Convolutions of more than two functions exist, and have a physical
meaning, in particular in optics (e.g., in interferometry, spectroscopy,
holography, etc.). For instance, the transmission function (i.e., instrument function) $W(\sigma)$ of a Fabry-Perot interferometer, as a function of
the wavenumber $\sigma = 1/\lambda$, with the wavelength λ in centimeters, is given
(according to Chabbal)[6] by an equation of the form

$$W(\sigma) = A(\sigma) \circledast D(\sigma) \circledast F(\sigma) \qquad (24)$$

in which $A(\sigma)$ is the so-called *Airy function*, characteristic of the Fabry-Perot transmission, according to physical optics; $D(\sigma)$ is a *diffusion function*, characteristic of the effects of imperfections in the Fabry-Perot mirrors; and $F(\sigma)$ is the function which characterizes the use of a source pinhole of a finite diameter rather than a point source, as assumed in the derivation of the Airy function $A(\sigma)$. We may recall that the Airy function $A(\sigma)$ can be obtained in the form

$$A(\sigma) = \frac{\text{emerging flux}}{\text{incident flux}} = \left(\frac{T}{1-R}\right)^2 \Big/ \left[1 + \frac{4R}{(1-R)^2} \sin^2(2\pi n l \cos i)\right] \quad (25)$$

in which R and T are the flux-reflection and flux-transmission coefficients of the Fabry-Perot mirrors, l is the separation of the mirrors, n is the refractive index of the medium between the mirrors, and i is the angle of the plane waves between the mirrors with the optical axis.

A function similar to that of (24) can be shown to characterize other spectroscopic instruments (e.g., grating spectrometers). Interferometric arrangements used in the recording of holograms and in the many applications of holography can also be characterized by equations comparable to (24).

2.3. Convolution Theorem (Fourier Transform of Convolution Integral)

The Fourier transform of a convolution integral of two functions will now be shown to be equal to the product of the Fourier transforms of the two functions. We shall also prove the inverse theorem: The Fourier transform of the product of two functions is equal to the convolution of the Fourier transforms of the two functions.

To prove the first form of the convolution theorem, consider the function $f(t)$:

$$f(t) = \int_{-\infty}^{+\infty} f_1(\tau) \, f_2(t-\tau) \, d\tau \quad (26)$$

Let

$$f(t) \longrightarrow F(u) \quad (27)$$

by Fourier transformation, that is,

$$F(u) = \int_{-\infty}^{+\infty} \exp(+2\pi i u t) \left[\int_{-\infty}^{+\infty} f_1(\tau) \, f_2(t-\tau) \, d\tau\right] dt \quad (28)$$

Now, let us interchange the order of integration in (28), according to the usual assumptions, and also introduce the exponential factor inside the bracket, and take $f_1(\tau)$ outside the bracket:

$$F(u) = \int_{-\infty}^{+\infty} f_1(\tau) \left[\int_{-\infty}^{+\infty} \exp(+2\pi iut) \, f_2(t-\tau) \, dt \right] d\tau \qquad (29)$$

We now apply to (29) the Fourier-transform translation theorem of (9a). We recognize, accordingly, that the bracket in (29) can be written

$$\int_{-\infty}^{+\infty} f_2(t-\tau) \exp(+2\pi iut) \, dt = F_2(u) \exp(2\pi iu\tau) \qquad (30)$$

where $F_2(u)$ is the Fourier transform of $f_2(t)$. Equation (29) can now be written

$$F(u) = \int_{-\infty}^{+\infty} f_1(\tau) \, F_2(u) \exp(2\pi iu\tau) d\tau \qquad (31)$$

Since $F_2(u)$ can be taken outside the integral, when the integration with respect to τ is being carried out, (31) finally becomes

$$F(u) = F_2(u) \int_{-\infty}^{+\infty} f_1(\tau) \exp(2\pi iu\tau) \, d\tau \qquad (32)$$

which is immediately recognized as

$$F(u) = F_1(u) \cdot F_2(u) \qquad (33)$$

We may therefore conclude that if

$$f_1(t) \longrightarrow F_1(u)$$

and

$$f_2(t) \longrightarrow F_2(u)$$

then

$$f(t) = \int_{-\infty}^{+\infty} f_1(\tau) \, f_2(t-\tau) \, d\tau \longrightarrow F(u) = F_1(u) \cdot F_2(u) \qquad (34a)$$

that is,

$$\boxed{f_1 \otimes f_2 \longrightarrow F_1 \cdot F_2} \qquad (34b)$$

Equation (34) is the first form of the convolution theorem, which states that Fourier transform of the convolution of two functions is equal to the product of the transforms of the two functions.

Next we demonstrate the second form of the convolution theorem. Let $f(t)$ be the convolution of two functions $f_1(t)$ and $f_2(t)$; that is,

$$f(t) = \int_{-\infty}^{+\infty} f_1(\tau)\, f_2(t - \tau)\, d\tau \tag{35}$$

and let $f_2(t)$ be the Fourier transform of a function $F_2(u)$; that is,

$$f_2(t) = \int_{-\infty}^{+\infty} F_2(u)\, \exp(+2\pi iut)\, du \tag{36}$$

It follows that

$$f_2(t - \tau) = \int_{-\infty}^{+\infty} F_2(u)\, \exp[+2\pi iu(t - \tau)]\, du \tag{37}$$

We now introduce (37) in (35) and obtain

$$f(t) = \int_{-\infty}^{+\infty} f_1(\tau) \int_{-\infty}^{+\infty} F_2(u)\, \exp(+2\pi iut)\, \exp(-2\pi iu\tau)\, du\, d\tau \tag{38}$$

Because of the uniform convergence of the second integral with respect to τ, we may invert the orders of integration in (38) and obtain

$$f(t) = \int_{-\infty}^{+\infty} F_2(u)\, \exp(+2\pi iut) \underbrace{\int_{-\infty}^{+\infty} f_1(\tau)\exp(-2\pi iut)d\tau}_{F_1(u)}\, du \tag{39}$$

The second integral is recognized as $F_1(u)$, the inverse Fourier transform of $f_1(\tau)$ and (39) can therefore be written, finally, as

$$f(t) = \int_{-\infty}^{+\infty} F_1(u)\, F_2(u)\, \exp(+2\pi iut)\, du \tag{40}$$

In comparing (40) with (35) we may, therefore, conclude that

$$\boxed{f_1 \otimes f_2 \longleftarrow F_1 \cdot F_2} \tag{41}$$

Equation (41) states that the Fourier transform of the product of two functions is equal to the convolution of the transforms of the two functions.

3. Correlation Functions

An important distinction must be made between convolution integrals, on one hand, and correlation functions, on the other, especially because they appear, at first sight, to have a comparable form.

Basically, a correlation function, rather than a convolution function, will arise as a suitable representation, whenever one of the two functions in the representation is a conjugate of some function.

3.1. Definition of Correlation Functions

By definition, an autocorrelation function $\varphi_{11}(\tau)$ is given by the expression

$$\varphi_{11}(\tau) \equiv \int_{-\infty}^{+\infty} f_1(t) f_1^*(t + \tau)\, dt \tag{42}$$

and a cross-correlation function φ_{12} is given by the expression

$$\varphi_{12}(\tau) = \int_{-\infty}^{+\infty} f_1(t) f_2^*(t + \tau)\, dt \tag{43}$$

Similarly, we have

$$\varphi_{21}(\tau) = \int_{-\infty}^{+\infty} f_1^*(t + \tau) f_2(t)\, dt \tag{44}$$

3.2. Transforms of Correlation Functions

Consider the product

$$F_1 F_2^* \tag{45}$$

where

$$F_1 \longrightarrow f_1 \tag{46}$$

and

$$F_2 \longrightarrow f_2 \tag{47}$$

by Fourier transformation. We wish to obtain an expression for the Fourier transform of the product $F_1 F_2^*$. We first note [see (41)] that (45) gives

$$F_1 F_2^* \longrightarrow f_1(t) \otimes f_2^*(-t) \tag{48}$$

At this point, we need to write out (48) completely. We have

$$f_1(t) \otimes f_2^*(-t) = \int_{-\infty}^{+\infty} f_1(\tau) f_2^*[t - (-\tau)]\, d\tau \tag{49}$$

that is,

$$f_1(t) \otimes f_2{}^*(-t) = \int_{-\infty}^{+\infty} f_1(\tau) f_2{}^*(t + \tau) \, d\tau \qquad (50)$$

Equation (50) is recognized, according to (43), as the correlation function $\varphi_{12}(\tau)$. We may therefore write

$$F_1 F_2{}^* \longrightarrow f_1(t) \otimes f_2{}^*(-t) \equiv \varphi_{12}(\tau) \qquad (51)$$

It is frequently helpful to write out the correlation function in a form comparable to the convolution integral. We define by the symbol $\star$ the correlation operation, according to the equation

$$\varphi_{12}(\tau) \equiv f_1(t) \star f_2{}^*(t) \qquad (52)$$

Finally, we may write (51) in the form

$$\boxed{F_1 F_2{}^* \longrightarrow f_1 \star f_2{}^*} \qquad (53)$$

Similarly, we find

$$F_1{}^* F_2 \longrightarrow f_1{}^* \star f_2 \qquad (54)$$

We may also note the expression for the transformations obtained under the condition

$$f_1 \longrightarrow F_1$$
$$f_2 \longrightarrow F_2 \qquad (55)$$

when we have

$$f_1 f_2{}^* \longrightarrow F_1 \star F_2{}^* \qquad (56)$$

and

$$f_1{}^* f_2 \longrightarrow F_1{}^* \star F_2 \qquad (57)$$

We may note, however, that care must be taken in recognizing that the $+$ and $-$ signs in $\exp(2\pi iux)$ and $\exp(-2\pi iux)$, respectively, lead one to a transform, and the other to an inverse transform. If this is taken into account, we may write a complete set of two-way transformations as: Given

$$g_1 \longrightarrow G_1$$
$$g_2 \longrightarrow G_2 \qquad (58a)$$

then [see the definition in (2)]

$$G_1 \longrightarrow g_1(-x)$$
$$G_2 \longrightarrow g_2(-x) \qquad (58b)$$

and the set of *transformation rules for correlation functions* becomes

$$g_1 g_2{}^* \longrightarrow G_1 \star G_2{}^*$$
$$g_1{}^* g_2 \longrightarrow G_1{}^* \star G_2 \tag{59}$$

and

$$G_1 G_2{}^* \longrightarrow (g_1 \star g_2{}^*)_{-x}$$
$$G_1{}^* G_1 \longrightarrow (g_1{}^* \star g_2)_{-x} \tag{60}$$

Finally, for the set of autocorrelation functions, we have the following *set of transformation rules for autocorrelation functions:* Given

$$f_1 \longrightarrow F_1$$
$$f_1{}^* \longrightarrow F_1{}^*(-u) \tag{61}$$

we have

$$f_1 f_1{}^* \longrightarrow F_1 \star F_1{}^*$$
$$f_1{}^* f_1 \longrightarrow F_1{}^* \star F_1 \tag{62}$$

We may note that (62) is known as *Parseval's theorem*, which can be taken, for example, as a statement of energy conservation. We may further note that

$$F_1 F_1{}^* \longrightarrow (f_1 \star f_1{}^*)_{-x}$$
$$F_1{}^* F_2 \longrightarrow (f_1{}^* \star f_2)_{-x} \tag{63}$$

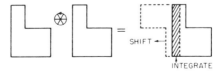

FIG. 2. Correlation and convolution functions: graphical illustration.

and also, by comparison of (63) with (51) that

$$f_1(t) \circledast f_1{}^*(-t) = [f_1 \star f_1{}^*]_{-t} \qquad (64)$$

A graphical representation of the convolution and correlation operations is given in Fig. 2. We may note, in particular, the significant difference between an autocorrelation and a convolution for functions (such as that illustrated) which have no twofold rotation symmetry. The same remark applies, of course, to cross-correlation functions! The remark is of particular significance in coherent and even in noncoherent optical image processing (see Section 7.1 in Chapter V and Section 4 in Chapter VI).

4. Distributions

The theory of distributions, as first described by Laurent Schwartz in 1950–1951, and developed by many authors (in particular, J. Arsac, A. Erdélyi, M. J. Lighthill, and G. Temple), is proving to be an increasingly powerful tool in modern optics and electro-optical science. The reader is referred to one of the new treatises for even the basic definitions. However, a few examples will be used to illustrate the power of the method.

4.1. Definitions

The "Dirac delta function" $\delta(x)$ may be defined as a *distribution* by the equation

$$\left< \delta(x-a), \, g(x) \right> = g(a) \qquad (65)$$

which, for $a = 0$, becomes

$$\left< \delta(x), \, g(x) \right> = g(0) \qquad (66)$$

where $g(x)$ belongs to a certain class of "testing functions" (e.g., the class of infinitely differentiable functions with finite support). In a broader sense, distributions belong to the class of linear "functionals" (i.e., functions of functions).

The notion of distributions enables one to define Fourier transforms "in the sense of distributions" for functions which have no Fourier transform in the strict sense of Eqs. (1) and (2). We may recall that a function can have a Fourier transform in the strictest sense only if it is absolutely integrable, that is if

$$\int_{-\infty}^{+\infty} |f(x)| \; dx < \infty \qquad (67)$$

For instance, the function $f(x) = 1$ does not have a transform in this sense. The form of Eq. (65) is also known as a *scalar product*. For the case of a *function* $f(x)$ a scalar product is defined by the integral

$$\int_{-\infty}^{+\infty} f(x) \; g(x) \; dx = \left< f(x), \, g(x) \right> \qquad (68)$$

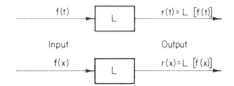

FIG. 3. Linear systems (t = temporal, x = spatial).

4.2. Impulse Response of a Linear System

Let L be a linear operator, operating on $f(t)$ or $f(x)$, shown in Fig. 3. We require the linear system to have the properties of:

Linearity:

$$L[a_1 f_1(t) + a_2 f_2(t)] = a_1 L[f_1(t)] + a_2 L[f_2(t)] \qquad (69)$$

with a_1 and a_2 = constants and

Time (or space) invariance:

If
$$L[f(t)] = r(t)$$

or
$$L[f(x)) = r(x)$$

then
$$L[f(t - t_1)] = r(t - t_1) \qquad (70)$$

or
$$L[f(x - x_1)] = r(x - x_1)$$

The impulse response of the linear system (optical system, electrical network, etc.), is represented by $h(t)$ or $h(x)$ (see Fig. 4).

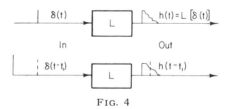

FIG. 4

For an *optical filter* (e.g., interference filter, optical network[5]) $h(t)$, the time impulse response, is in order because the component waves of different temporal frequencies are delayed by different amounts by propagation through the filter. For an *optical image-forming system*, $h(x)$, the space impulse response is in order. (The space impulse response $h(x)$ is nothing but what is ordinarily called the diffraction pattern.)

4.3. Response of a Linear System to an Arbitrary Input Function

Now let the input function $f(t)$ be formed of a *sum of unit impulses:*

$$f(t) = \int_{-\infty}^{+\infty} f(t_1)\,\delta(t_1 - t)\,dt \qquad (71)$$

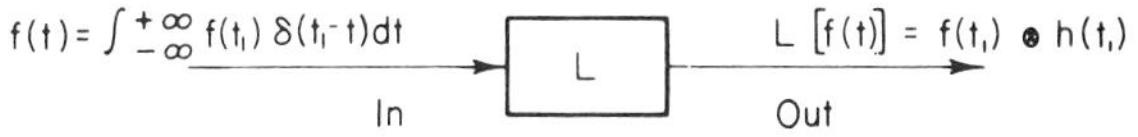

$$f(t) = \int_{-\infty}^{+\infty} f(t_1)\,\delta(t_1-t)\,dt \qquad \text{In} \qquad L \qquad \text{Out} \qquad L[f(t)] = f(t_1) \circledast h(t_1)$$

FIG. 5

Recalling that $<\delta(t_1-t), f(t_1)> = f(t)$ according to (65), we have the relation between input and output shown in Fig. 5, where, $h(t_1)$ is the impulse response of the system L.

One immediately finds for the output function

$$L[f(t)] = L\left[\int_{-\infty}^{+\infty} f(t_1)\,\delta(t_1-t)\,dt_1\right]$$
$$= \int_{-\infty}^{+\infty} f(t_1)\,L[\delta(t_1-t)]\,dt_1 \tag{72}$$

because of the fact that L is a linear operator. It follows that

$$L[f(t)] = r(t) = \int_{-\infty}^{+\infty} f(t_1)h(t-t_1)\,dt_1 \tag{73}$$

showing the remarkable fact that the output $r(t)$ is equal to the *convolution* of the impulse response function with the input function.

In case of noncoherent image formation, for which the diffraction pattern or on-axis spread function is

$$h(x) = s(x)$$

one has, for an input $f(x)$ equal to the geometric-optics image of the object, that is, for $f(x) = O(x)$, the output $r(x) = I(x)$, where

$$I(x) = O(x) \circledast s(x) \tag{74}$$

as already indicated in III.20.

5. Spectral Analysis[†]

Let $\Phi(\omega)$ be the spectral energy distribution in the input of a linear system (Fig. 6). The output $R(\omega) = L[\Phi(\omega)]$ is as shown in Fig. 6. $R(\omega)$ may be the spectrum as recorded in a grating spectrometer, as a function of ω or λ, or the response in an optical spectrum analyzer using heterodyning detection.

Let the impulse response of the spectrometer be $h(\omega)$. Also, let the input $\Phi(\omega)$ be given by the equation

$$\Phi(\omega) = \int_{-\infty}^{+\infty} \Phi(\omega_0)\delta(\omega_0-\omega)\,d\omega_0 \tag{75}$$

[†]See also Chapter IV.

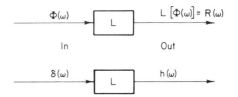

FIG. 6. Spectroscopic (linear) systems. $h(\omega)$ represents the spectral "image" at the frequency ω.

One has for the output, according to (73), the function

$$R(\omega) = \int_{-\infty}^{+\infty} \Phi(\omega_0) h(\omega_0 - \omega) d\omega_0 \qquad (76)$$

that is,

$$R(\omega) = \Phi(\omega) \otimes h(\omega) \qquad (77)$$

which is recognized as the basic equation governing the behavior of spectrometers. $R(\omega)$ is the quantity actually recorded in the spectrometer. One has by Fourier transformation of (77) [see (34b)],

$$T[R(\omega)] = T[\Phi(\omega)] \cdot T[h(\omega)] \qquad (78)$$

We recall that (see, for example III.23) for optical as well as for electrical network systems,

$$T[h(\omega)] = \text{frequency response of system} \qquad (79)$$

References†

1. J. Arsac, "Transformations de Fourier et théorie des distributions." Dunod, Paris, 1960.
2. R. C. Jennison, "Fourier Transforms and Convolutions for the Experimentalist." Pergamon Press, New York, 1961.
3. M. J. Lighthill, "An Introduction to Fourier Analysis and Generalized Functions." Cambridge University Press, London and New York, 1958.
4. A. A. Kharkevich, "Spectra and Analysis," translated from Russian. Consultants Bureau, New York, 1960.
5. See, for example, G. W. Stroke, Proc. IEEE 52, 855 (1964).
6. R. Chabbal, Rev. Opt. 37, pp. 49, 366, 501 (1958).
7. E. L. O'Neill, "An Introduction to Statistical Optics," Addison-Wesley, Reading, Massachusetts, 1963.

†See also *Mathematical References* on page 15, at end of Chapter I.

APPENDIX: APPLICATIONS OF HOLOGRAPHY[†] by I. P. NALIMOV

1. Introduction

The interest in holography is not exhausted by the fact that it brought about the possibility of the introduction of the third dimension into optics. Holography touched upon almost all of the traditional areas of applied optics and caused the review of the previously formed boundaries between them. The principle of holography enabled one to reconsider certain areas of infrared techniques, super-high-frequency techniques, acoustics, and x-ray and electron microscopy. In other words, this principle affects all those areas where wave interference plays a role.

However, when one considers the present state of holography, it is convenient to classify its applications according to the utilized properties of the holograms and not according to the types of radiations utilized in order to obtain the holograms. The proposed classification of holography applications (see Table I) does not pretend to be universal, but it allows one to systematize over a hundred papers published mainly after the appearance of Stroke's monograph. A number of earlier papers were also used, since Stroke's book describes the questions of application very briefly. Following this principle, it is possible to distinguish ten main directions.

The most obvious application of holography is the three-dimensional photograph. Here one uses a combination of the photographic and locational properties of the hologram which allow one not only to register the image, but also to determine the distance to each of its points. In this category, one can also mention such processes as sound-, radio-, and infrared-vision, which all possess a common first stage—the recording of the hologram. In order to transfer the image into the visible region, one uses one additional property of holograms: By changing the scale of the interference picture proportionally to the change in the wavelength, it is possible, when reconstructing the image in visible light, to conserve its three-dimensionality.

Image recognition is one of the most interesting and promising applications of holography. It is based on the ability of the hologram to pick out from a group of objects only those whose "images" are recorded in it. What determines the most interesting property of holograms in the area of pattern recognition, which almost brings them close to ideally matched filters? Let us consider the most general holographic scheme (Fig. 1). Let A and B be coherently radiating objects; in particular they may be any objects illuminated by a sufficiently coherent laser. If the

[†] Translated from the Russian by permission. (Edited by GWS.)

TABLE I
HOLOGRAPHY APPLICATIONS

Three-Dimensional Photography	Technology
Image photography Photogrammetry Contour photography Pulsed-laser photography of moving objects Underwater photography Sound vision Radio vision Microwave antenna modeling	Surface application of complicated microimages Microfinishing **Imaging through distorting media** Observation of the walls of incorrect shape Image-coding Observations in a turbulent atmosphere
Image Recognition	
Reading of prints and manuscripts Three-dimensional object recognition Aerial photograph analysis Associative (correlative) search	**Microscopy** Three-dimensional observation of living micro-objects X-ray microscopy Electron microscopy
Volume Holograms	**Cinematography**
Wave photography Memory systems of high capacity with an associative choice	Three-dimensional projection systems
Interferometry	**Television**
Measurement of complex-surface vibrations Measurement of unfinished complex-surface deformations Three-dimensional phase objects. Aerohydrodynamics Interferometric measurements Nondestructive testing	Transmission of holograms over distances **Optics** Compensation of lens aberrations Lensless optics Combined lens-holographic aberrationless systems

radiation is capable of creating, in a certain plane, an interference picture—a system of standing waves—then one may place there a photographic plate and obtain a hologram. By illuminating the hologram $A + B$ with an ideal copy of the initial wavefront of one object (for instance the object B), we shall obtain a faithfully reconstructed wavefront capable of faithfully imaging the other object. The hologram will transmit

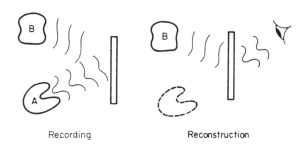

Recording   Reconstruction

FIG. 1. The most general holography scheme.

only that part of the spatial spectrum which is close to the spectrum recorded on it. In other words, it will "respond" only to the image of one of "its own" objects, with the condition that the object be placed in the appropriate position. †If, for instance, the object B is a combination of point sources comprising the code of the letter, then the hologram $A + B$ illuminated by the letter A permits one to reconstruct (i.e., to extract from the hologram) the code B. Using recordings with various angles on the same hologram, one can record many letters. Thus one finds a new channel of communications between man and the computer which will liberate the operator from the manual introduction of data. Moreover, it will probable be possible, using this method, to resolve a very important problem of computational technology—the recognition of multidimensional images.

Holographic memory aroused great interest in view of its extremely high capacity (the theoretical limit is 10^{12} bits/cm^3), and also because of its associative properties, due to which it resembles human memory. ‡ The mechanism of holographic memory is related to the interference recording of information in a photosensitive volume. Too few investigations

---

†Usually one takes a plane or a spherical wavefront for B, since it is easier to form such a front and to aim it at the hologram. [In reality, this may be achieved in all rigor only under certain "correlation" conditions (see pp. 127 - 137, dealing with "Resolution-retrieving compensation of source effects in holography.") (Note by GWS.)]

‡For a mathematical analysis of the analogies between holographic recording and the human memory, see e.g. D. Gabor.[34] See also Longuet-Higgins,[87] and Van Heerden.[128] (Note by GWS.)

were conducted along these lines for us to be confident about the possibilities of this memory. The delay lies mainly in the development of
photosensitive materials of high optical quality and sensitivity. [†]

It is expected that quite soon industry will have a new method for
nondestructive surface control which will be based on the use of holograms. [‡] This application of holography is a particular case of differential
interferometry. In the same manner that the hologram "recognizes" objects
which are recorded on it, it also can be made to display changes of shape
or displacements of the object. This may be achieved by successively
recording, in the same hologram, the "component" holograms corresponding to the two "states" of the object. Usually the quantitative indicators
characterizing these changes can be found from the structure and the
density of the interference fringes covering the reconstructed image and
formed by the superposition of the waves from the object and the waves
of the same object after its change, as reconstructed by means of the
same hologram. Eventually this line of research will enable us to have
contactless control of complex unfinished surfaces (their vibrations, deformations, cracks, and changes in their reflective properties). The
factors which have slowed down a general industrial development of this
method are mainly of a technical rather than a theoretical nature. For
instance, in holographic interferometry, just as in other forms of holography, one has to guarantee total stability of the optical elements which
yield the interference picture.

The technological applications of holography—the utilization of the
real image for purposes of testing, processing and manufacture—are just
beginning to be developed. Nevertheless, they possess a great future.
The advantages of holographic processing (or machining) of materials
vis-à-vis the usual laser processing are related to the possibility of
contactless superposition of the most complicated designs and apertures
on surfaces of the most involved form, as well as to the absence of
lenses. Using a hologram, it is possible to obtain, within the limits of
the field, approximately an order more of resolved elements than by
means of the best lens. The reason for this is that the lens acts as an
ideal lens only in the vicinity of the axis, while its field resolution

---

[†] The use of alkali-halide crystals (e.g. KBr and KCl), such as those under
study in a number of laboratories (e.g. Carson Laboratories, Bristol, Connecticut)
as well as of photochromic glasses such as those under study (e.g. by Corning
Research Laboratories, Raleigh, North Carolina) appears to have shown good
initial promise for such holographic storage purposes. (Note by GWS.)

[‡] In the U.S. several companies are actively working on the problem (see,
for example, the striking results demonstrated by GC-OPTRONICS of Ann
Arbor, Michigan, Figs. 5A and 5B). (Note by GWS.)

falls off near the edges. A hologram, on the other hand, has a more uniform distribution of resolution over the field. Holographic technology
may acquire unusual qualities when one combines the focusing, recognition, and inspection properties of holograms. The development in
this direction requires the increase in power and coherence of laser
beams. [†]

Holographic inspection not only widens the sensitivity of the human
eye for all laser frequencies, but also opens the possibility of three-
dimensional observation of objects located behind opaque walls of
arbitrary form, as well as in inhomogeneous and scattering media. In this
case, the hologram is registered in IR- or UV-laser radiation. The
change of the three-dimensional image into the visibile spectrum may
be accomplished in the same manner as in sound vision and microscopy
using scale transitions, i.e. a decrease or increase of the interferogram
(hologram) proportional to the change in the wavelength. What amounts,
in effect, to a "removal" of the wall is accomplished through its
"compensation" during the reconstruction of the real image. The lack
of developments in this direction is mainly due to the absence of
multi-element IR receivers with high resolution. [†]

The creation of a lensless holographic microscope will allow biol-
ogists and physicians to observe three-dimensional images of living
tissues and **microorganisms**.[§] The use of x-ray radiation brings about
the possibility of large magnifications (up to 10^6) with the retention of

---

[†]Many of the remarks with regard to the status of industrial technology in
many of the fields pertain, no doubt, more to the Soviet Union than, for example,
to the U.S. (Note by GWS.)

[†]It would appear that these concepts, rather vaguely expressed, might be of
a particular interest also in acoustical holography. (Note by GWS.)

[§]It appears increasingly, as a result of recent investigations, that perhaps a
most promising form of holography, in view of three-dimensional microscopy, is
that based on the method of "white-light reflection holography," first described
by G.Stroke and A. Labeyrie.[119] This appears to be the only method, so far, capable of simultaneously storing, in the same hologram, very widely extended
lateral fields (normal to the optical axis) in addition to the great depth of field
(along the optical axis), and significantly without lenses, thus preserving the
high-resolution capability, which characterizes "lensless" holographic imaging.
This method is now being developed with regard to cinematography of blood circulation in capillaries *in vivo* by Stroke and Burke, notably for multicolor
imaging. A recent adaptation of these principles is a method by K.A. Stetson,[134]
holography with total internally reflected light, which materializes these advantages in cases where reflection of the reference beam from the emulsion itself
may be acceptable. A new type of high-resolution wide-field lens capable of

resolution. The principle of the microscope action, which is well described in Stroke's book, is based on scale transformations and geometric magnification in divergent beams. The difficulties of realization of high resolution microscopy are due to the absence of x-ray lasers.† For the microscopy of a living cell, it is necessary to improve the coherence and power of the existing ultraviolet lasers.

Holographic movies will bring about the possibility of projecting and observing three-dimensional dynamical images. One of the possible realizations of three-dimensional cinematography may be the following. Using different angles, one registers on the hologram various moments of the scene. During reconstruction the hologram (or the illuminating beam) is rotated, creating the effect of motion. The viewer looks at the hologram as in a window, behind which the events take place. However, the method of creating large holograms the size of a cinema screen through which many people might be observing the events, is still not clear.‡ Moreover, we still do not have the method for magnification of three-dimensional images, since a simple projection is unsuitable here. All this limits, at least for the near future, the future of holographic cinema to demonstrations, or special applications (for instance, on-board devices for "blind" landings of aircraft).

imaging a volume 4.2 mm in diameter and 120 μ deep with a resolution in excess of 1,100 lines/mm recently developed by J. Wilczynski and R.E. Tibbetts [135] of IBM and independently by E. Hugues of CERCO, France may perhaps prove to be the most important new development in this direction by permitting one to premagnify the field before its holographic recording, and thus to surmount certain emulsion shrinkage and observation difficulties, among others. (Note by GWS.)

† In contrast with a general belief, Stroke showed as early as 1966 that no x-ray "lasers" are in fact required for x-ray holography! This may be readily understood if one recalls that it is a property of diffracting crystals, when diffracting comparatively incoherent (spatially and temporally incoherent) x-ray radiation, to make the emergent "diffracted" beams precisely sufficiently monochromatic (both spatially and temporally) to produce the observed diffraction patterns by Bragg "interference" of the beams from different crystal "layers"! If this were not true, there would in fact be no x-ray diffraction. In simple words, it may be said that a crystal acts with respect to incident x-ray beams as a spatial and temporal "coherence" filter! The importance of this conclusion cannot be overestimated, of course. (Note by GWS.)

‡ Such holograms have now been demonstrated by the Conductron Corp. of Ann Arbor, Michigan, notably, most recently, by the recording of holograms of living persons (coherence "volume" in excess of 1 cubic meter), using a new type of high-coherence pulsed ruby laser system L.D. Siebert. [133]

Three-dimensional television, using the holographic principle, may be, at present, realized only in its simplest form. Its creation is hampered by the absence of dynamical (erasable) holograms and by the absence of methods for reading such holograms, [†] as well as by the impossibility of transmission of large volume information. Undoubtedly, quantum optics and later technology are moving in the direction which is of interest to three-dimensional television. Holography, one way or another, utilizes systems of super-wideband optical communication, modulation, and scanning of light beams developed for other purposes. Even now the holographic principle will, possibly, guarantee an increase in transmission reliability and may perhaps be used for the coding of images.

Finally, the hologram will be a serious competitor to the lens. While possessing as good a resolution in the focal spot, holograms, as distinguished from lenses, do not violate the three-dimensionality of the images. Furthermore, they are simple in preparation and are lighter. Thus, point holograms—Fresnel's zone plates—may be used for focusing of wide beams of up to 10 m in diameter, while the creation of lenses or mirrors of such dimension is simply unreasonable. The utilization of holograms in conjunction with lenses will enable us to create uncomplicated aberrationless systems.

2. Three-Dimensional Photography

The creation of the laser is, first of all, the creation of a new source of visible radiation. As such, the laser opens up unique possibilities for image projection. "It is specifically in the field of visual applications that the laser will find one of its most important applications" wrote the pioneer of laser technique, Shawlow. [107]

The properties of laser holograms and the ways to obtain them for three-dimensional image applications were thoroughly discussed in a series of reviews. [3,77,91,102,111,112] Let us note here two main advantages of holographic representations vs. the usual photographic ones: (1) When "holographing" an object, one need not focus the rays; therefore there is no danger of obtaining a blurred image because of a

[†] Several systems using moving lens elements for television purposes have been described (e.g. Phillips, Eindhoven). A microscope, using a superposition of sections in the same microphotograph for some special applications has also been recently reported. [109]

lack of focusing; (2) the hologram registers the whole object equally sharply throughout its depth. This cannot be achieved by any sophistication of photographic technique. [†] Another advantage of holography lies in the fact that the nonlinearity of photoregistration does not influence the quality of "tone" transmission. Even using the most contrasting emulsion for holographic recording, one may still reconstruct a continuous-tone image.

The quality of the three-dimensional representations obtained by means of a laser is already so high that Gabor had the right to make the semijocular statement[22]: "In order to obtain such pictures with the help of a mercury-arc lamp, I would have to collimate its radiation in such a manner that the exposure would have taken me as much time as has elapsed from the moment of the discovery of holography in 1947 until today."

The new development in this direction proceeded along the lines of the improvement of already well-known methods,[9,63] especially by the introduction of pulsed-laser holography using ruby lasers.[4,5,50] Moreover, within this development a number of new applications of holography were demonstrated: thus, it was used for underwater pictures,[39,54] for preparation of three-dimensional contour maps,[45,46] and for the reconstruction of directivity patterns of microwave antennas.[2]

Historically the first practical application of holography should be considered the disdrometer. This device[21,100] was designed for the investigation of rapidly moving particles suspended in the atmosphere, such as rain drops or fog particles, snow flakes, ice crystals, or aerosols of the size of 3 to 3000 μ. Usual photography does not allow one to maintain in focus each moving particle for the duration of time necessary for the exposure. The photographic method is unable to register all the particles within a certain volume simultaneously and with the same sharpness. The disdrometer eliminates these difficulties. A high-quality modulated ruby laser, of 10 mW power, illuminates the moving particles in a volume of up to 5000 cm^3 for the duration of 20 nsec. The three-dimensional distribution of particles is "frozen" on the hologram, and may be later viewed, consecutively, using a continuous-wave laser, for instance, a helium-neon laser.

A portable version of the installation[65] which will allow registering the cloud structure directly from an airplane is under development. This will make possible forecasting fogs and artificially influencing them.

[†] Significant progress may perhaps result from the use of thermoplastic recording aids in holography. For a general background, see e.g. Glenn.36a,37,37a (Note by GWS.)

It was not originally expected that the disdrometer might find many more uses than were supposed initially. The company "Technical Operation," which prepared the first three samples of the disdrometer, received over 400 inquiries about this installation.[66] Among the host of possible applications, one may list air pollution control, control of turbulent low-density streams, the study of aerosol dynamics, analysis of lubricating compounds, observation of processes occurring in the inner volumes of steam turbines, the study of the growth and the composition of fog droplets, control and analysis of medical aerosols, and pictures of tracks in bubble-chambers.

The last application will expedite the work of the experimentalists who study tracks in bubble chambers and the Wilson chamber. At present, in order to reconstruct the geometry of the trajectories and to determine the kinematics of high-energy processes, one uses stereophotographs and computers. The a posteriori analysis of three-dimensional holographic images will enable us to reduce the volume of computation.

Moreover, holography will permit an increase in the focal depth of bubble-chamber photographs.[130] However, at this point a number of difficulties arise for the researchers. First of all, since the angles go up to 45°, one has to use photographic emulsions of high resolution, which may be rather insensitive. Secondly, the process of holographic photography is complicated in a Wilson chamber by the 30 cps frequency width. [†] Thirdly, the strong magnetic field of the chamber causes a Faraday rotation of the plane of polarization which decreases the contrasts of the interference fringes. Finally, the brightness and the contrast of the reconstructed tracks are not large.

In order to overcome this last complication[62] the use of image reconstruction from phase holograms was proposed. It was demonstrated experimentally that it is possible to reconstruct images of phase objects (of the type of air bubbles in glass) in volumes which are relatively extended along the line of vision. It is possible to use Schlieren-photography for observation of three-dimensional phase objects reconstructed from a hologram.[116]

If one talks about "holographing" small transparent (as in the track chamber) or opaque objects contained in a large volume, then one may use Fraunhofer holograms.[19] The photographic plate is placed in the near zone of a large aperture, which is simultaneously the far zone for the small particles. Then the hologram registers the interference picture of the Fraunhofer diffraction of the particles. When the image of these holograms is reconstructed in the far field of the particles, the virtual

---

[†]Unclear in original.(Translator's note.)

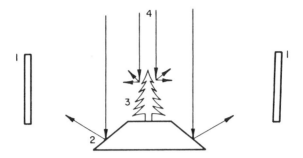

FIG. 2. A holographic installation with an object view angle of 360°. (1) Circular hologram, (2) conical surface of the mirror, (3) object, (4) coherent illumination.

image is absent.

For the purposes of demonstration of large-volume images, a number of holographic devices have been developed [47,51,121] with an observation angle of the object reaching 360°. The reference beam is produced by a convex conical (Fig. 2) or a spherical mirror. After "holographing" the object, the object is removed and in its place the observer views, through a film, the reconstructed image, which may be considered under any angle. The difficulties here are related to the manufacturing of a non-planar mirror and to adjustments. Since this is where the advantages of three-dimensional holographic representation are most obvious, devices of this type will find wide uses in advertising and for the demonstration of illustrations in lectures.

Using powerful laser pulses of short duration, it is possible to "freeze" rapidly changing processes throughout the volume under investigation and then to study, as long as necessary, the three-dimensional representation reconstructed by a continuous laser. Since ruby lasers are still insufficiently coherent, † one has to adopt special measures: optical-path equalization, i.e. the equalization of the optical distances of the reference and object beams, [4,5] and the selection of longitudinal modes. [50] In order to record holographic representations of large objects on small holograms by means of a laser of moderate power one may use lenses placed between the object and the hologram. It should be noted that the problem of pulsed-laser holography of large

†They now are sufficiently coherent. See, e.g., Ref. 133. (Note by GWS.)

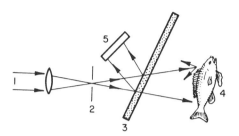

FIG. 3. Underwater holography scheme. (1) Laser light, (2) diaphragm, (3) aquarium wall which gives the reference beam, (4) object, (5) hologram.

objects (above 5000 cm^3) and large homogeneous surfaces is still not solved, since the lasers necessary for this purpose have still not been developed.

Under usual conditions, the human eye is an irreplaceable instrument for visual investigation of three-dimensional objects. Where the conditions for visual observation are absent, e.g. under water, inside complicated devices, and so on, holography may be of significant help. Experiments were conducted for making holograms of microorganisms in aquaria filled with sea water.[54] The recording was done using a ruby laser with a 60 nsec exposure. In the reconstruction, which used a helium-neon laser, the microscope was focused on different planes of the three-dimensional scene. If one were to prepare a number of holograms in the deep sea, then it would be possible to study the realistic behavior of deep-sea organisms in their natural habitat; for instance, their relative illumination.

Another form of underwater holography has been described.[39] The object of investigation is the analysis of acoustical vibrations of underwater objects. The experimental scheme is shown in Fig. 3. It is curious to note that the reflections from the walls of the aquarium were used for the creation of the reference beam.

Holography offers an unusual opportunity for sound vision[40] and radiowave vision.[20,55,56,59][†] The simple notion of such visualization

---

[†]We do not mention here x-ray, electron-beam, and other waves of that range, since their wavelength is significantly (more than 10^4 times) shorter than the light wavelength. This makes the visual observation of a three-dimensional image more difficult.

is based on the fact that the interference pictures of various wave fields
are identical, if their wavelengths are identical. However, generally
speaking, it is not necessary to have equal wavelengths in order to observe a true image. If the recording is performed using radiation with a
large wavelength λ_1 and the reconstruction is done using light of small
wavelength λ_2, then the scale transition—the shrinking of the initial
hologram by $M = \lambda_1/\lambda_2$ times—will also allow the preservation of three-
dimensionality in the reconstruction. † Sound and radio vision, in this
version, requires multielement receivers of high resolution, ‡ and the
recording of the resulting "electronic" holograms in the form of photographic holograms. These may then be illuminated by a laser and the
images first "seen" by sounds or radio waves may thus be made
optically observable. It is possible that for ultrasound one may use a
piezo-ceramic mosaic as such a receiver (for instance, a ceramic formed
by barium titanate) with an electron beam reader and an indication on a
television screen. Radio waves require special multielement antenna
grids.

It is possible, in some cases, to avoid using the photographic fixation
of a sound hologram. An original method[86] of sound vision involves the
immersion of ultrasound sources and the object in water (Fig. 4). Ripples
will appear on the surface of the water as a result of the interference between the direct and the object beams of the ultrasound. By illuminating
this interference pattern picture by a laser, one may immediately reconstruct the image and object. The representations of three-dimensional
objects will be distorted due to the inequality of light wave and ultrasound wavelengths, but by regulating the focal distance of the telescope,
one may view consecutively various cross sections of the three-dimensional
object.

This method will be of use for observations in nontransparent water
and in other oceanographic investigations. There is hope, moreover, that
the investigators will be able, in the future, to see the inner organisms of
living beings. Using the same method, creation of an operations control

---

†Unfortunately, it may readily be shown that for holographic imaging, just as
for lenses, the axial-magnification $M_{AX} = M^2$, where M is the lateral magnification. Accordingly a significant axial "compression" characterizes the images in
this case. A similar relative axial "extension" characterizes images in holographic microscopy. Fortunately, however, the relative distortion is only
$M_{AX}/M = M$, and the effect generally quite tolerable. (Note by GWS.)

‡ In reality [see e.g. Cutrona *et al.*13b] a small, *low-resolution* antenna is
the desirable element for use in the side-looking radar scheme, using this scheme
of holographic image reconstruction. (Note by GWS.)

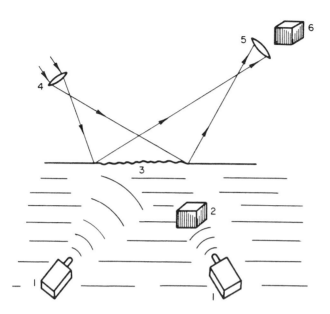

FIG. 4. A scheme for three-dimensional underwater sound vision with a holographic visualization. (1) Ultrasound sources submerged in water, (2) object, (3) ultrasound hologram on the surface of the water, (4) laser light, (5) telescopic receiver, (6) reconstructed image.

for the inside of hermetically closed blocks of electronic equipment has been proposed. It is necessary, however, in all these cases to place the object under investigation in a liquid. The main difficulty present is the creation of an ultrasound reference beam of sufficient sphericity.

Radio vision utilizing "synthetic" holograms has been accomplished.[20] † The Fresnel picture was created gradually using mechanical scanning in the plane of a Gabor-type hologram. A diminished photograph of the oscillogram reconstructed the original object under a helium-neon laser radiation . Since λ_1 = 3 cm and λ_2 = 0.63 μ the reconstructed image was extremely small and no parallax was present during the observation. In order to magnify this image, it has been proposed to "glue" together many small

† The reader's attention is again drawn to the high degree of perfection of the microwave holograms, as recorded in the side-looking radar scheme, and described by Cutrona et al.[13b] See also Leith and Ingalls[75a] and Kock.[58a] (Note by GWS.)

holograms. Then the object will be visible, as if it was observed through a multiplicity of small openings in a screen. The author supposes that holographic television in radio waves may be operated on this principle.

Another interesting application of radio vision has been described.[125] A holographic principle is proposed for the purpose of mapping a surface of a planet from a spaceship circumnavigating this planet. The planet and the spaceship would be illuminated by a surface-based microwave source. Along the track of the spaceship, which plays the role of a one-dimensional hologram, the interference between the direct wave and the wave scattered by the planet in the direction of the flight is registered. Using these radar maps, one may obtain surface structure information which is not given by the usual radio-wave or optical methods. For instance, making a radar map of Venus is probably the only way to investigate it. Only a landing may give more information. A system of this type is being developed by NASA for cosmic investigations.[68] Analogous on-board systems designed for military aircraft of the tactical-reconnaisance type are being developed by the Perkin-Elmer Co.[67] This method will also use radar maps recorded on a photographic film and later, one can observe the reconstructed image in a coherent light. For fast image reconstruction, "holographed" with "invisible" radiation, this firm is developing a two-dimensional spatial light modulator. This modulator will permit visual observation of images recorded by means of infrared or microwave holograms, transmitted through radio channels.

The original method for preparation of three-dimensional contour maps and models[45,46] is illuminating the object either by two collimated coherent beams with a small angle between them, or by a single beam containing the radiation of two slightly differing frequencies (for instance, an argon laser with $\Delta \lambda = 65$ Å). Using a single frequency beam, one may reconstruct the image from such a hologram. The contour lines of the image will give information on the distance of various details (if the hologram is at right angles to the beam).

The measurement of the directivity pattern of a microwave antenna (especially of a large one) causes difficulties since, in order to make such measurements, one has to move significant distances away from the antenna. Instead of this, measuring the distribution of the microwave field at relatively small distances from the antenna and preparing, using holography, an optical model of this field has been suggested.[2] The optical image reconstructed from this hologram may be transformed by means of a lens into the directivity pattern in a certain plane. The article establishes the possibility of scale transitions from microwave holograms into the optical region. Results of measurement of directivity patterns in the 3 cm range, using diapason ranging and holographic methods, are presented.

Holography opens up the possibility of creation of three-dimensional representations of objects which were not successfully observed, and also of "synthetic" objects.[85]† For instance, it is possible to consider the x-ray of a unknown complicated protein molecule in such a manner that one obtains its hologram rather than the image which can give only planar information. Then the binary hologram—a collection of black-white lines—may be drawn on a piece of paper and decreased photographically. Now one can illuminate such a synthetic hologram by a laser and reconstruct the volume representation of the molecule. Experiments which obtain synthetic holograms,[36] and the method of preparation of synthetic holograms in order to imitate three-dimensional objects[113] have been described. A luminous fiber is mechanically moved in space, and for each position its representation is registered on a hologram, thereby creating a cube consisting of 120 luminescent points after the reconstruction. The problems of creation and analysis of holograms by computers have also been considered.[48]

3. Image Recognition

The recognition properties of holograms found applications in a number of practical systems. The most familiar one is the holographic reader.[44,111,112] The device is designed for the reading of microfilms and for the introduction of the text into a computer. The hologram recognizes 100 letters which are recorded on it under various angles (the Lohmann θ - modulation principle). The recognition manifests itself through finding of the correlation maximum when the spectrum of the unknown letter is compared with a set of known letters, which includes the desired letter. In other words, the hologram, acting as a matched filter, will allow through only that spatial spectrum which is close to the spectrum recorded on it.

The advantage of this method consists in the small dimension of the recognition element—the hologram, which contains all possible variants. Gabor[32] calculated that one hologram may contain up to 30 letters in 30

---

†Optically synthesized holograms for such applications may be obtained according to the method described by Stroke et al.[120e]. See also Rogers[104a], and Spitz and Werts[113], who verified the generality of the image-synthesis method first proposed in 1965 by Gabor and Stroke et al.[34b]. The Gabor and Stroke et al. paper may also be taken as providing the theoretical basis for the several forms of holographic interferometry as subsequently developed by Burch, Stetson, Powell, Brooks, Hefflinger, Wuerker, and others (see pp. 90-96 of the present volume). (Note by GWS.)

variants in combination with the machine code for every letter. This means that it is possible to introduce manuscripts into the machine. Recognition will be indicated by the appearance, behind the hologram, of the maximum signal in a form of a set of bright points—the machine code of the given letter. An analogous hologram will allow[64] printing automatically or exhibiting on a videoscreen the values furnished by the computer. [†]

In order to prevent the changes in the hologram of the letter, whenever the letter is translated, Gabor[32] proposed the utilization of the Fourier hologram, i.e. a hologram obtained in the focus of a lens. Another method[86] guarantees the invariance with respect to translation through a comparison of the text with a Fraunhofer hologram prepared in the far field.

Of great interest for the military is the location of necessary objects on aerial photographs. Knowing the geographical coordinates of the object, it will be possible to reconstruct from the aerial photograph the map of the locality. Automatic discovery of chromosomes of known forms will simplify genetic investigations. Holographic systems required for this purpose are being developed in the United States by a number of companies.[67,71]

A more general direction in recognition is related to the obtaining of phantom or "ghost" images. By illuminating a Fourier hologram by a fragment of the initial image, it was possible to obtain a phantom image of the whole object.[101] [‡] For instance, using a woman's hat, it was possible to reconstruct the image of the whole face and head. This property of holograms may find application for recognition and determination of a location of a fragment of an object, for instance, the search of a page using a known line, and in other analogous cases. It is also possible to search for a three-dimensional object using some known detail.[12]

[†]See the discussion and the illustration of this method on pp. 79-86 of this book. (Note by GWS.)

[‡]These methods were first recognized in some aspects by Denisyuk,[14] and somewhat later, apparently independently, by Van Heerden.[128a] They were described in a general form by Stroke[120b,c] and independently, specifically for character (i.e. image) recognition, by Gabor.[32] This work was followed by the extensions of Pennington and Collier[12,101] and others. For a general theory, see pp. 127-137, this book. The most recent developments include Lanzl et al.,[64a] as well as new extensions to holographic image deblurring by holographic Fourier-transform division, as first proposed by Stroke and Zech,[120a] and now verified, for "black-on-white" two-dimensional objects by Lohmann and Werlich,[86a] and for a continous-tone three-dimensional object by Stroke et al.[120f] See also Stroke[117b]. (Note by GWS.)

The mechanism of such search was described in Section 1. It is only necessary to consider in this case that the object B is a fragment of the object A.[†] Practically, indeed, the image reconstructed through compensation of the source extent is nothing else than a phantom (i.e. "ghost") image.

The difference between some experiments[101] and Stroke's work on compensation of the source extent, consists in the use of a diffusion lighting of the transparency.[‡] The phantom image of a three-dimensional diffusion reflecting object was considered by Collier and Pennington.[12] Naturally, as distinguished from the experiments of Pennington and Collier[101] the hologram no longer registers the Fourier image of the object. Let us now explain the appearance of the phantom image. Each fragment scatters light diffusely throughout the hologram and, consequently, it registers the result of the interference between one fragment and the other parts of the object. The amplitude and phase of all waves emanating from the object were determined relative to the corresponding characteristics of fragment waves. Therefore, by illuminating the hologram by one fragment, one obtained the image of the whole object—its phantom image.

In the experiments,[101] it was possible to move the illuminating object relative to its Fourier hologram and the phantom image followed it, retaining its sharpness. Here, however, whenever the illuminating object was moving, the phantom image vanished. The reason for this distinction is clear: The Fourier spectrum is one of those rare cases when a diffraction spectrum is independent of the location of the point on the plane.

[†] The mathematical description of this phenomenon is given on pages 127 - 137 of this book. (Note by I. Nalimov.)

[‡] This often-repeated statement is not in fact, correct. A diffusor was used in the experiments described by Stroke et al.[120b,c] in 1965, and the recorded Fourier-transform hologram was reconstructed by Fourier transformation. In these terms, the Stroke et al.[120b] are examples of what has now come to be called "ghost" imaging (or "phantom" imaging). (Note by GWS.)

4. Interference Memories ("Volume" Holograms) †

As distinguished from the previous applications, which utilize thin emulsions, it is convenient to use thick holograms for memory devices. Gabor[33] investigated the scheme of holographic recording in a three-dimensional medium. The recording consists of the following: The volume hologram registers a set of standing waves created by the object and reference beams. During reconstruction such a hologram operates as a volume diffraction grating, i.e. a resonant structure which yields a diffraction picture for certain wavelengths and angles of incidence. In other words, the reconstruction occurs only when the reconstructing beam is analogous to the recording beam in the angle and the wavelength. The prospects of using holograms in recording devices is determined by the following two properties. First of all, the recording of each object point turns out to be uniformly distributed throughout the recording volume. Therefore, even significant damages in the hologram are not important. These will just degrade the signal-to-noise ratio. This very property enhances the reliability of reconstruction of the recorded information. Secondly, each point of the hologram contributes to the reconstructed image. This means that the memory possesses associative properties, i.e. the choice of the necessary information is made using a certain sign rather than the address of the cell in which it is stored, as is done in the usual memory devices. An additional advantage of holographic memory is the absence of lenses during the recording. Thanks to this, it is possible to utilize to the

---

† The use of thick (i.e. "volume") media for holographic recording was first proposed by Denisyuk[14] and (as we noted above) somewhat later, and apparently independently by Van Heerden.[128a] The first holograms in which "volume" effects were deliberately used to achieve certain advantages in multi-color holography were described by Pennington and Lin[101a] using lasers for the image reconstruction. "Volume" holograms, capable of reconstructing color images with *white light* (rather than laser light) were first described by Stroke and Labeyrie,[119] where they stressed, significantly, that they considered their new method an extension of the above-mentioned work of Denisyuk, in that they first introduced *white-light* reconstruction into holography. See also Lin *et al.*,[84] Leith *et al.*,[80] and Stroke.[116a] The new type of "planar focused-image hologram" recorded with "in-line" background, invented by Stroke[117] in 1966, is probably the most luminous of all holograms using white-light reconstruction, in that it requires *no filter* whatsoever, as a result of the low dispersion of the hologram recorded in this special way. As now shown by Nassenstein,[93] this type of hologram may be, moreover, *copied* in ordinary white light! The theory of "deep" (i.e. volume) holograms is given by a number of authors; see, for example, Kogelnik[61a] and most recently Gabor and Stroke.[34a] (Note by GWS.)

limit (in so far as the limiting wavelength of light allows) the resolving properties of the photographic material.

On the other hand, the fact that the visual image almost does not suffer when only a part of the hologram is used, convinces us that the hologram possesses a great amount of information which is superfluous for the eye. In this sense, holographic recording is not economical and investigations on the basis of information theory are necessary. These will enable us to find optimal methods of holographic recording for various conditions.

The most often utilized recording media are the thick layered photoemulsions, basic-halogen crystals, and photochromic glasses. Thick-layered photoemulsions have been investigated.[30,31,80,81,83] It was shown that the emulsion operates as a three-dimensional hologram if its thickness exceeds the separation between the interference fringes. Multicolored images were recorded on a thick layered plate using helium-neon (6328 Å) and argon (4880 Å and 5145 Å) lasers.[31] Even though the multicolored beams were directed at the hologram with the same angle, each color created its own system of interference surfaces in the emulsion. Therefore, when this was reconstructed at the Bragg angle, a three-dimensional color image was created.

The optical properties of three-dimensional holograms have been thoroughly investigated theoretically.[80] † Experiments with thick-layered emulsions have shown the sensitivity of the reconstructed image to the angle and the wavelength of the reading beam. Using a rotation of the hologram after each exposure, holograms of moving objects were prepared. Rotating the hologram in a laser beam, one could observe the moving image under the Bragg angle.

The theory of reconstruction of images from thick holograms has been presented.[8] The rigorous solution of the diffraction problem for circular and linear polarization was obtained. Numerical estimates were given for a number of cases which are typical for holography, and it was shown that the maximum intensity is observed for the Bragg angle.

Experiments with photochromic glasses have been described.[53] Up to now, one obtained a 60 lines/mm resolution. In order to ensure that the image does not vanish, exposures of no less than 1.5 hours using an argon laser of 25 mW were needed.

Thirty different images[23] of dimension $2.5 \times 2.5 \times 0.2$ cm^3 were recorded in a KBr crystal, even though theoretically it is possible to record

---

†See introductory footnote at beginning of this section.

500,000 such images. The crystal was heated up to 80°C for these recordings. When this crystal was irradiated by a laser, the color centers were whitened, and, as a result of this, an interference picture was registered. Subsequently, the crystal was cooled to 0°C, and the image was reconstructed when the radiation was passed at the same angle. The basic halogen crystals possess significant advantages compared to thick emulsions:

(1) Simple cooling is sufficient for the "development" of such photoemulsion.

(2) Color-spoiling contraction of the emulsion does not occur.

(3) Experiments are easily reproduced, since the preparation of crystals of the same thickness is not difficult.

(4) The sensitivity of crystals does not change from sample to sample, as it does for film.

In 1966, everyone's attention was directed to three-dimensional holograms reconstructed in white light.[†] Even though the applications of Denisyuk's[14-18] "wave photographs" are not restricted to memory systems, it is appropriate that we consider these and other works of other authors in this section of the review, since the mechanism of recording in a light-sensitive volume is the same here as in a memory system.

Denisyuk was the first to note the similarity of holography to the Lippmann process of color photography. In his arrangement, a coherent beam, which was passed through a Lippmann plate, was reflected by an object located on the other side of the plate. The interference of the direct and reflected beams created "wave photography," i.e. the recording of information concerning the optical properties of the object. Whenever the hologram was illuminated by white light from a source of a sufficiently small angular dimension, the color image of the object, a concave spherical mirror, was created. A characteristic peculiarity of Denisyuk's wave photographs was the fact that the reference beam was introduced from the reverse side of the hologram, and the interference planes (layers) were created almost parallel to the surface (and not perpendicular to the surface, as is the case in the later "volume"

[†] Stroke and Labeyrie[119] [reproduced, *in extenso*, in the Russian translation of this book].

holography experiments with usual photoemulsions[30,31,80,81,83] [†]).The distance between the planes was very small ($\sim\lambda/2$). During the reconstruction, the illuminating beam passed through many layers (several tens) of emulsion. Only a small portion of the beam was diffracted while thereby forming the image, and the remaining part was transmitted and lost.

Since Denisyuk's first papers have become classics, it is useful to quote his own definition[16] which formulates very concisely the idea of these papers: "For Rayleigh scattering of the radiation by the object, the intensity of the wave field in the space around the object represents 'as a model', with sufficient degree of accuracy, the 'optical scattering operator' of this object. This property of the radiation enables us, by fixing this field in a material medium, to obtain a spatial structure whose optical properties coincide with the optical properties of the object."

Stroke's works[118] and page 213 of the Russian translation of this book[‡] differ in the following: It became possible, thanks to lasers, to separate the reference and the object beams and to obtain high-quality images of complicated extended objects. Besides memory systems, the wave photographs of Denisyuk, and especially their later versions developed by Stroke and his co-workers[84,118] and also by other authors[126] may find applications in the following areas:

(a) Representation (projection) techniques, which will create a total illusion of reality of the projected images; for instance, three-dimensional portraits reconstructed by sunshine;

(b) Hydro-location, radio-location, and ultrasonic inspection methods;

(c) Preparation of dispersing elements of the volume diffraction grating type. Modeling of three-dimensional gratings in crystallographic investigations.

[†]See footnote at beginning of this section.

[‡]Stroke's article, which is the first experimental endeavor in this direction is on p. 213 of the Russian translation of this book. This article describes the realization of Denisyuk's idea using a laser. (Note by I. Nalimov.) (It should also be remembered that white light can reconstruct any hologram, not necessarily three-dimensional ones only. It suffices to place an interference filter between it and the source.) (Note by I. Nalimov.)

5. Holographic Interferometry

Thanks to the invention of holography, the area in which interferometry may be utilized has expanded to such an extent that it is now
possible to investigate not only simple idealized surfaces, but also diffusion scattering objects of arbitrary shape. Until now, the interferometric method was utilized only for the determination of the quality of
highly polished mirror-like surfaces, lenses, and polished objects of
regular shape. [†] A holographic interferometer does not require specular
reflection. The reference wave front does not require an analytical expression, and the obstacles in the optical path may be quite substantial.
Moreover, it turns out that it is possible to create an interferogram of
images which either exist at different times or are recorded using radiation with differing wavelengths. The analysis of the problems of holographic interferometry has been described most thoroughly. [1,41-43,115]

Prior to the consideration of various applications, let us pause to
discuss the main interferometric schemes. The simplest method for ex-
perimentation is the double exposure one. Using a laser and a fine-
grain photographic plate, it is possible to record on the hologram each
surface twice, provided the surface is not absolutely black. If, in the
interval between the exposures, we were to deform the surface slightly
at certain places, then the reconstructed image would be covered by interference fringes at these places. [‡] The advantages of this method lie
in the fact that it does not require perfect optics and precise adjustments. Moreover, if the film is exposed at two different instants of
time, it will allow us to study the stationary, as well as the non-
stationary processes. The attraction of this method also lies in the
ease with which one may prepare differential interferograms which register small changes in the optical paths of the rays or the location of
objects of complex shape.

If the surface is vibrating, the hologram will be exposed, so to speak,
many times, recording a multiplicity of images in a certain range of positions. The wavefronts which are reflected by the vibrating surface are
time-averaged on the hologram. The reconstructed image possesses a
system of interference bands which determines the nodes and the con-

[†] The only exceptions were the Schlieren-method and similar applications
with transparent objects (e.g. in wind tunnels, microscopy, and so on). (Note
by GWS.)

[‡] The fringes may be located in the interobject space, as well as "inside"
the object [1,42]

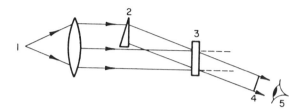

FIG. 5. The interference scheme of "live" holographic interferometry. (1) Laser, (2) prism, (3) hologram on which the image of the prism is recorded, (4) coinciding "object" and "reconstructed" wavefronts, (5) the observer.

tours of the spots with a constant vibration amplitude.

These two preceding methods allow one to study the total displacement or deformation and time-averaged vibrations. A "live" (instantaneous) interferometric study of motions and vibrations is also possible. It can be made to yield immediate information on the changes. If the developed hologram is placed into its recording position (as originally used for the object) while the object itself is also irradiated by a laser, then interference between the reconstructed and object wave fronts will arise. For instance, in the scheme depicted in Fig. 5, we have two plane waves going in the direction of the observing eye (or camera): One is a direct wave which passed through the prism, and the other one (reconstructing an image of the prism) is reconstructed from the hologram. These waves will precisely coincide. If, however, we change slightly the angle of the prism, the observer will see a system of parallel interference bands.

It should be noted that the quality of the optics is not essential in this as well as in the other methods, since this quality is the same for both stages. However, one does require here a precise adjustment since after the development, the hologram should be placed in the initial position, otherwise the interference fringes will be determined not by the changes of the object, but by the fact that the object and its image will not coincide. Certainly, visual control yields fewer details of the object under investigation than photographic control (for instance, with a double exposure), but one gains time with a visual control.

The applications of holographic interferometry may be divided into four main areas of investigation: measurements of vibrations, of deformations, of phase objects (i.e. such objects which alter only the phase of the passing wave, leaving its intensity unchanged), and interferometric measurements.

The interferometric analysis of vibrations of diffusely scattering objects has been considered.[104,114] The observation was conducted using the method of multiple exposure, as well as instantaneous interferometry. In the latter case, one prepared the hologram of the motionless object and then the vibrating object was viewed through it.

The application of holography to vibration analysis allowed one to undertake an investigation of a number of problems which were previously not solvable. For instance, it is proposed[70] to study oscillations of membranes in microphones and loudspeakers as well as surface waves on different substances. This opens the possibility of detection of vibrations of objects in a vacuum even when there is no medium which transmits the sound. An original method for detection of defects in thick and thin metallic devices of complicated form, which cannot be studied using x-ray methods, has been proposed. The use of a powerful pulse of an IR laser to create sound waves in a metal and then detect these waves from the surface using a holographic method is suggested.

A demonstration[73] of the high efficiency of holographic vibration analysis, which is used for nondestructive ultrasonic control, includes detection and study of flaws, imperfections, and fissures in solid bodies. Two plates of cold rolled steel were prepared with one of them having a crack. Then using a solenoid, vibrations of frequencies of 110 to 617 Mc were induced in the plates. The interference pictures for these plates differed strongly from each other and showed different frequency responses. In the place where the crack was located, one always observed swellings, since the metal was weakened there and the oscillations had their largest amplitude at this point.

Holographic control of unfinished surfaces may be used in a variety of situations. All these applications are based on the interference method of comparison of several states, using a double exposure of a single hologram. Thus it is possible to investigate all changes in solid bodies which are due to the form and the quality of their surfaces.[93] The changes may be caused by heating, pressure, or swelling. For instance, it is possible to investigate air bubbles and irregularities in the walls of hollow vessels.[95] The heating of air inside the vessel causes the expansion of the wall, and the regions with higher heat conductivity expand more than the normal regions. The interference band picture will allow one to find these places. Similarly, one may investigate vessels under pressure. Densely-spaced interference fringes will correspond to the weakened places. A study of creep of materials has been proposed.[13] (Certainly the creep should not exceed a wavelength during the time of exposure.) For instance, the deformation of a loaded piece of channel iron was carried out in this way, by holographic interferometry.[7]

All the above mentioned applications utilized a comparison of different states of the same object. Of great interest for industry is the possibility of comparison of differing objects, for instance, of the object which one desires to control with a standard. The comparison may be conducted[6] using oblique illumination. In this case, the decoding of the interference picture depends not only on the deviation of the object from the standard, but also on the accuracy of its placement in an assigned position with respect to the hologram. †

The holographic interferometry of dynamical phase objects in hydro- and aerodynamics of rarified streams, gaseous flows, etc., requires the application of special methods which are closely related to the usual interferometry of these objects.[90,122-124] However the methods used in holographic interferometry differ from the previously considered methods by the absence of focusing on the photographic plate. Because of this, the amount of information recorded on the photographic plate is increased. The experimental scheme in this application of holographic interferometry is essentially the same as in the method of double exposure. The appearance of interference fringes is related to the fact that the light, after passing the denser regions, for instance the front of the shock wave, is retarded and falls behind in phase compared to light, which passed the same path, in a homogeneous medium.

The work[52,96,132] done permitted one to obtain and investigate the reconstructed image of the "phase topology" of another phase object—a plasma. (In order to obtain a plasma, the radiation of a ruby laser was focused on a certain point of the medium.) The unfocused laser radiation was used[96] to obtain Gabor-type holograms. Then one[52] was able to photograph a laser spark throughout the three stages of its development, using a two-beam scheme (e.g. in the form used by Leith and Upatnieks). From the holographic measurement, one was able to count the density of electrons at the various stages. The Schlieren method and a helium-neon laser at 6328 Å were used for the observation of the reconstructed image of the spark.[132]

Three examples of the high degree of perfection and of the capabilities of holographic interferometry are shown in Figs. 5A, 5B, and 5C. ‡

---

†A general theory of holographic interferometry is only now being published. [See, e.g., Brown, Grant, and Stroke.[5a]] As noted by Dr. L.M. Soroko, the editor of the Russian translation of this book, in his "Foreword," the theoretical basis for the theory of hologram interferometry was first given by Gabor and Stroke et al[34b] See also Powell and Stetson,[104] Heflinger et al.,[43] and Tsujiuchi and Tsuruta.[124a] See also Shiotake et al.[136] and Tsuruta et al.[137]

‡(Note and figures added by GWS.)

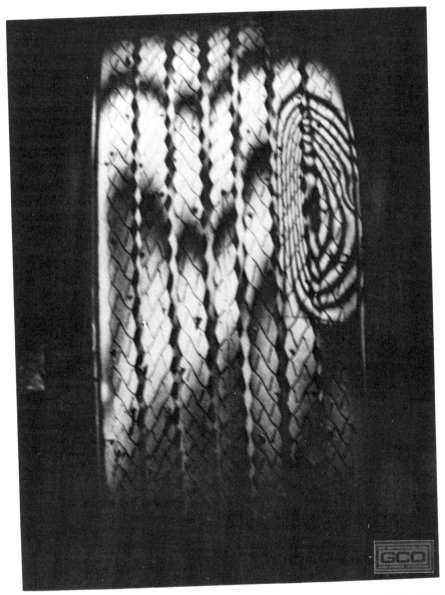

FIG. 5A. "Real time" (live) holographic interferogram of a large (8.25 x 14) tubeless automobile tire showing defect (separation between tread and outer ply of a two-ply bias) using natural creep of normally inflated tire, photographed during two successive states (i.e. "original" tire state, used to produce holo-gram, and "successive" state, existing during observation). [*Courtesy Dr. Ralph M. Grant, GC-OPTRONICS, Ann Arbor, Michigan.*]

FIG. 5B. "Double-exposed" holographic interferogram of a large (8.25 x 14) tubeless automobile tire showing deformation as a result of thermal (heat lamp) loading. [*Courtesy Dr. Ralph M. Grant, GC-OPTRONICS, Ann Arbor, Michigan.*]

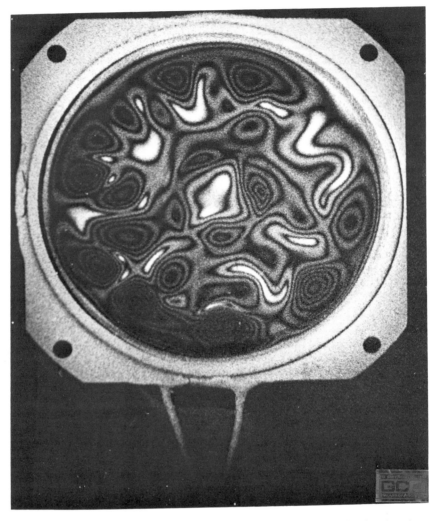

FIG. 5C. "Time-averaged" holographic interferogram, directly showing the topology of deformation of a vibrating high-fidelity loudspeaker, approximately 6 in. long, vibrating at about 2000 cps. [*Courtesy Dr. Ralph M. Grant, GC-OPTRONICS, Ann Arbor, Michigan.*]

6. Holography in Technology

The combination of high-power laser radiation and of the astounding ability of holograms to focus light into complicated three-dimensional figures, allows one to hope for a widespread use of holography in technological surface-finishing processes. Holography will be especially useful in cases when the patterns imprinted on the surface are required to have a high resolution— of the order of 1 μ —over extended portions of the field. Thus the best lenses are capable of producing at most 3000 linearly resolved elements within the field, while the holograms easily yield 10,000 resolved elements. †

---

†A "resolved element" may be considered as corresponding to the width, at half intensity, of the diffraction pattern (see, e.g. Abbe and Rayleigh resolution criteria, pp. 74-77 of this book, and also the discussion of resolution by Gabor, pp. 266 ff., this book). In this sense, at least in theory it may be readily shown that a perfectly corrected lens and a perfectly processed hologram, both of the same dimensions, could be made to "resolve" the *same* number of elements in a given field! However, in addition to their ability to store image information, with equal quality (i.e. with the maximum possible resolution) over essentially unlimited depths of field (rather than only for a single surface, which is the case for an ideal lens) the practical advantages of *lenslessly* recorded holograms, in comparison to lenses, reside in the following facts: (1) It is possible to achieve aberrationless wide-field imaging (and this from a single-element hologram, provided of course that the wavefront, used in the reconstruction, duplicates the reference wavefront used in the recording with great rigor, as shown in the analysis on pp. 127-137 of this book), (2) It is possible to realize, in general, holograms of much greater dimensions than appears readily possible for complicated, multi-element lens systems of comparable performance, and thus to correspondingly gain in the number of "resolved" elements. However, it may be in order to note that these important advantages of holographic imaging may only be achieved if the spurious effects, characteristic of recording with *coherent* light, are suitably minimized or otherwise eliminated. Among these spurious effects, the most notable one, for diffusely reflecting objects, is the "speckle," the worm-like interference patterns formed near the surface of objects illuminated in laser light. Fortunately, it has been found by many authors (see, e.g. Stroke,[117c]; see also Fig. 32c and d, p. 142 of this book) that the image-degrading "speckle," which would result in very considerable loss of actually obtained resolution, may be made essentially undetectable by (a) either using the large hologram to form a real image (Fig. 32c), or else (b) looking through the hologram with a sufficiently large aperture (for instance $f/10$ or larger, where f is the distance of the reconstructed virtual image from the lens—eye or camera—used to view the image). This last condition is of considerable importance in the attainment of high-resolution images in microscopy, e.g. in that based on the scheme first described by Stroke and Labeyrie[119]). With these remarks in mind, it becomes clear that the enormous superiority of holographic imaging, in comparison with conventional imaging, does not reside primarily in a comparison of the resolving "power" of a hologram with that of even the most

In distinction to the majority of other applications of holography, technology uses the reconstructed *real* image. In order to achieve maximum resolution, it is necessary to focus the real image as closely as possible to the hologram. Then the number of Fresnel zones contributing to each point of the image will be a maximum (for given dimensions of the hologram). If the number of Fresnel zones is small, then the sharpness decreases. On the other hand, if the hologram is located excessively close to the image, the direct laser beam will be incident upon the surface and the contrast will be worsened, which is not always desirable. Therefore, the choice of the optimal hologram-surface distance is very important.

Apparently the first technological application of holography is the contactless printing of microcircuits[28,72] (Fig. 6).

The real image is made to expose the desired regions in the layer of "photo-resist." Upon the usual washing away of the unexposed parts, the desired pattern remains imprinted. This is a protective layer which protects the film from exposure.

Besides the absence of lenses and the possibility of high resolution, holographic printing of microcircuits possesses a number of specific advantages. The absence of a contact with the surface removes the possibility of scratching and dust settling and also facilitates the automation of printing. Contact printing interferes with the possibility of obtaining ultrahigh resolutions since it requires ideally plane surfaces. In contrast to the use of contact-printing masks, which are capable of giving about 100 prints, the hologram may serve as long as necessary. Finally, the dust particles found in the space between the hologram and the photoresistor are no longer dangerous, since the registered information about every point of the object is uniformly distributed throughout the hologram. The dust particle will only lower the brightness of a point by a certain magnitude.

Holographic technology requires a substantial increase in the power output of a laser and simultaneously the increase in its coherent properties. The larger the power is, the larger is the possible area of

---

perfect lens system. Rather, the enormous superiority of holographic imaging resides in the capability of storing in a single photographic hologram intrinsically resolved information for a field extending over a depth and width which could not, in the present state of the art, be stored in comparably extractable form in a single photograph recorded with an image-forming lens or mirror system. It should be noted, however, that "image-restoration" or "image-deblurring" schemes, either holographic, or digital, or indeed electronic, may require detailed examination of possible advantages, specifically for each application. (Note by GWS.)

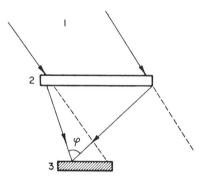

FIG. 6. Contactless printing of microcircuits using holograms. (1) Laser light, (2) holographic (photographic) mask, (3) microcircuit.

the image. A significant shortcoming of focusing holograms is their low efficiency. Usually, only several per cent of the incident light energy participates in the creation of the real image. The remaining radiation passes through without diffraction and is lost. In order to increase the brightness of the real image, it is proposed to place the holograms inside the laser resonator.[103] If the metalized hologram (Fig. 7) is used in place of one of the resonator mirrors, and the other mirror is made nonreflecting, then up to 50% of the laser power may be used to form the real image. For transmitting phase holograms, the efficiency may reach 25% since two real and two imaginary images will arise.

Further improvement of holographic technology will probably lead to a combination of the focusing properties of the holograms and their unusual recognition and introscopic "capabilities." $S + \partial \ell$

7. Holographic Imaging through Turbulent, Distorting, or Opaque Media

Image transmission through transparent walls of irregular form (for instance, through ribbed glass) usually causes astigmatic distortions of the image. Images transmitted through an inhomogeneous or scattering medium (for example, a medium with random or turbulent fluctuations of the index of refraction) lose their sharpness and become "washed out." A unique possibility of avoiding these distortions as offered[61] by the holographic method of recording and reconstruction of the wavefront. Which property of the hologram ensures this possibility? Let us assume that the image of the object along with the "image" of the wall (or of

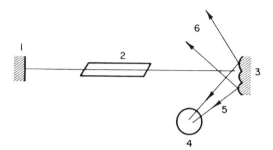

FIG. 7. A metallized hologram in a laser resonator. (1) Nonreflecting mirror of the resonator, (2) the active medium of the laser, (3) the metallized hologram, (4) the working surface, (5) the real image, (6) the imaginary image.

the scattering medium) is recorded on the hologram (Fig. 8a). If, during the reconstruction, the hologram is simply irradiated by a laser beam, then the image which arises as the result of this will be totally devoid of meaning (Fig. 8b). However, if we simply place the distorting wall, retaining its original position relative to the hologram, into the path of the rays, as shown, then, the *real* image will be reconstructed, on the other side of the wall *without* distortion (Fig. 8c). Since the real image is "pseudoscopic" (inverted relief) as far as visual observation is concerned, it is possible to turn it "inside out" and to obtain a visually more pleasing image.[106] To achieve this end, we shall record the real image on a second hologram. Then, when the latter is viewed in laser light (Fig. 9), the observer will perceive the three-dimensional image, precisely similar to the original object, "suspended" in space between him and the hologram. The wall was successfully removed as if it never existed.

It should be noted that the operation of "turning the real image inside out" (i.e. of making the "pseudoscopic" image "nonpseudoscopic") would not be necessary at all, had we been able to compensate immediately for the imaginary image. This, however, is apparently unachievable for the imaginary image. The mechanism of compensation consists in the following: All phase shifts acquired in the forward direction are compensated by opposite-in-sign phase shifts in the reverse path. In other words, if a wave S is propagating in the forward direction, then in the reverse direction we should have the wave $S*$–the complex conjugate wave; i.e. only the real image may be compensated.

The theoretical basis for this remarkable property of holograms—the

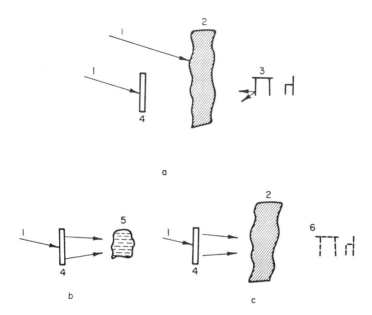

FIG. 8. Holographic observation through walls of irregular form. The image is "washed out" when the reconstruction is done without compensation. (a) Recording, (b) reconstruction without compensation, (c) reconstruction with compensation. (1) Laser light, (2) irregularly shaped wall, (3) the object, (4) hologram, (5) the "washed out" (uncompensated) image, (6) compensated image.

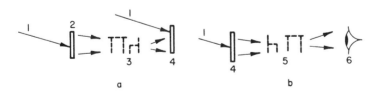

FIG. 9. The method for turning the real image "inside out." (a) Real image reconstructed from the hologram 2 is recorded onto another hologram 4; (b) when the latter is illuminated the observer views the copy of the original object suspended in the air between him and the hologram. (1) Laser light, (2) hologram, (3) real image, (4) hologram, (5) the copy of the original object, (6) the observer.

reconstruction of undistorted images upon recording through inhomogeneous media—lies in the reciprocity theorem. The latter follows from the main property of the Green's function—the possibility of interchanging the source and the observation point. In its general form, this property is formulated in the following manner: Let the antenna A located at the point O_1 be the radiator, and the antenna B located at the point O_2 be the receiver. Let us now allow the B antenna to radiate from the point O_2, creating the same field as previously. Then, according to the property of reciprocity, antenna A shall possess the same field as the antenna B in the first case, regardless of the properties of the medium and the form of the antenna. It is important to note that the theorem is valid regardless of the inhomogeneity of the medium.

For simplicity, let us consider a medium without absorption. Let the index of refraction $n(r)$ vary along the path according to a certain definite or according to some random law. The propagation of a light wave in such a medium is described by the wave equation

$$\nabla^2 u + k^2 n^2(r) u = 0$$

where $k = 2\pi/\lambda$ is the wave number in free-space and u is the component of the electric field vector. We shall assume that the dimensions of the inhomogeneity are much larger than the wavelength of the radiation.

The wave propagating in an approximately direct line may be described by the following solution of the equation:

$$u = S(r) \exp[-iknz],$$

where $S(r)$ is a function which varies slowly with r.
However, this equation also possesses another solution

$$u = S*$$

which describes a wave propagating in the reverse direction.

Therefore, in order to reconstruct the image degraded by the irregularity of the medium, three conditions must be satisfied: (1) The index of refraction along the reverse path should be able to duplicate $u(r)$ in six coordinates which determine the position and the orientation of the inhomogeneity in space; (2) a complex conjugate wave (to the forward one) must be propagating in the reverse direction; (3) the divergence of the beam illuminating the hologram during the reconstruction should be equal to the divergence of the reference beam during the recording so

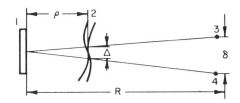

FIG. 10. Holographic transmission of nondistorted images through a turbulent layer *without* compensation. (1) Hologram, (2) wavefronts of the reference
and the subject waves in the turbulent medium, (3) the source, (4) the object.

that the magnification of the wall's (medium's) image should be equal
to unity.

Further investigations have shown that in a number of cases it is
possible to dispense with the compensation[38]; for instance, in the case
of a "thin" turbulent medium (layer). It is only necessary (Fig. 10) [†] to
place the object and the reference source sufficiently close to each
other such that the distance between them is

$$\delta < \frac{R}{\rho} \Delta$$

where R is the object-hologram distance; ρ —the wall to hologram distance; Δ —the dimension of the inhomogeneity.

When this condition is satisfied, the object and reference waves suffer an almost equal phase delay, and, therefore, the hologram records an
interference picture identical to the one which is recorded in the absence
of the turbulent layer. Consequently, the layer is no longer necessary
for the reconstruction of the copy. It is especially easy to preserve resolution if the hologram is located right next to the layer. Then the object
and the source may be located at any distances, or the medium may be
in the form of a thick layer. The shortcoming of the method lies in the
fact that the coherent source has to be located side-by-side with the
"concealed" object.

It is assumed that a system based on these principles will allow
the transmission of undistorted images from satellites which are in
orbit.[49,74,75,108] For the usual observation methods, the image of a

[†] The arrangement of Fig. 10 is the "lensless Fourier-transform hologram"
arrangement (Stroke[116a]). The use of this arrangement has been recognized by
Goodman *et al.*[38] as having this additional significant advantage for holographic
imaging through turbulent media, and has been verified by them for the imaging
of 2-D objects, as shown. In keeping with the principles of "lensless Fourier-
transform holography" of 3-D objects (see Fig. 21, p. 119, Chapter VI), the
method of imaging through "distorting" media may be more generally also used
(with suitable precautions, see Fig. 10A) for the holographic improvement of
imaging of 3-D objects, as compared to a direct photograph taken through the
"distorting" medium. (Note by GWS.)

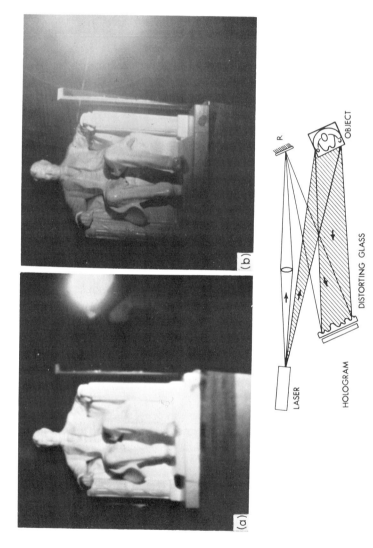

FIG. 10A. Imaging through a distorting medium using "lensless Fourier-transform holography." (a) Direct photograph through distorting glass. (b) Holographic photograph recorded through distorting glass. The holographic image-quality compensation is achieved by having the phase distortion exp $[i\phi_D]$ of the waves from the various object points canceled by the distortion exp $[-i\phi_D]$ of the reference wave from R. Compensation is good when $[\phi_D{}' - \phi_D] < \pi/4$. (After Stroke, Puech, and Indebetouw, $J.\ Opt.\ Soc.\ Am.$ [submitted 22 May 1968].)

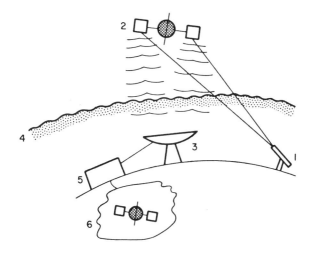

FIG. 11. A scheme for holographic transmission of images beyond the limits of the troposphere. (1) Laser, (2) space ship, (3) receiver, (4) turbulent atmosphere, (5) converter, (6) image

cosmonaut leaving the space-ship is distorted because of the atmospheric turbulences in the lower layers. A powerful Earth-based laser (Fig. 11) illuminates the satellite. The radiation reflected from the two parts of the satellite is registered by an Earth-based receiver. The received hologram is transformed (it is not reported in what manner) into a visible image. The influence of the turbulence on the quality of the hologram is eliminated because the wavefronts of the two parts of the satellite have identical path lengths and their phase difference is preserved.

However, in our opinion, caution need be applied to the possibility of practical realization of this system, since the required diameter of the telescope and the required laser power are far too large.

In many practical cases, the wall possesses regular surfaces (Fig. 8) which facilitate its installation. Regular surfaces may be created artificially by placing the tested material into a vessel with liquid.

It is curious that the compensation method allows one to obtain undistorted images through transparent surfaces of irregular form, for instance, through frosted glass and a wavy water surface. It is known, of course, that observations through these surfaces are practically impossible when one uses ordinary optics or the unaided eye.

Wide possibilities for holographic imaging through "turbulent" media open up in connection with the utilization of the principle of visualization of invisible images which was discussed by Leith et al.[78] and realized by Dooley.[20] In order to transfer the image into the visible region, the use of scale change is proposed—an increase and a decrease in the size of the hologram. A combination of scale change with the compensation method[92] allows one, in principle, to observe undistorted three-dimensional objects behind walls of arbitrary form, opaque to light! Numerous technical difficulties must be resolved in order to realize this possibility. One requires sources of a sufficiently coherent and powerful radiation. Of certain interest in this context is the CO_2 laser with the wavelength of $10.6\,\mu$.

Besides such imaging, the compensation method may find use in secret document coding and the study of wave propagation through inhomogeneous media. Possibly, it will be feasible to establish laser communications in the turbulent atmosphere.[†] This communications net will not be subject to interference by the fluctuations of the index of refraction in the air.

Instead of compensating for the wall, it is possible to record its image on a separate hologram.[118] If we now illuminate both holograms simultaneously, then we will be able to reconstruct the image of the object which is located behind the wall.

8. Holographic Microscopy

Holographic microscopy began its development mainly due to the efforts of the Stroke and Leith groups in the University of Michigan in the United States. Stroke's work, which aimed at the achievement of resolutions of the order of 1 Å (which corresponds to the dimensions of molecules of living tissue) is described in detail in the main chapters of this monograph. Stroke[117] continued the investigation of Gabor's scheme of a holographic microscope. The reference and object beams were directed along one axis and the image of the object was focused on the hologram. Practically, the reference and the object waves possessed not only the same direction but also the same curvature. Since the recording was performed in the $17\,\mu$ thickness of Kodak 649F emulsion layer, the opportunity was taken to reconstruct the image in

[†] L.J. Porcello pointed out that the forerunner of the current work on holographic imaging through turbulent media, as well as in fact the forerunner of microwave holography (e.g. as used in synthetic-aperture radar) is the work of Rogers.[104b] (Note by GWS.)

a white light (using a carbon arc). Although the reconstructed object (a model of an atom of 15 cm dimension) could be observed only under a certain angle, the observer could focus on different planes of the image, thereby extracting information about the third dimension of the object.†

An experiment[131] in optical modeling of an x-ray holographic microscope (of the Gabor type) which operates on the basis of a double Fourier transform was conducted without the use of a laser. The reference wave was created by a point opening in the plane of the object. During reconstruction, it was possible to obtain sharp images of periodic structures.

In the articles of Leith and his co-workers,[76,78] microscopy schemes in divergent laser beams are described, magnification formulas are given, and conditions for which the reconstructed image is aberration-less are determined. A more detailed theory of angular and longitudinal magnification has been presented[88] where one finds expressions for third-order aberrations arising in the reconstruction of point objects as well the ways for their elimination.

The first high-quality three-dimensional images of micro-objects have been obtained.[10,11,69,129] Micro-images were obtained using the bright-field method as well as the polarization method.[10,11,69] The magnification was effected in two steps. First, using an ordinary microscope, the magnified image was projected onto the hologram. Subsequently, during the reconstruction with the help of a lens, additional

---

†Significantly, in the paper, "White-Light Reconstruction of Holographic Images Using Transmission Holograms Recorded with Conventionally-Focused Images and 'In-Line' Background," Stroke[117] showed that it was literally possible to "restore" the third dimension information in the recording of conventionally focused photographs, in a conventional photographic arrangement, provided (1) that the recording is carried out in coherent light from a laser and (2) that a coherent background wave is added to an otherwise conventionally focused image! The white-light reconstruction, in this case, is made possible as a result of the very low "dispersion" of the hologram (grating), rather than of any emulsion thickness effect. Because all of the spectral components are usable simultaneously in this case (rather than only a very narrow "filtered" part of the spectrum, usable in the case of the "white-light reflection holograms of the Stroke and Labeyrie[119] type) these "in-line focused-image transmission holograms" are perhaps in practice the most "efficient" holograms to date. Moreover, as shown by Nassenstein, ordinary white light (rather than laser light) is sufficient[93] for the production of copies (replicas) of this type of 'focused-image' hologram. Other aspects of focused-image holography, together with historical references to related work are given by Kock et al.[60a] under the title "Focused-Image Holography—A Method for Restoring the Third Dimension in the Recording of Conventionally-Focused Photographs." (Note by GWS.)

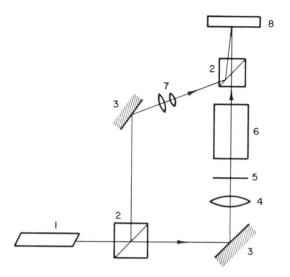

FIG. 12. A holographic microscope which allows one to observe the image at various depths. (1) Laser, (2) beam-splitting prism, (3) mirror, (4) condenser, (5) object, (6) microscope, (7) microscopic lenses, (8) hologram.

magnification was achieved. Even though this method destroys three-dimensionality (in its usual meaning), it was possible to view various planes in depth in the reconstructed image.

One of the possible schemes of such a microscope[129] is shown in Fig. 12. Using this microscope, colored holograms of the neuron chain were prepared with the help of a helium-neon laser. By photographing the reconstructed quasi-three-dimensional image, the experimenters succeeded in obtaining photographs of two planes separated from each other by 40 μ, while the dimensions of the smallest details reached 1 μ. Using the usual holographic method (without the microscope), the authors succeeded in reaching a resolution of only 12 μ, and, only after much exertion, 5 μ.[†]

[†] It is now generally understood that a fundamental advantage expected from the use of holograms in microscopy, in addition to their depth-of-field property (along the optical axis), is their ability to store, in a single hologram, also high-resolution pictorial information over vastly more extended lateral fields (orthogonally to the optical axis) than is possible by imaging through a microscope or other high-resolution lens. It becomes readily apparent, especially by comparing Fig. 12 with the arrangement used for the recording of the "white-light reflection holograms" (according to Stroke and Labeyrie [118]) that it is this last method which permits one to fulfill simultaneously the two conditions re-

quired for the holographic recording of high-resolution image-information with great depth of field as well as with great lateral field extents: (1) Have the hologram be as close to the object (or its image) as required for the best-possible high-resolution microscope lens necessary for the attainment of the desired resolution, (2) have the hologram plate be as large as the desired lateral field. Clearly, when the very restricted lateral field of high-resolution microscope lenses are taken into account, it becomes evident that the "lensless" recording, with the hologram close to the object, and the reference beam incident from the back side of the hologram, provides an obvious approach to the solution of this problem. Moreover, the "white-light reflection holograms" may be recorded with laser wavelengths incident onto the hologram simultaneously from the same direction, and the "white" light used in the reconstruction (whether originating from a multi-color "white-light" laser, or from an incoherent white light source) also has all of its component wavelengths incident from the same direction. Indeed, unlike the transmission holograms, which are characterized by a spectral color dispersion of the different color component images, when white light is used in the reconstruction (which results in complicated color "cross-talk," as a result of the recording of the holograms in the form of planar gratings), the "white-light reflection holograms" can be readily shown to have the remarkable property of reconstructing all of the color component images in the same direction without spectral color dispersion (Stroke [117a]). Indeed, when the hologram is recorded in the wavelengths λ_1, λ_2,...,λ_n, with the object beam E_0 and the reference beam E_R incident onto the holographic plate from opposite sides, normal to the plate, so that each of the spectral components is described by

$$\widetilde{E}_0 \, (x_1 \, y_1 \, z_1 \, t) = \overline{E}_0 \, (x_1 \, y) \, e^{i \, (\omega t \, - \, kz)}$$

$$\widetilde{E}_R \, (x_1 \, y_1 \, z_1 \, t) = \overline{E}_R \, (x_1 \, y) \, e^{i \, (\omega t \, + \, kz)}$$

and the corresponding hologram by

$$I(x_1 \, y_1 \, z) = |\overline{E}_0|^2 + |\overline{E}_R|^2 + |\overline{E}_0| \, |\overline{E}_R| \, 2 \cos 2 \, kz$$

it is seen that the spacings of the "Lippmann-Bragg" layers along the z direction (parallel to the surface of the plate) are

$$d_1 = \lambda_1 / 2, \quad d_2 = \lambda_2 / 2, ..., d_n = \lambda_n / 2, \quad \text{i.e. } \lambda/d = 2 = \text{constant!}$$

Accordingly, when the images are reconstructed from this hologram by "Bragg" reflection, using illumination under the angle i with respect to the Lippmann-Bragg planes, the angle of diffraction i' is given by

$$\sin i' = \frac{\lambda}{a} - \sin i \quad (i' = i).$$

Clearly, all of the spectral image components will be diffracted in the same direction (noticing that $\sin i = $ constant by illumination from the same direction and $\lambda/a = $ constant by the particular nature of the recording!). (Note by GWS.)

Another advantage of microscopic holography lies in the fact that it preserves the phase information about the object. Moreover, the possibility of lensless magnification enhances, in a number of cases, the resolving capability of this system as compared to the usual lens-microscope. The proposed applications of the holographic microscope are related to the observations of cancer-affected tissues as well as mobile or short-lived micro-objects, for instance, amoebas and blood cells. It will be possible to register these objects very fast and then to view them in depth.

9. Holographic Cinematography

Holographic methods allow one, in principle, to reproduce a moving object from a stationary screen. † To achieve this, it is necessary to record on one hologram the consecutive positions of the object using different angles.[27,82,98,99] During the reconstruction the hologram is consecutively illuminated by the same sources and with the same angles. This results in the motion effect.

Other types of holographic cinematography are also possible, for instance, the panoramic stereoscopic visual device[94] or sequential rotation of the source or the hologram itself. The techniques of modern cinematography—rotating drums and mirrors—may also be utilized. Problems arise from the fact that the superposition of images recorded in neighboring time instances cause undesirable distortions. Up to now success has been obtained in recording up to seven images with different angles.[82] The problems of preparing holograms the size of a movie screen are still very great. Nonetheless, this is apparently the only way to create a mass display, since the problem of magnifying three-dimensional small holograms for large audiences is unlikely to be found. ‡

A more realistic problem is presented by the development of systems for three-dimensional indication designed for individual use, such as a system for blind (instrument) landing. The creation of a large number of holograms using a model of the airport with the landing runway is proposed; every hologram shall display the view of the landing strip as

---

† Such a system, using a large concave mirror system for holographic cinematography has been demonstrated by the Conductron Corporation (Ann Arbor, Michigan).

‡ Besides this, the image may be three-dimensional and colored (even though the film itself is gray) and will not require any additional visual aid devices for its viewing.

seen by a landing pilot. The radiolocator will pinpoint the position of the plane in relation to the strip, and a computer will choose the appropriate hologram. The pilot will see the three-dimensional landing strip as if it were real and will therefore be capable of landing the plane in the midst of a dense fog, using only the holograms. Only a laboratory model of such an installation is in existence.

10. Holographic Television

The general interest expressed in the possibility of realization of three-dimensional television is rather understandable. Such television will bring about the optimum similarity of the art and technology of television with real conditions and will allow one to create an almost complete sense of presence.

Although the first three-dimensional television image (based on another principle) was demonstrated by Shmakov 17 years ago,[110] the future prospects of three-dimensional television are certainly associated with holography. The development of such systems is being discussed rather intensely and, apparently is making some progress. Thus, it has been mentioned that in 1967 there will be shown a model of a commercial holographic system which will transmit three-dimensional images and will not require a band width exceeding 6 MHz.[25]

The idea of holographic television was apparently stated for the first time by Rogers[105] in his 1958 patent—before the invention of the laser. The most detailed discussion about the requirements on the holographic system of three-dimensional television[79] shows that such a system will require a bandwidth of about 10^{11} cps (with a screen resolution of 700 lines/mm) [†] which exceeds by four orders of magnitude the bandwidth of the modern television channels. Consequently, at present, the transmission of three-dimensional images over the usual television channels is restricted to simple objects or may be done in the slow scanning regime.

If one prepares large-structure holograms and chooses a small angle between the object and reference beams, then it will be possible to transmit them immediately on television. The first successful telecast[26] of such holograms has already been conducted. However, this method is useful only for small two-dimensional objects such as slides. Compared to the usual telecast, this method has the following advantages: The information about the image is transmitted in a coded form and such a tele-

---

[†] At present the laboratory models of television systems "EIDOPHOR" have a resolution which does not exceed 100 lines/mm. (Unclear in original—GWS.)

cast possesses extremely high noise stability. Even if up to 90% of the information is lost (for instance the connection was broken for 9 out of 10 min. because of the noise), it is still possible to reconstruct the various contours of the whole initial image.

Another possible way is television in the microwave range.[56] Multi-element antenna grids may be utilized in the capacity of the microwave hologram. The quantity of information contained in the hologram obtained in the millimeter range is not very large and it may therefore be transmitted using the usual means. Observation at the receiving end of the television channel may be realized, according to some proposals, through illumination of the reduced hologram by a laser. However, such holograms are very small and do not exhibit a noticeable parallax. If a set of such holograms is glued together, the object will appear as if it was being observed through a multiplicity of small openings in the screen.

Future advances in holographic television will apparently proceed from several directions. Firstly, improvements in television technology will allow an increase in the speed of the transmission.[†] and in the quality of three-dimensionality of holographic images. Furthermore, the development in laser technology will assure the creation of super-wide-band optical communication links and a corresponding system for modulation and scanning of light beams. Apparently, the utilization of the laser beam is the only way to transmit the colossal volume of information contained in a hologram.

A third direction is related to the development of dynamical image receivers and faster reacting screens with heightened resolving abilities. Today, the most promising materials seem to be the photochromatic materials and thermoplastics.[‡] The first has its resolution on the molecular level with, however, still small sensitivity. The latter are fast-acting materials. Even now, the preparation of a hologram takes only several seconds, and even this time may be lowered to fractions of a second.

Moreover, holographic television must find the means for economy of the passband. It is possible, for instance, to lower, without any sig-

---

[†] Since a 25×25 cm^2 hologram contains 6×10^{10} elements (with a resolution of 1000 lines/mm), it is not difficult to compute that, using today's technology, it would take about 2 hours to transmit it over a standard channel with a bandwidth of 8×10^6 cycles.

[‡] Thermoplastic recording, invented by W.E. Glenn, was first described by him.[37] Applications to color image projection and television are described by Glenn,[36a] and Glenn and Wolfe.[37a] (Note by GWS.)

nificant ill-effect, the field of vision in the vertical direction. It is necessary, besides, to utilize the fact that the consecutive images hardly differ one from another. By creating at the rim of the hologram a divergent reference beam, it is possible to increase the smallest element of the hologram. The darkening of nonessential details of the image and other optical tricks are also possible. Finally, not all of the information recorded on the hologram is required for the image reconstruction, and it will be necessary to learn how to control this property of redundancy.

11. Holographic Optics

Relatively few papers as yet deal with the creation of lensless optics and the use of holograms in systems with lenses. Apparently, these applications will be developed intensively when the necessary holographic technology has been further developed. Furthermore, a certain inertia of the interested investigators, which is due to loyalty to traditional methods, must be overcome. This is especially characteristic of the area lens optics, where the technology has achieved perfection, and a newcomer, such as holography, has a difficult time breaking through. Nevertheless, it is already clear that the hologram will prove to be useful in some cases and may even phase out the lens. For instance, a totally lensless photocamera has been developed in which the focusing is done by Fresnel zone plates.[57] It is desirable to establish *large* telescopes on the satellites in order to perform astronomical experiments. A glass lens with a diameter of 3 m weighs 3 tons. A zone plate of the same dimensions made out of thin plastic is almost weightless. Moreover, it may be folded up, and later, in orbit, unfolded into a large sheet.[60] To prepare zone plates, it is sufficient to photograph the interference picture. This is considerably simpler than preparing and handling a lens.

Zone plates may find application in laser communications.[57] At the transmitting end, they will allow a sharp collimation of the transmitted signal. At the receiving end, the zone plates will focus wide beams of up to 10 m diameter.

It should not be forgotten, however, that chromatism characterizes most types of holograms and, in particular, the zone plates. This chromatism of holograms is due to their diffraction ancestry. Quite probably, the main application of holographic optics will be found in coherent systems, even though methods for achromatization of holograms, which would allow their use with nonmonochromatic light, are known.[58,97]

Let us note also the low efficiency of holograms, which usually does not exceed 10-20%. Methods for obtaining holograms with a high efficiency are still being developed. [†]

Holograms find application also in combination with lenses. This is done for elimination of the spherical aberrations in the lenses,[127] for studies on the transmitting functions of optical systems,[29] and for preparation of high-resolution diffraction gratings.[35] [‡]

[†] The theoretical efficiency of various types of holograms has been investigated by a number of authors. See, for example, Kogelnik,[61a] including a complete list of references. See also Bergstein and Kermisch[2a]; Gabor and Stroke.[34a]

[‡] In addition to these types of holographic optics, Stroke[138] has recently proposed and verified the principle of a new class of two-step optical imaging systems, usable for incoherent light, and achieving "aperture synthesis" by *a posteriori* holographic Fourier-transform processing of images recorded in incoherent light with very "imperfect" imaging systems (e.g., pinhole cameras, Fresnel-zone systems, etc.). Paradoxically, in fact, the new principle will tend to work best with imperfect rather than "perfect" focusing systems, provided notably that the autocorrelation function of the spread function h (x, y) be very sharply peaked (approximating a "delta" function). This will be the case, for example, for a Fresnel-zone plate and for a "pinhole" camera, consisting simply of an aperture pierced with a randomly-disposed array of holes placed in front of a photographic plate, and for x-ray astronomy applications, among others. The method is to some extent closely related to Stroke's method of holographic image "deblurring" using *a posteriori* "extended-source lensless Fourier-transform holography" compensation.[139] The principle of the new method may be readily described in simple terms. Let $f(x, y)$ be the desired image and $h(x, y)$ the "spread function" (i.e. "image" of a point) produced by the system: Significantly, the spread function need not be a simple "conventional" diffraction pattern, but may consist, for example, of the multiplicity of images formed by a pinhole camera. The photograph recorded in incoherent light is $g(x, y)$. To extract the desired image $f(x, y)$ from $g(x, y)$, we form the "lensless Fourier-transform" hologram of $g(x, y)$ in the form $I = 1 + |\overline{G}|^2 + \overline{F} \ \overline{H} + \overline{F}^* \ \overline{H}^*$. If we now replace the hologram I into its recording position and illuminate it with the light from the spread function $h(x, y)$, placed into the place of the "point" (delta") function used to record the hologram, the imaging-wave of interest transmitted through the hologram is $\overline{F}^* \overline{H} * \overline{H}$ giving by Fourier transformation (e.g., in the focal plane of a lens "looking" through the hologram) $f^* \circledast h * h^* = f^*$, provided that we do in fact have the condition $h * h^* = \delta$ as stated. Alternatively, we may record the "extended-source" lensless Fourier-transform hologram of $g(x, y)$ (or the corresponding Fourier-transform hologram using a lens) in the form $(\overline{G+H})(\overline{G+H})^* = |\overline{G}|^2 + |\overline{H}|^2 + \overline{G}\overline{H}^* + \overline{G}^*\overline{H} = |\overline{G}|^2 + |\overline{H}|^2 + \overline{F} \ \overline{H} \ \overline{H}^* + \overline{F}^*\overline{H}^*\overline{H}$ and then, this time, illuminate the hologram with the "point" function placed into the position of the "spread" function h (x, y) used to record the hologram, in order to again obtain the identical "deconvolution" in the form of f^*, the conjugate image of f, or of $f!$ An example of holographic "aperture synthesis"[138] is shown in Fig. 13. [For further details, see also *Photographic Appendix*, Fig. 14]. (Note by GWS.)

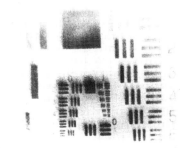

b

c

Blurred image formed by
multiple pinhole array

Single image holographically synthesized
from superposed multiple images

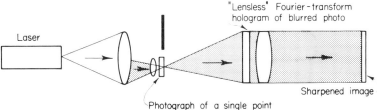

d

$$ h \longrightarrow \bar{H}\left[1 + |\bar{G}|^2 + \bar{F}\bar{H} + \bar{F}^*\bar{H}^*\right] \longrightarrow f $$

FIG. 13. *A posteriori* holographic "aperture synthesis" using extended-source Fourier-transform holography. (a) Random-array 9,000-point pinhole aperture generated by computer.[140,141] This also represents the point-spread function, notably at x-ray wavelengths where holes are large compared to wavelength.[142] (b) Blurred image which would be formed by using pinhole-aperture in place of a lens for image formation. Because of diffraction effects at optical wavelengths, this image was reconstructed by point-source illumination of the hologram of the test chart, recorded by using pinhole aperture as the "extended source" in place of the point source as a model experiment, to verify the principles. (c) Image reconstructed by illuminating the "extended-source" Fourier-transform hologram with the point-source aperture (spread function), according to the arrangement illustrated in (d). (See also footnote on p. 225 and Fig. 14, p. 253.) (After G. W. Stroke.[138,140])

APPENDIX REFERENCES †

1. *Alexandrov, E.B., and Bonch-Bruevich, A.M., Zh. Tekhn. Fiz.* **37**, 360-369 (1967).
2. Bakhrakh, L.D., and Kurochkin, A.P., *Dokl. Akad. Nauk SSSR* **171**, 1309-1312 (1966).
2a. Bergstein, L., and Kermisch, D., *Proc. Symp. Mod. Opt., Brooklyn Polytech., Brooklyn, New York, 1967*, J. Fox, ed. pp. 655-680. Brooklyn Polytech. Press, Brooklyn, New York, 1967.
2b. Boersch, H., Geiger, J., and Raith, H., *Z. Physik* **160**, 66 (1960).
3. Borodina, S.B., *Radioelektron. za Rubezhom* (27-28), 78-84 (1966).
4. Brooks, R.E., *et al., Appl. Phys. Letters* **7**, 92 (1965).
5. Brooks, R.E., Heflinger, L.O., and Wuerker, R.F., *IEEE J. Quantum Electron.* **2**, 275-279 (1966).
5a. Brown, G.M., Grant, R.M., and Stroke, G.W., *J. Acoust. Soc. Am.* (1968).
6. Burch, J.M., *Prod. Eng.* **44**, 436, 444 (1965).
7. Burch, J.M., Ennos, A.E., and Wilton, R.J., *Nature* **209**, 1015 (1966).
8. Burckhardt, C.B., *J. Opt. Soc. Am.* **56**, 1502-1509 (1966).
9. Carcel, J.T., *et al., Appl. Opt.* **5**, 1199-1201 (1966).
10. Carter, W.H., and Dougal, A.A., *Electronics* **39**, 204 (1966).
11. Carter, W.H., Engeling, P., and Dougal, A.A., *IEEE J. Quantum Electron.* **2**, 44 (1966).
12. Collier, R.J., and Pennington, K.S., *Appl. Phys. Letters* **8**, 44 (1966).
13. Collier, R.J., Doherty, E.T., and Pennington, K.S., *Appl. Phys. Letters* **7**, 223-225 (1965).
13a. Cutrona, L.J., Leith, E.N., Palermo, C.J., and Porcello, L.J., *IRE Trans. Inform. Theory* **6**, 386-400 (1960).
13b. Cutrona, L.J., Leith, E.N., Porcello, L.J., and Vivian, W.E., *Proc. IEEE* **54**, 1026-1032 (1966).
14. Denisyuk, Yu. N., *Dokl. Akad. Nauk SSSR* **144**, 1275 (1962).
15. Denisyuk, Yu. N., *Opt. i Spektroskopiya* **15**, 522-532 (1963).
16. Denisyuk, Yu. N., Ob otobrazhenii opitcheskikh svoistv ob'ekta v volnovom pole rasseyannogo im isluchemiia. Dissertatsiia na Soiskanie Uchenoi Stepeni Kandidata Fiziko-Matematicheskikh Nauk, 1963. 133 pp. 34 illust. bibl.
17. Denisyuk, Yu. N., *Opt. i Spektroskopiya* **18**, 275-283 (1965).
18. Denisyuk, Yu. N., *Usp. Nauchn. Fotogr. i Kinematogr.* **11**, 46-56 (1966).
19. De Velis, J.B., Parrent, G.B., and Thompson, B.J., *J. Opt. Soc. Am.* **56**, 423-427 (1966).
20. Dooley, R.P., *Proc. IEEE* **53**, 1733-1735 (1965).
21. *Elec. News* (536), 26 (1966).
22. *Elec. Weekly* (298), 6 (1966).
23. *Electronics* **39**, 35 (1966).
*24. *Electronics* **39**, 94 (1966).
25. *Electronics* **39**, 25 (1966).
26. Enloe, L.H., Murphy, J.A., and Rubinstein, C.B., *Bell System Tech. J.* **45**, 335-339 (1966).
27. Erdos, P., *IBM Tech. Discl. Bull.* **9**, 291 (1966).
28. Field, R.K., *Electron. Design* (15) 17-18 (1966).
29. Françon, M., Lowenthal, S., May, M., and Prat, R., *Compt. Rend.* **263**, 237-240 (1966).

† (Updating references added by GWS.)

228

30. Friesem, A.A., *Appl. Phys. Letters* 7, 102 (1965).
31. Friesem, A.A., and Fedorowicz, R.J., *J. Appl. Opt.* 5, 1085, 1086 (1966).
31a. Gabor, D., *Nature* 161, 777-778 (1948).
32. Gabor, D., *Nature* 208, 422, 423 (1965).
33. Gabor, D., *Electronics* 39, 94 (1966).
34. Gabor, D., *Nature* 217, 584 (1968).
34a. Gabor, D., and Stroke, G.W., *Proc. Roy. Soc.* A304, 275–289 (1968).
34b. Gabor, D., Stroke, G.W., *et al., Phys. Letters* 18, 116-118 (1965).
35. George, N., and Matthews, J.W., *Appl. Phys. Letters* 9, 212-219 (1966).
36. Givens, M.P., and Siemens-Wapniarsky, W.J., *J. Opt. Soc. Am.* 56, 537 (1966).
36a. Glenn, W.E., *J. Opt. Soc. Am.* 48, 841-843 (1958).
37. Glenn, W.E., *J. Appl. Phys.* 30, 1870-1873 (1959).
37a. Glenn, W.E., and Wolfe, J.E., *Intern. Sci. Technol.* pp. 28-35, etc. (1962).
38. Goodman, J.W., Huntley, Jr., W.H., Jackson, D.W., and Lehmann, M., *Appl. Phys. Letters* 8, 311 (1966).
39. Grant, R.M., Lillie, R.L., and Barnett, N.E., *J. Opt. Soc. Am.* 56, 1142 (1966).
40. Greguss, P., *J. Phot. Sci.* 14, 329-332 (1966).
41. Haines, K.A., and Hildebrand, B.P., *IEEE Trans. Instr. Meas.* 15, 595-602 (1966).
42. Haines, K.A., and Hildebrand, B.P., *Appl. Opt.* 5, (1966).
43. Heflinger, L.O., Wuerker, R.F., and Brooks, R.E., *J. Appl. Phys.* 37, 642-649 (1966).
44. Hersch, P., *Elec. News* (4955), (1965).
45. Hildebrand, B.P., and Haines, K.A., *J. Opt. Soc. Am.* 56, 537 (1966).
46. Hildebrand, B.P., and Haines, K.A., *Phys. Letters* 21, 422, 423 (1966).
47. Hioki, R., and Suzuki, T., *Japan. J. Appl. Phys.* 4, 816 (1965).
48. Huang, T.S., and Prasada, B., Quart. Progr. Rept. No. 81, pp. 199-205. Mass. Inst. Technol., Cambridge, Massachusetts, 1966.
49. *IEEE Spectrum* 3, 114 (1966).
50. Jacobson, A.D., and Mc Clung, F.J., *Appl. Opt.* 4, 19 1509 (1965).
51. Jeong, T.H., Rudolf, P., and Luckett, A., *J. Opt. Soc. Am.* 56, 1263, 1264 (1966).
52. Kakos, A., Ostrowskaya, G.H., Ostrovskii, Yu. I., and Zaidel, A.N., *Phys. Letters* 23, 81-83 (1966).
53. Kirk, J.P., *Appl. Opt.* 5, 1684, 1685 (1966).
54. Knox, C., *Science* 153, 989 (1966).
55. Kock, W.E., "Sound Waves and Light Waves." New York, 1965.
56. Kock, W.E., *Proc. IEEE* 54, 331 (1966).
57. Kock, W.E., *Elec. News* (567), 32 (1966).
58. Kock, W.E., *Proc. IEEE* 54, 1610, 1611 (1966).
58a. Kock, W.E., *Proc. IEEE* 56, 238-239 (1968).
59. Kock, W.E., and Harvey, F.K., *Bell System Tech. J.* 20, 564-587 (1951).
60. Kock, W.E., Rosen, L., and Rendeiro, J., *Proc. IEEE* 54, 1599, 1600 (1966).
60a. Kock, W.E., Rosen, L., and Stroke, G.W., *Proc. IEEE* 55, 80-81 (1967).
61. Kogelnik, H., *Bell System Tech. J.* 44, 2451-2454 (1965).
61a. Kogelnik, H., *Proc. Symp. Mod. Opt., Brooklyn Polytech., Brooklyn, New York, 1967*, J. Fox, ed., pp. 605-617, Brooklyn Polytech. Press, Brooklyn, New York, 1967.

62. Komar, A.P., Stabnikov, M.B., and Turukhano, B.G., *Dokl. Akad. Nauk SSSR* **169**, 1052-1053 (1966).
63. Landry, J., *J. Opt. Soc. Am.* **56**, 1133 (1966).
64. Lanza, C., *IBM Tech. Discl. Bull.* **8**, 1559, 1560 (1966).
64a. Lanzl, F., Mager, J., and Waidelich, Z. *Angew. Phys.* **24**, 157-159 (1968).
65. *Laser Focus* **1**, (6), (1965).
66. *Laser Focus* **2**, (6), 5 (1966).
67. *Laser Focus* **2**, (7), 4, 5 (1966).
68. *Laser Focus* **2**, (10), 10 (1966).
69. *Laser Focus* **2**, (11), 8 (1966).
70. *Laser Focus* **2**, (13), 7 (1966).
71. *Laser Focus* **2**, (13), 15 (1966).
72. *Laser Focus* **2**, (14), 4 (1966).
73. *Laser Focus* **2**, (17), 30-32 (1966).
74. *Laser Focus* **2**, (19), 11 (1966).
75. *Laser Focus* **2**, (20), (1966).
75a. Leith, E.N., and Ingalls, A., *Appl. Opt.* **7**, 539-544 (1968).
75b. Leith, E.N., and Upatnieks, J., *J., Opt. Soc. Am.* **52**, 1123 (1962).
76. Leith, E.N., and Upatnieks, J., *J. Opt. Soc. Am.* **55**, 569, 570 (1965).
77. Leith, E., and Upatnieks, Yu., *Usp. Fiz. Nauk* **87**, 521-538 (1965); also see *Nauka i Zhizn* (11), 23-31 (1965).
78. Leith, E. N., Upatnieks, J., and Haines, K.A., *J. Opt. Soc. Am.* **55**, 981-986 (1965).
79. Leith, E.N., Upatnieks, J., Hildebrand, B.P., and Haines, K.A., *J. Soc. Motion Picture Television Engrs.* **74**, 893-896 (1965).
80. Leith, E.N., *et al.*, *Appl. Opt.* **5**, 1303-1312 (1966).
81. Leith, E.N., Upatnieks, J., Kozma, A., and Massey, M., *J. Opt. Soc. Am.* **56**, 536 (1966).
82. Leith, E.N., Upatnieks, J., Kozma, A., and Massey, N., *J. Soc. Motion Picture Television Engrs.* **75**, 323 (1966).
83. Lin, L.H., and Pennington, K.S., *Bell Lab. Record* **43**, 416 (1965).
84. Lin, L., Pennington, K.S., Stroke, G.W., and Labeyrie, A.E., *Bell System Tech. J.* **45**, 659, 660 (1966).
85. Lohmann, A.W., *J. Opt. Soc. Am.* **56**, 537 (1966).
86. Lohmann, A.W., and Armitage, J., *J. Opt. Soc. Am.* **54**, 1404 (1964).
86a. Lohmann, A.W., and Werlich, H.W., *Phys. Letters* **25A**, 570 (1967).
87. Longuet-Higgins, H.C., *Nature* **217**, 104 (1968).
88. Meier, R.W., *J. Opt. Soc. Am.* **55**, 987-992 (1965).
89. Mueller, R.K., and Sheridon, N.K., *Appl. Phys. Letters* **9**, 328, 329 (1966); also see *Electronics* **39**, 37, 38 (1966).
90. Mustafin, K.S., Seleznev, B.A., and Shtyrkov, E.Il, *Opt. i Spektroskopiya* **22**, 319-320 (1967).
91. Nalimov, I.P., *Zarubezhnaya Radioelektron.* (2), 3-28 (1966).
92. Nalimov, I.P., Author's certificate, entry No. 1062342/26-25 of 3/12/66. issued 11/5/66.
93. Nassenstein, H., *Phys. Letters* **21**, 290, 291 (1966).
94. Nassimbene, E.G., *IBM Tech. Discl. Bull.* **8**, 1397, 1398 (1966).
95. *New Scientist* **31**, 18 (1966).
96. Ostrovskaya, G.V., and Ostrovskii, Yu. I., *Zh. Eksperim. i Teor. Fiz. Pis'ma v Redaktsiyu* **4**, 121-123 (1966).

97. Paques, H., *Proc. IEEE* 54, 1195 (1966).

98. Paques, H., and Smigielski, P., *Compt. Rend.* 260, 6562-6564 (1965).

99. Paques, H., and Smigielski, P., *Opt. Acta* 12, 359-378 (1965).

100. Parrent, G.B., and Thompson, B.J. *Opt. Acta* 11, 183-192 (1964).

101. Pennington, K.S., and Collier, R.J., *Appl. Phys. Letters* 8, 14-16 (1966).

101a. Pennington, K.S., and Lin, L.H., *Appl. Phys. Letters* 7, 56 (1965).

102. Platonenko, V.T., *Usp. Fiz. Nauk* 87, 575-580 (1965).

103. Pole, R.V., Wieder, H., and Myers, R.A., *Appl. Phys. Letters* 8, 229-231 (1966).

104. Powell, R.L., and Stetson, K.A., *J. Opt. Soc. Am.* 55, 1593-1598 (1965).

104a. Rogers, G.L., *Proc. Roy. Soc. Edinburgh* A63, 319-325 (1951).

104b. Rogers, G.L., *J. Atmospheric Terrest. Phys.* 11, 51-53 (1957).

105. Rogers, G.L., *J. Sci. Instr.* 43, 677 (1966).

106. Rotz, F.B., and Friesem, A.A., *Appl. Phys. Letters* 8, 146-148 (1966); see also revision *Appl. Phys Letters* 8, 240 (1966).

107. Schawlow, A., *Microwaves* 5, 16 (1966).

108. *Sci. News* 90, 349 (1966).

109. *Sci. Res.* p. 65 (1968).

110. Shmakov, P.V., *Tsvetn. i Ob'emnoe Televidnie* (1955).

111. Soroko, L.M., "Lektsii po Golografii." Preprint. O. I. Ya. I., March 1965.

112. Soroko, L.M., *Usp. Fiz. Nauk*, 90, 1-46 (1966).

113. Spitz, E., and Werts, A., *Compt. Rend.* 262, 758-760 (1966).

114. Stetson, K.A., and Powell, R.L., *J. Opt. Soc. Am.* 55, 1694, 1695 (1965).

115. Stetson, K.A., and Powell, R.L., *J. Opt. Soc. Am.* 56, 1161-1168 (1966).

116. Story, J.B., Ballard, G.S., and Gibbons, R.H., *J. Appl. Phys.* 37, 2183 (1966).

116a. Stroke, G.W., *Appl. Phys. Letters* 6, 201 (1965).

117. Stroke, G.W., *Phys. Letters* 23, 325-327 (1966).

117a. Stroke, G.W., *J. Phys. (Paris)* 28, 196-203 (1967).

117b. Stroke, G.W., *Phot. Korr.* 104, 82-85 (1968).

117c. Stroke, G.W., *Proc. Symp. Mod. Opt., Brooklyn Polytech., Brooklyn, New York, 1967*, J. Fox, ed. Brooklyn Polytech. Press, Brooklyn, New York, 1967.

118. Stroke, G.W., and Labeyrie, A.E., *Phys. Letters* 20, 157-159 (1966).

119. Stroke, G.W. and Labeyrie, A.E., *Phys. Letters* 20, 368 (1966).

120. Stroke, G.W., and Zech, R.G., *Appl. Phys. Letters* 9, 215-217 (1966).

120a. Stroke, G.W. and Zech, R.G., *Phys. Letters* 25A, 89 (1967).

120b. Stroke, G.W. et al., *Phys. Letters* 18, 274 (1965).

120c. Stroke, G.W. et al., *Appl. Phys. Letters* 7, 178 (1965).

120d. Stroke, G.W. et al., *Brit. J. Appl. Phys.* 17, 497-500, 2 plates (1966).

120e. Stroke, G.W., Westervelt, F.H., and Zech, R.G., *Proc. IEEE* 55, 109-111 (1967).

120f. Stroke, G.W., Puech, C., and Indebetouw, G., *Phys. Letters* 26A, 443-444 (1968).

121. Supertzi, E.P., and Rigler, A.K., *J. Opt. Soc. Am.* 56, 524 (1966).

122. Tanner, L.H., *J. Sci. Instr.* 43, 81-83 (1966).

123. Tanner, L.H., *J. Sci. Instr.* 43, 346 (1966).

124. Tanner, L.H., *J. Sci. Instr.* 43, 878-887 (1966).

124a. Tsujiuchi, J., and Tsuruta, T., *Japan. J. Appl. Phys.* 36, 232(76)-239(83) (1967).

125. Tyler, G.L., *J. Geophys. Res.* 71, 1559-1567 (1966).

126. Upatnieks, J., Marks, J., and Fedorowicz, R.J., *Appl. Phys. Letters* 8, 286 (1966).

127. Upatnieks, J., Vander Lugt, A., and Leith, E.N., *Appl. Opt.* **5**, 589-593 (1966).
128. Van Heerden, P.J., *Appl. Opt.* **2**, 393 (1963).
128a. Van Heerden, P.J., *Appl. Opt.* **2**, 387 (1963).
129. Van Ligten, R.F., and Osterberg, H., *Nature* **211**, 282 (1966).
130. Welford, W.T., *Appl. Opt.* **5**, 872 (1966).
131. Winthrop, J.T., and Worthington, C.R., *Phys. Letters* **21**, 413 (1966).
132. Zaidel, A.N., Ostrovskaya, G.V., Ostrovskii, Yu. I., and Cheslidze, T. Ya., *Zh. Tekhn. Fiz.* **36**, 2208-2210 (1966).
133. L.D. Siebert, *Appl. Phys. Letters* **11**, 326-328 (1967); *Proc. IEEE* **56**, pp. 1242-1243 (1968).
134. K.A. Stetson, *Appl Phys. Letters* **11**, 225-226 (1967).
135. J. Wilczynski and R.E. Tibbetts, private communication from JW to GWS.
136. N. Shiotake, T. Tsuruta, Y. Itoh, N. Takey, and K. Matsuda, *Japan, J. Appl. Phys.* **7**, 904-909 (1968).
137. T. Tsuruta, N. Shiotake, and Y. Itoh, *Japan. J. Appl. Phys.* **7**, 1092-1100 (1968).
138. G.W. Stroke, *Phys. Letters* **28A**, 252-253 (November 1968).
139. G.W. Stroke, *Phys. Letters* **27A**, 405-406 (1968).
140. G.W. Stroke, J.H. Underwood, and R.B. Hoover, *Astrophys. J.* (in press).
141. R.B. Hoover, *J. Opt. Soc. Am.* **58**, 721 (1968).
142. R.H. Dicke, *Astrophys. J.* **153**, L101-L106 (August 1968).

Bibliography †
Surveys

1. Dickinson, A., and Dye, M.S., Principles and practice of holography. *Wireless World* **73**, 56-61 (1967).
2. Elliott, E.R., Lasers—where they are now. *Elec. Weekly* (334), **7**, 34 (1967).
3. Ennos, A.E., Holography and its applications. *Contemp. Phys.* **8**, 153-170 (1967).
4. Françon, M., La coherence en Optique. *Atomes* (241), 162-166 (1967).
5. Gabor, D., Les transformations de l'information en Optique. *Opt. Acta* **13**, 299-310 (1966).
6. Maréchal, A., Theory and practice of image formation. *J. Opt. Soc. Am.* **56**, 1645-1648 (1966).
7. Marquet, M., Performances en holographie. *Rev. Opt.* **45**, 404-416 (1966).
8. Ramberg, E.G., The hologram—properties and applications. *RCA Rev.* **27**, 467-500 (1966).
9. Smith, A.B., Direct-view 3-D images!, *Radioelectron.* **38**, 46-49 (1967).
10. Yates, J.M., Wavefront image evaluation, *Brit. J. Phot.* **133**, 328-330, 343 (1966).

†Also see p. 189, Editor's remarks. Among other useful lists of holography publications, the reader may wish to consult that published by J.N. Lattia in *J. Soc. Motion Picture Television Engrs*, **77**, 1-22 (April 1968) under the title "A classified bibliography on holography and related fields."

Physical Optics

1. Champagne, E.B., Nonparaxial imaging, magnification, and aberration properties in holography. *J. Opt. Soc. Am.* **57**, 51-55 (1967).

2. Konjaev, K.V., Interference method of two-dimensional Fourier transform with spatially incoherent illumination. *Phys. Letters* **24A**, 490-491 (1967).

3. Pavlenko, Yu. G., Vosstanovlenie protsessa uprugogo rasseyaniia v kogerentnom svete. *Zh. Eksperim. i Teor. Fiz.* **52**, 699-701 (1967).

4. Reynolds, G.O., and De Velis, J.B., Hologram coherence effects. *IEEE Trans. Antennas Propagation* **15**, 41-48 (1967).

5. Roig, J., Taravellier, R., and Mas, G., Interferences entre une onde coherente Σ' emise par un laser et une onde Σ'' diffractee par un object de faibles dimensions, dans le cas d'un faisceau forment astigmatique. *Compt. Rend.* **263**, 1014-1017 (1966).

Technique

1. Bolstad, J.O., Holograms and spatial filters processed and copied in position. *Appl. Opt.* **6**, 170 (1967).

2. Brumm, D.B., Double images in copy holograms. *Appl. Opt.* **6**, 588 (1967).

3. Carpenter, R.L., and Clifford, K.I., Simple inexpensive hologram viewer. *J. Opt. Soc. Am.* **57**, 276 (1967).

4. Complete holographic "camera" system. *Microwaves* **6**, 62, 64 (1967).

5. De, M., and Sevigny, L., Polarisation holography. *J. Opt. Soc. Am.* **57**, 110-111 (1967).

6. De Bitetto, D.J., White-light viewing of surface holograms by simple dispersion compensation. *Appl. Phys. Letters* **9**, 417-418 (1966).

7. Froehly, C., Etude holographique du second harmonique de l'onde emise par un laser a rubis. *Compt. Rend.* **263**, 1304-1307 (1966).

8. Kock, W.E., Rosen, L., and Rendeiro, J., Realism of lens action in holograms. *Proc. IEEE* **54**, 1985 (1966).

9. Kock, W.E., Rosen, L., and Stroke, G.W., Focused-image holography—a method for restoring the third dimension in the recording of conventionally-focused photographs. *Proc. IEEE* **55**, 80-81 (1967).

10. Martienssen, W., and Spiller, S., Holographic reconstruction without granulation. *Phys. Letters* **24A**, 126-127 (1967).

11. Mikayelyan, A.L., Razumov, L.N., Sakharova, N.A., and Turkov, Yu. G., O poluchenii gologramm fur'e s Pomoshch yu impul'snogo rubin ovogo lazera. *Zh. Eksperim. i. Teor. Fiz. Pis'ma v Redaktsiyu* **5**, 148-150 (1967).

12. Rosen, L., Apparent rotation of hologram virtual images. *J. Opt. Soc. Am.* **57**, 278-279 (1967).

13. Rosen, L., Holograms of the aerial image of a lens. *Proc. IEEE* **55**, 79-80 (1967).

14. Rosen, L., The pseudoscopic inversion of holograms. *Proc. IEEE* **55**, 118 (1967).

15. Stroke, G.W., Funkhouser, A., Leonard, C., Indebetouw, G., and Zech, R.G., Hand-held holography. *J. Opt. Soc. Am.* **57**, 110 (1967).

16. Suzuki, T., and Hioki, R., Frequency response of photographic emulsion in holography. *Japan J. Appl. Phys.* **5**, 1257-1259 (1966).

17. Vandewarker, R., and Snow, K., Low spatial frequency holograms of solid objects. *Appl. Phys. Letters* **10**, 35 (1967).

18. Yoshihara K., and Kitade A., Holographic spectra using a triangle path interferometer. *Japan J. Appl. Phys.* **6**, 116 (1967).

Three-Dimensional Photography

1. Fingerprints in 3-D. *Electronics* **40**, 52 (1967).
2. Friesem, A.A., and Fedorowicz, R.J., Multicolor wavefront reconstruction. *Appl. Opt.* **6**, 529-538 (1967).
3. Greguss, P., Ultrasonoholography.*Sci.J.(London)* **2**, 83 (1966).
4. Hildebrand, B.P., and Haines, K.A., Multiple-wavelength and multiple-source holography applied to contour generation. *J. Opt. Soc. Am.* **57**, 155-162 (1967).
5. Marom, E., Color Imagery by wavefront reconstruction. *J. Opt. Soc. Am.* **57**, 101-102 (1967).
6. Pole, R.V., 3-D imagery and holograms of objects illuminated in white-light. *Appl. Phys. Letters* **10**, 20-22 (1967).
7. Stroke, G.W., Westervelt, F.H., and Zech, R.G., Holographic synthesis of computer-generated holograms. *Proc. IEEE*, **55**, 109-111 (1967).
8. Thompson, B.J., and Ward, J.H., Particle sizing—the first direct use of holography. *Sci. Res.* **1**, 37 (1966).
9. Thompson, B.J., Ward, J.H., and Zinky, W.R., Application of hologram techniques for particle size analysis. *Appl. Opt.* **6**, 519-528 (1967).
10. Tricoles, G., and Rope, E.L., Wavefront reconstruction with centimeter waves. *J. Opt. Soc. Am.* **56**, 542 (1966).
11. Tricoles, G., and Rope, E.L., Reconstructions of visible images from reduced-scale replicas of microwave holograms. *J. Opt. Soc. Am.* **57**, 97-99 (1967).
12. Ward, J.H., and Thompson, B.J., In-line hologram system for bubble-chamber recording. *J. Opt. Soc. Am.* **57**, 275-276 (1967).
13. Waters, J.P., Holographic image synthesis utilizing theoretical methods. *Appl. Phys. Letters* **9**, 405-407 (1966).

Recognition of Images

1. Marchant, M., and Knight, D., Multiple recording of holograms. *Opti. Acta* **14**, 199-201 (1967).
2. Marquet, M., Bourgeon, M.H., and Saget, J.C., Quelques applications de l'holographie. *Compt. Rend.* **264**, 35-37 (1967).
3. Vienot, J. Ch., and Bulabois, J., Differenciation spectrale et filtrage par hologramme de signaux optiques faiblement decorrelees. *Opt. Acta* **14**, 57'70 (1967).

Interference Memory

1. Brinton, Jr., J.B., Remember with microwaves or lasers. The two technologies are vying for future computer-memory applications. *Microwaves* **6**, 10, 12, 116 (1967).
2. Burckhardt, C.B., Display of holograms in white-light. *Bell System Tech. J.* **45**, 1841 (1966).
3. Is memory holographic? *New Scientist* (531), 221 (1967).
4. Nikityuk, N.M., "Svet, Kvanty i Vychislitel'naya Tekhnika." Znanie Press, 1967.

Interferometry

1. Wolfe, R., and Doherty, E.T., Holographic interferometry of the distortion of thermoelectric cooling modules. *J. Appl. Phys.* 37, 5008-5009 (1966).

Restoration of Atmospherically Degraded Images

1. Morgan, S.P., Restoring atmospherically degraded images. *Nature* 213, 465-469 (1967).

Holographic Microscopy

1. Carter, W.H., and Dougal, A.A., Field range and resolution in holography. *J. Opt. Soc. Am.* 56, 1754-1759 (1966).
2. De M., and Sevigny, L., Three-beam holography. *Appl. Phys. Letters* 10, 78-79 (1967).
3. Ellis, G.W., Holomicrography: transformation of image during reconstruction a posteriori. *Science* 154, 1195-1197 (1966).
4. Holograms to get IC picture. *Electronics* 40, 26 (1967).

Holographic Motion Pictures

1. Hologram magnification technique developed. *Laser Focus* 2, 4 (1966).
2. Ross, R.M., Ring hologram for 3-D display-single concept motion picture. *IBM Tech. Discl. Bull.* 9, 390 (1966).
3. Shchekochikhin, V., Stereokino na printsipakh lazernoi golografii. *Kinomekhanik* 4, pp. 41-43 (1967).

Supplemental Bibliography †

Surveys

1. Platonenko, V.T., Golografiia, fotografirovanie i vosstanovlenie volnovogo fronta. *Usp. Fiz. Nauk* 90, 199-201 (1966).
2. Soroko, L.M., Golografiia i interferentsionnaia obrabotka informatsii. *Usp. Fiz. Nauk* 90, 1-46 (1966). Bibliography, 98 titles.
3. Soroko, L.M., Bezlinzovaia optika. Fotografirovanie v luchakh lazera. *Priroda* 55, 37-48 (1966).
4. Blum, J., Holography: the picture looks good. *Electronics* 39, 139-143 (1966).
5. Closets, F., Des ≪objets-fantômes≫ en plein jour. *Sci. Avenir* (239), 8-12, 70 (1967).
6. Collier, R.J., Some current views on holography. *IEEE Spectrum* 3, 67-74 (1966).

†Here has been collected supplementary literature for 1965-1966 for which there are no references in the monographs nor in the surveys. For a bibliography covering preceding years see: *J. Soc. Motion Picture Television Engrs.* 75, 373-434, 759-809 (1966).

7. Davy, J., Breakthrough in three dimensions. *Management Today* p. 21, 22 (1966).
8. Di Pentima, A., Holography. *AAAS Meeting, 133rd,* Washington, D.C., December 1966; *Science* 154, 1363, 1364 (1966).
9. Eaglesfield, C.C., Holograms: what uses have they? *Discovery* 27, 23-26 (1966).
10. Gabor, D., Holography, or the whole picture. *New Scientist* 29, 74-78 (1966).
11. Gabor, D., Holography—the reconstruction of wavefronts. *Electron. and Power* 12, 230-239 (1966).
12. Kogelnik, H.W., Holography—fundamentals and demonstrations. *The Scanner* 16, 1 (1966).
13. Leith, E.N., Holography's practical dimension. *Electronics* 39, 88-94 (1966).
14. Leith, E.N., Holography—lensless 3D photography. *Ind. Res.* 8, 40-44 (1966).
15. Leith, E.N., and Upatnieks, J., Wavefront reconstruction photography. *Phys. Today* 18, 26-32 (1965).
16. Leith, E.N., and Upatnieks, J., True 3D image from laser photography. *Electron. World* p. 34, 35 (1965).
17. Leith, E.N., and Upatnieks, J., Holograms: their properties and uses. *SPIE J.* 4, 3-6 (1965).
18. Leith, E.N., and Upatnieks, J., Photography by laser. *Sci. Am.* 212, 24 (1965).
19. Morris, R.E., Physical principles of holography. *J. Phot. Sci.* 14, 291-296 (1966).
20. Nassenstein, H., Abbildungsverfahren mit Reconstruction des Wellenfeldes (Holographie). *Z. Angew. Phys.* 22, 37-50 (1966).
21. Ose, T., Holography. *Oyo Butsuri* 35, (1966).
22. Paques, H., and Smigielski, P., Holographie. *Opt. Acta* 12, 359-378 (1965).
23. Spiller, E., Optische Nachrichtenübertragung durch Holographie. *Umschau Wiss. Tech.* 66, 288-292, 315-321 (1966).
24. Thompson, B.J., and Parrent, Jr., J.B., Holography. *Sci. J. (London)* 3, 42-49 (1967).

Physical Optics

1. Mikayelyan, A.L., and Bobrinyev, V.I., Shumovye ogranicheniia pri poluchenii ob'emnikh izobrazhenii. *Zh. Eksperim. i Teor. Fiz. Pis'ma v Redaktsiyu* 44, 172-174 (1966).
2. Pistol'kors, A.A., O razreshayushchei sposobnosti gologrammy. *Dokl. Akad. Nauk SSSR* 172, 334-337 (1967).
3. Burch, J.M., Gates, J.W., Hall, R.J., and Tanner, L., Holography with a scatter-plate as a beam-splitter and a pulsed ruby laser as a light source. *Nature* 212, (5068), 1347 (1966).
4. Cathey, Jr., W.D., Spatial phase modulation of wavefronts in spatial filtering and holography. *J. Opt. Soc. Am.* 56, 1167-1171 (1966).
5. Delingat, E., The Fourier spectra of photographic objects. Part 1. Statement of the problem and apparatus for research. *Optik* 23, 177-188 (1965/1966).

6. Gabor, D., Improvements in and relating to optical apparatus for producing multiple interference patterns. U.S. Patent No. 2770166, November 1956.

7. Helström, C.W., Image luminace and ray tracing in holography. *J. Opt. Soc. Am.* **56**, 433-441 (1966).

8. Kock, W.E., and Rendeiro, J., Some curious properties of holograms. *Proc. IEEE* **53**, 1787 (1965).

9. Kopylov, G.I., On some possible properties of holograms. *Phys. Letters* **21**, 645, 646 (1966).

10. Lohmann, A.W., Wavefront reconstruction for incoherent objects. *J. Opt. Soc. Am.* **55**, 1555, 1556 (1965).

11. Lohmann, A.W., Reconstruction of vectorial wavefronts. *Appl. Opt.* **4**, 1667 (1965).

12. Lurie, M., Effects of partial coherence on holography with diffuse illumination. *J. Opt. Soc. Am.* **56**, 1369-1372 (1966).

13. Mandel, L., Color imagery by wavefront reconstruction. *J. Opt. Soc. Am.* **55**, 1697, 1698 (1965).

14. Marquet, M., Fortunato, G., and Royer, H., Etude theorique de la correspondance objet-image dans l'holographie. *Compt. Rend.* **261**, 3553-3555 (1965).

15. Meier, R.W., Depth of focus and depth of field in holography. *J. Opt. Soc. Am.* **55**, 1693 (1965).

16. Meier, R.W., Cardinal points and the novel imaging properties of a holographic system. *J. Opt. Soc. Am.* **56**, 219-223 (1966).

17. Parrent, G.B., and Reynolds, G.O., Space-band-width theorem for holograms. *J. Opt. Soc. Am.* **56**, 1400 (1966).

18. Peters, P.J., Incoherent holograms with mercury light source. *Appl. Phys. Letters* **8**, 209 (1966).

19. Rosen, R., Focused-image holography with extended sources. *Appl. Phys. Letters* **9**, 337-339 (1966).

20. Royer, H., Contribution a l'etude de l'information en holographie. *Compt. Rend.* **261**, 4003 (1965).

21. Stroke, G.W., Brumm, D., Funkhouser, A., Labeyrie, A., and Restrick, R.C., On the absence of phase-recording or ≪twin-image≫ separation problems in Gabor (in-line) holography. *Brit. J. Appl. Phys.* **17**, 497-500 (1966).

22. Suzuki, T., and Hioki, R., Speckled diffraction pattern and source effect on resolution limit in holography. *Japan J. Appl. Phys.* **5**, 814-818 (1966).

23. Winthrop, J.T., and Worthington, C.R., Fresnel-transform representation of holograms and hologram classification. *J. Opt. Soc. Am.* **56**, 1362-1369 (1966).

24. Worthington, C.R., Production of holograms with incoherent illumination. *J. Opt. Soc. Am.* **56**, 1397 (1966).

Fresnel Zone Plates

1. Einighammer, H.J., Zur Abbildung von Röntgensternen mit dem Hologramm-Teleskop. *Optik* **23**, 627-641 (1965/1966).

2. Horman, M.H., and Chau, H.H.M., Zone plate theory based on holography. *Appl. Opt.* **6**, 317-322 (1967).

3. Kock, W.E., Rosen, L., and Rendeiro, J., Holograms and zone plates. *Proc. IEEE* **54**, 1599, 1600 (1966).

4. Kock, W.E., Three-color hologram zone plates. *Proc. IEEE* **54**, 1610, 1611 (1966).

5. Paques, H., Achromatization of holograms. *Proc. IEEE* 54, 1195 (1966).
6. Roig, J., Taravellier, R., and Mas, G., Interferences entre les ondes dif-
    fractees par des ecrans de faibles dimensions et un fond coherent fourni
    par un laser. *Compt. Rend.* 263, 608-611 (1966).
7. Waldman, G.S., Variation on the Fresnel zone plate. *J. Opt. Soc. Am.* 56,
    215-218 (1966).

Technique

1. Zaidel', A.N., Konstantinov, V.B., and Ostrovskii, Yu. I., Lazernaia
    Rezol' vometriia, *Zh. Nauchn. i Prikl. Fotogr. i Kinematogr.* 11, 381
    (1966).
2. Klimenko, I.S., and Pukman, G.I., Gaborovskoe vosstanovlenie volnovogo
    fronta s pomoshch 'yu lazera. *Opt. i Spektroskopia* 21, 751-752 (1966).
3. Komar, A.P., Stabnikov, N.V., and Turukhano, B.G., Golografiia s
    priamym i opornym puchkom. *Dokl. Akad. Nauk SSSR* 173, 1059-1061 (1967).
4. Kosourov, G.I., Kalinkina, I.N., and Golovei, M.P., Vosstanovlenie
    izobrazheniia po gologramme v nemonokhromaticheskom svete. *Zh.
    Eksperim. i Teor. Fiz. Pis'ma v Redaktsiyu* 4, 84-86 (1966).
5. Konstantinov, B.P., Zaidel', A.N., Konstantinov, V.B., and Ostrovskii,
    Yu. I., Fotografirovanie v kogerentnom svete, eksperimental 'naia
    tekhnika i razreshayushchaia sposobnost' metoda. *Zh. Tekhn. Fiz.* 36,
    1718-1721 (1966).
6. Beddoes, M.P., and Aktar, S.A., Some techniques for producing holograms
    with lasers. *Electron. and Commun.* 13, 61 (1965).
7. Carcel, J.T., Rodemann, A.H., Florman, E., and Domeshec, S., Simplifi-
    cation of holographic procedures. *Appl. Opt.* 5, 1199-1201 (1966).
8. Corcoran, V.J., Herron, R.W., and Jaramillo, J.G., Generation of a holo-
    gram from a moving target. *Appl. Opt.* 5, 668, 669 (1966).
9. De Bitetto, D.J., On the use of moving scatterers in conventional holo-
    graphy. *Appl. Phys. Letters* 8, 78-80 (1966).
10. Falconer, D.G., Role of the photographic process in holography. *Phot.
    Sci. Eng.* 10, 133-135 (1966).
11. Harris, F.S., Sherman, G.C., and Billings, B.H., Copying holograms.
    Appl. Opt. 5, 665, 666 (1966).
12. Hoffman, A.S., Doidge, J.G., and Mooney, D.G., Inverted reference-
    beam hologram. *J. Opt. Soc. Am.* 55, 1559 (1965).
13. Kozma, A., Photographic recording of spatially modulated coherent light.
    J. Opt. Soc. Am. 56, 428-434 (1966).
14. Landry, J., Coffee-table holography. *J. Opt. Soc. Am.* 56, 1133 (1966).
15. Landry, M.J., Copying holograms. *Appl. Phys. Letters* 9, 303, 304
    (1966).
16. Leith, E.N., Photographic film as an element of a coherent optical sys-
    tem. *Phot. Sci. Eng.* 6, 75-80 (1962).
17. Leith, E.N., Kozma, A., and Upatnieks, J., Requirements for hologram
    construction. *Laser Letters* 3, 2-10 (1966).
18. Nassimbene, E.G., and Ross R.M., Reducing noise in holograms. *IBM
    Tech. Discl. Bull.* 8, 1396 (1966).
19. Rigler, A.K., Wavefront reconstruction by reflection. *J. Opt. Soc. Am.*
    55, 1693 (1965).
20. Rogers, G., The design of experiments for recording and reconstructing

three-dimensional objects in coherent light (holography). *J. Sci. Instr.* **43**, 677 (1966).

21. Ross, R.M., Making holograms using Brewster's angle. *IBM Tech. Discl. Bull.* **8**, 1404 (1966).

22. Royer, H., Study of the recording of high frequencies by photographic emulsions. *Compt. Rend.* **261**, 5024-5027 (1965).

23. Slaymaker, F.H., The elimination of building vibration in an optical laboratory. *Appl. Opt.* **5**, 1766-1768 (1966).

24. Van Ligten, R.F., Influence of photographic film on wavefront reconstruction. I. Plane wavefronts. *J. Opt. Soc. Am.* **56**, 1-9 (1966).

25. Urbach, J.C., and Meier, R.W., Thermoplastic xerographic holography. *Appl. Opt.* **5**, 666, 667 (1966).

26. Urbach, J.C., The role of screening in thermoplastic xerography. *Phot. Sci. Eng.* **10**, 287-297 (1966).

27. Walles, S., Magnification and observation of a holographic interference pattern. *Opt. Acta* **13**, 241-246 (1966).

28. Wilmot, D.W., Schinelier, E.R., and Heuman, R.W., Hologram illumination with a flashlight. *Proc. IEEE* **54**, 690, 691 (1966).

FIG. 1. Hologram recording arrangement, L, 100 mW helium-neon laser (6328 Å); B, beam splitting prism; D, diffusor; O, object; R, reference mirror; H, hologram. (After Stroke. [1])

FIG. 2. View of three-dimensional image (O) reconstructed by bleached "phase" hologram (H) upon illumination with helium-neon laser (L). Note "speckled" (granular) appearance of image when photographed with small aperture (required to show great depth of field).

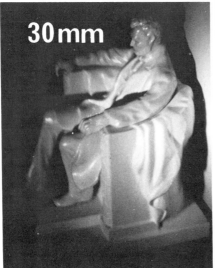

FIG. 3. Holographic images photographed through bleached "phase" hologram (see Fig. 2). Note "speckled" (granular) appearance of image when it is
photographed with 3 mm diameter pupil (corresponding to pupil of eye). The
speckle is caused by interference field pattern formed near object by spherical
wavelets originating from neighboring points of diffusing object. The speckle is
not observable when image is observed with a relative aperture D/f greater than
$1/10$ (D, diameter of pupil of camera, microscope lens, eye, etc.; f, distance of
pupil from image). Image on left was photographed through hologram with $D = 30$
mm and $f = 300$ mm. Note importance of this remark in view of high-resolution
microscopy: No image-degrading (resolution-loss) effects will be observed in
holographic reconstruction schemes with observation using objectives such that
D/f exceeds $1/10$. (After Stroke.[1])

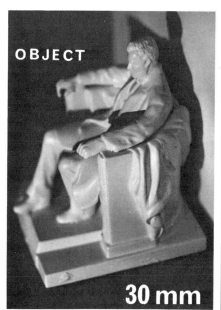 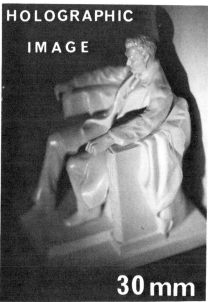

FIG. 4. Direct photograph (left) and photograph of virtual image of same object through hologram, both in laser light. Note high degree of perfection in holographic imaging attainable with contemporary techniques. Note also that *real* holographic images, reconstructed with large-area holograms, i.e. with D/f greater than $1/10$, are also equally perfect, without speckle (see, e.g., Fig. 32c, p. 142).

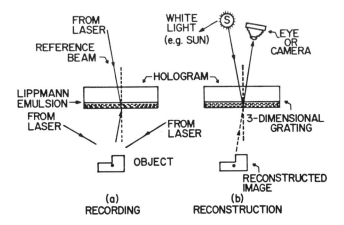

FIG. 5. White-light reconstruction of holographic images using the Lippmann-Bragg effect. (After Stroke and Labeyrie. [2])

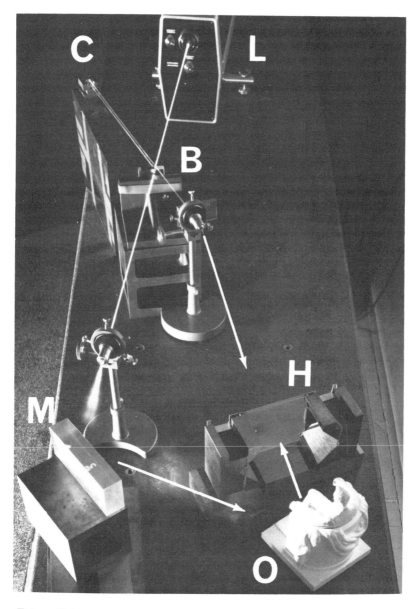

FIG. 6. Holographic recording of "white-light reflection holograms" according to Fig. 5. L, 100 mW continuous-wave helium-neon laser (6328 Å); B, beam-splitting mirror; C, corner-cube mirror used in coherence-length path equalization; M, mirror; O, object; H, hologram (recorded here on 15–μ thick "conventional" Kodak 649F high-resolution emulsion, coated on "microflat" glass plate). Agfa 8E75 emulsions have recently appeared to be particularly suitable for the recording of very luminous "white-light reflection holograms."[20]

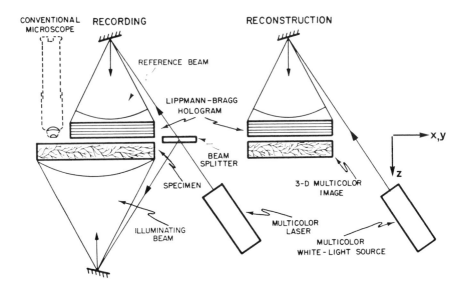

FIG. 7. Multicolor holographic microscopy using "white-light Lippmann-Bragg reflection holography." Notice considerable double advantage of holographic microscopy, using this method, compared to conventional microscopic photography (or using holographic recording through a microscope lens). In addition to the capability of storing "in focus" imaging information over a considerable *axial* depth along z, characteristic of all holograms, this type of hologram also additionally stores in a single photograph high-resolution microscopic information over an immense *lateral* field, along x, y, in comparison to an ordinary high-resolution photograph or an ordinary high-resolution hologram recorded through the microscope lens.

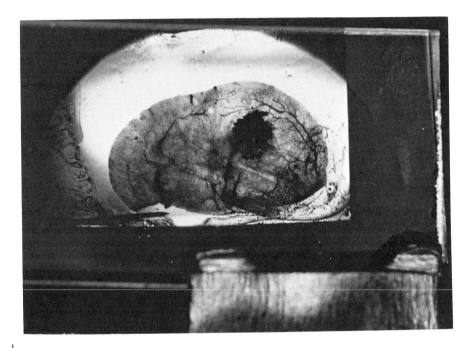

FIG. 8. Multicolor image reconstructed from "white-light reflection hologram" using method of Figs. 5 and 9, and displaying wide-field and large-depth advantages described in Fig. 9. The photograph shows holographically reconstructed image of 30-mm wide (along x, y), 3-mm deep (along z) section (whole-mount, cleared in oil of wintergreen) of guinea-pig ear. This early remarkable result shows normal blood vessels as well as some injection of india ink into lymphatic. Method is being developed for micro-cinematographic holographic photography of blood circulation in capillaries *in vivo*. (After Stroke and Burke.[3])

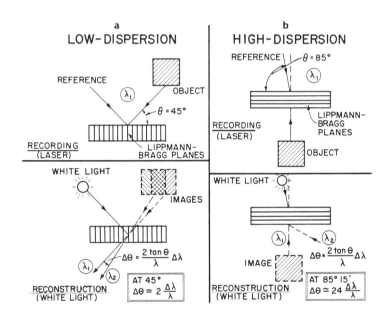

FIG. 9. Schematic diagram illustrating comparison between (a) conventional low-dispersion transmission hologram and (b) high-dispersion "white-light reflection" hologram. White-light holographic image reconstruction is readily obtained without any color "cross-talk" in high-dispersion hologram (b).

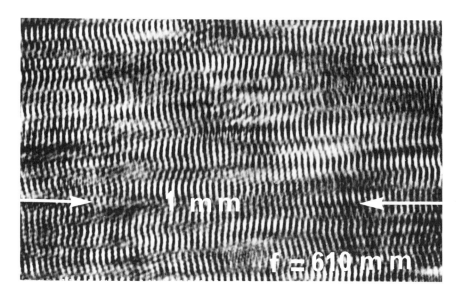

FIG. 10. Lensless Fourier-transform hologram of word "HOLOGRAPHY," illustrating diffraction-grating-like nature of holograms (see pp. 8-11, 116-127). Distance of word from reference point, 17.5 mm; distance from hologram, 610 mm. (For additional details on the theory of information storage in holograms see, e.g., Gabor and Stroke,[4] where it is shown that holograms recorded in diffused illumination, such as that shown, contain the information in the form of the auto-correlation function of densities or scattering powers between different space-elements in the photographic emulsion.)

FIG. 11. Imaging through a "diffusing" medium. The arrangement shown permits one to holographically obtain an image of a completely transparent "phase" object *(P)* placed *behind* a diffusor; not only is the "phase" object unobservable by the naked eye (or camera), but it is additionally "hidden" by the ground-glass diffusor *(D)*! The holographic compensation is achieved by first recording in a hologram H_1 the hologram of the diffusor *(D)* and the object *(P)* behind it, and next, in a second hologram H_2 the hologram of the diffusor *(D)* alone. (Suitable care is taken to have the emulsion of H_1 face the position of the emulsion of H_2, as shown.) The decoded image (Fig. 12) is obtained "interferometrically" by carefully aligning the two holograms in the reconstruction. (After Stroke and Labeyrie.[6])

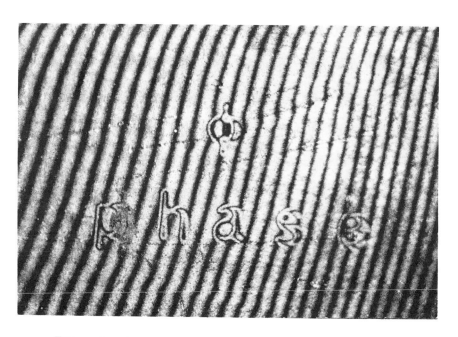

FIG. 12. Interferometrically decoded image of a phase-object placed *behind* a ground-glass diffusor (see Fig. 11).

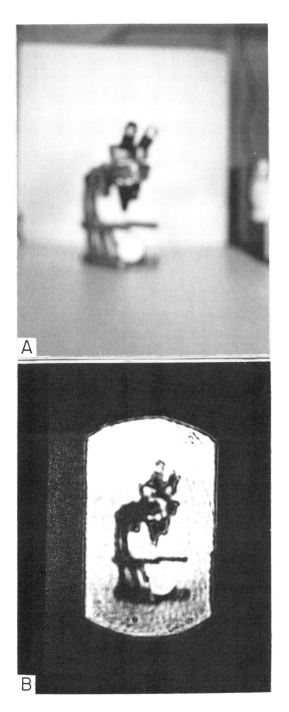

FIG. 13. Holographic sharpness restoration of blurred photographs using
Fourier-transform division. Early example showing a "sharp-focus" image (b)

holographically restored from the ordinary blurred photograph (a) of a three-dimensional object photographed with an ordinary camera in ordinary white-light (Stroke *et al.*[7]) according to the principles first described by Stroke and Zech.[8] Compensated sharpness restoration may thus be achieved for blurring caused by defocusing, image motion, atmospheric turbulence, and instrumental imaging defects (including aberrations), among others, including imaging defects at ultrasonic, microwave, and x-ray wavelengths. The holographic compensation realizes with the aid of holograms the "spatial filtering" principles first described by Maréchal and Croce.[9] The blurred photograph exposure $g(x,y)$ may be written in the form $g(x,y) = f(x,y) \circledast h(x,y)$, where $f(x,y)$ is the desired "gaussian" image, $h(x,y)$ is the "blurred" (e.g. out-of-focus) spread function (i.e. the image of a single "point" in object space), and is a spatial convolution integral.† This expression may be written, in the spatial Fourier-transform domain, as $G = F\,H$, where $\bar{G}$, $\bar{F}$, and $\bar{H}$ are the spatial Fourier transforms of $g, f,$ and h, respectively. The required deconvolution is thus carried out in the spatial frequency domain by *dividing* $\bar{G}$ by $\bar{H}$, to obtain $\bar{G}/\bar{H} = \bar{F}$ and from it, by a Fourier transformation, the desired function f. As first proposed by Stroke and Zech,[8] the complex Fourier-transform division filter $\bar{H}^{-1}$ may be synthesized in the form of a two-component filter, in the form $\bar{H}^{-1} = \bar{H}* / |\bar{H}|^2$. (For details and extensive references to related image-restoration work using nonholographic methods, e.g. digital computers, electronics, and separately realized amplitude and phase filters for the division, see cited references, and also e.g. Stroke.[10] See also Tsujiuchi.[11] See Harris,[12] for a description of one form of digital-computer image restoration. The advantages of holographic computing are referred to in Fig. 14.

†Strictly speaking, $f(x,y)$ is the "diffraction-limited" image of the object (assimilated to the "gaussian" image). A detailed theory of holographic image restoration methods is given in Ref. 21.

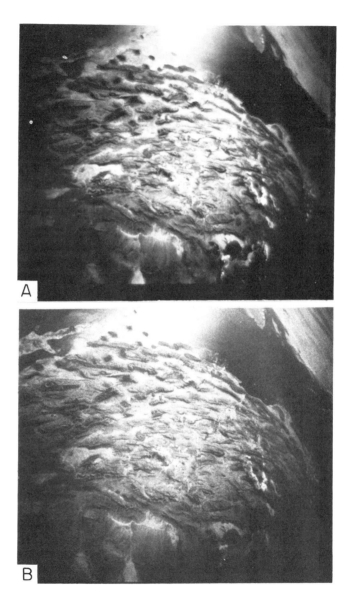

FIG. 14. Holographic image-sharpness restoration of blurred photographs using one-step "lensless fourier-transform holography" compensation with "extended sources" (after G.W. Stroke.[13]) (a) Accidentally blurred NASA photograph of Earth (portion of Iran) taken by astronauts Lovell and Aldrin from

Gemini XII satellite with hand-held $f = 38$ mm, $90°$ field, Hasselblad camera (Zeiss Biogon lens). [Photograph courtesy Herbert A. Tiedemann, NASA Manned Spacecraft Center, Houston. The original photograph was a 55×55 mm color photograph.] (b) Greatly enhanced sharpened image extracted from (a) by new holographic image-deblurring method.

Very satisfactory image-sharpening compensation, such as that shown in this early example may thus be achieved by using the method of "extended-source" lensless Fourier-transform holography of Stroke *et al.* (1965) [see, e.g., pp. 127-137, this volume], in extension of related independent work of D. Gabor.[14] Indeed, by forming a lensless Fourier-transform hologram of $g(x,y)$, the blurred photograph, in the form $I = 1 + |\overline{G}|^2 + \overline{G} + \overline{G}*$, and by noting that $\overline{G} = \overline{F} \, \overline{H}$, it is sufficient to illuminate the hologram with the wave $\overline{H}$, made to originate from the place of the point-reference source used to record the hologram, by placing there the image of the spread function $h(x,y)$ (e.g. as recorded on a fine-grain photographic transparency) to extract from the hologram the restored image f. The restoration is "perfect" to the extent that $\overline{H} \, \overline{H}* = 1$, i.e. that $h \maltese h* =$ "delta function," where $\maltese$ indicates a spatial correlation, noting that the Fourier transform of $\overline{F} \, \overline{H}* \, \overline{H}$ is $f \circledast (h \maltese h*)$. In the example shown, the spread function had a pentagonal shape (related to the iris diaphragm) with 0.2 mm maximum dimension. The image sharpening obtained here may be considered as due to the change of an approximately rectangular function to the corresponding triangular function obtained by autocorrelation. Clearly the new method may be used in conjunction with previous methods, when appropriate, as well as to sort out rapidly photographs for which digital-computer refinements may be in order.

A new extension of this method (Stroke *et al.*[17]) has now permitted us to accomplish the necessary *a posteriori* image synthesis of the multiple images recorded when a scatter-hole camera (with random pinholes) rather than a much more costly (and frequently less perfect) system (e.g., a grazing-incidence mirror system) is used for x-rays and gamma-rays imaging, notably in space applications, according to R.H. Dicke[18] and J.H. Underwood[19]. The original image is recorded in "incoherent" x-ray radiation in the form of a convolution of the random-array "impulse-response" (spread function) h with the desired image of the sources (stars, galaxies, nebulae, sun, etc.) f; that is, in the form $h \circledast f$. The image synthesis is accomplished, *a posteriori*, by first recording the "lensless Fourier-transform" hologram of this recording, using a point-source reference. The hologram $I = 1 + |\overline{G}|^2 + \overline{F} \, \overline{H} + \overline{F}*\overline{H}*$ is then replaced into its recording position and illuminated with the wave $\overline{H}$, originating from h, (the photograph of the "spread function") placed into the position of the point-source used in the recording. Because of the fact that the function h must in this case be chosen to be deliberately random (consisting of a computer-generated random array of pinholes), the autocorrelation $h \maltese h*$ can be made to be extremely close to a peaked (delta) function, and the image synthesis, obtained by Fourier transformation of $\overline{F}*\overline{H}*\overline{H}$ and equal to $f* \circledast (h \maltese h*)$ indeed results in bringing the multiplicity of images of f together into a single image $f*$ (identical to f!) as desired, providing the "solution" to the "problem" proposed by Dicke[18] and Underwood.[19] Experimental verifications[17] have been carried out in an optical analogue experiment, using an array of 9,000 pinholes of 0.046 mm diameter, randomly distributed according to an IBM 7094 computer and photographed on Kodak SP720 film (17 x 17 mm) from a Stromberg-Carlson 4020 plotter.[17] Complete success has been achieved in this synthesis, using for f both a single point (star) source and a test-bar pattern, as well as 3-D objects! The "extended-source lensless Fourier-transform" holographic image restoration (synthesis) method may thus be considered as providing the foundation for the design of a new class of *two-step image-forming systems*[8,10,13,17,18,19] (for "incoherent" light imaging), in which the first-step image may be recorded with a suitable focusing system, itself quite "imperfect" when used alone, and from which the "perfect" image may then be extracted by a second-step holographic image-restoration ("aperture synthesis"), such as that described here. Results obtained with this new method of "aperture synthesis"[17a] are shown in Fig. 13 (p. 227).

The increasing interest in the possibilities of "optical-computing" image restoration methods, such as those illustrated in Figs. 13 and 14 (and cited optical methods compared to digital-computer Fourier-transform processing (even of the type using the powerful Cooley-Tukey algorithm for fast Fourier transformation, permitting a reduction of computational time of the required Fourier transforms of a $n \cdot m$ matrix from $n^2 m^2$ to $n \log_2 n \cdot m \log_2 m$). For instance, according to D. Ansley,[15] in extension of a computation presented by J. Goodman, digital-computer image restoration on the fastest available computer would require 15 hr for a single photograph (100×100 mm with 150 lines/mm resolution, i.e. about 2.25×10^8 image elements thus processed "individually," compared to the few minutes required for holographic processing of all image elements simultaneously, as shown here!).

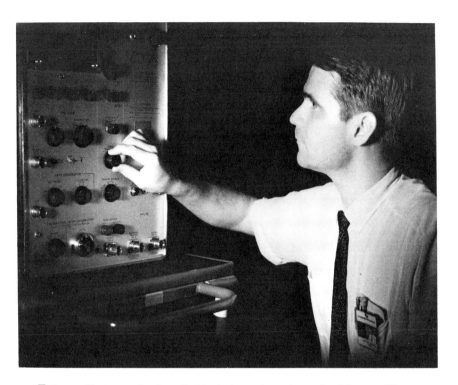

FIG. 15. High-speed "deep-field" holography using pulsed lasers. First image of a whole living person successfully reconstructed from a pulsed-laser hologram, recorded with the aid of a high-coherence pulsed-ruby laser system (L.D. Siebert [16]). The pulsed-laser system used had a measured coherence length in excess of 1 m and an output energy (at 6934 Å) of about 250 mJ (recording time about 30 nsec on Agfa "Scientia" 10E75 emulsion on an 8 x 10 in. plate). The importance of this achievement may be further appreciated when one recalls that a motion of only $\lambda/2$ (here about 5 millionths of an inch) of the subject towards the plate would have resulted in a complete holographic "black-out" of the image. [Courtesy D. Ansley and L. Siebert, Conductron Corp., Ann Arbor, Michigan.]

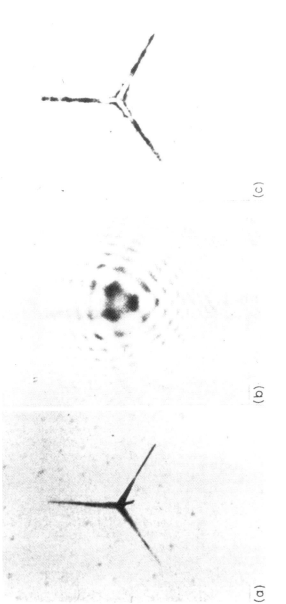

FIG. 16. Optical reconstruction of an image from a Fraunhofer electron-beam hologram. (a) Electron microscope image. (b) Hologram. (c) Holographically reconstructed image. The specimen is a smoke crystal of zinc oxide on carbon film (here the contrast γ of the hologram is about 4, permitting the reversal to be seen at the center of the reconstructed image of the crystal). [After A. Tonomura, A. Fukuhara, H. Watanabe, and T. Komoda, *Japan. J. Appl. Phys.* **7**, 295 (1968)]. [*Courtesy A. Tonomura, Hitachi Ltd., Tokyo, Japan*]

References for Photographic Appendix

1. G.W. Stroke, *Science Teacher* 34, No. 7 (October 1967).
2. G.W. Stroke and A.E. Labeyrie, *Phys. Letters* 20, 368-370 (1 March 1966).
3. G.W. Stroke and J.F. Burke, *Digest of San Diego Biomed. Symp.*, pp. 25-29 (1967).
4. D. Gabor and G.W. Stroke, *Proc. Roy. Soc.* A304, 275-289 (1968).
5. J.B. De Velis and G.O. Reynolds, *"Theory and Applications of Holography."* Addison-Wesley, Reading, Massachusetts.
6. G.W. Stroke and A.E. Labeyrie, *Phys. Letters* 20, 157-158 (1 February 1966).
7. G.W. Stroke, G. Indebetouw, and C. Puech, *Phys. Letters* 26A, 443-444 (1968).
8. G.W. Stroke and R.G. Zech, *Phys. Letters* 25A, 89-90 (31 July 1967).
9. A. Maréchal and P. Croce, *Compt. Rend.* 237, 607 (1953).
10. G.W. Stroke, *Phot. Korr.* 104, 82-85 (April 1968).
11. J. Tsujiuchi, in *"Progress in Optics"* (E. Wolf, ed.), Vol. II, pp. 133-182. North-Holland Publ., Amsterdam, 1963.
12. J.L. Harris, *J. Opt. Soc. Am.* 56, 569-574 (1966).
13. G.W. Stroke, *Phys. Letters* 27A 405-406 (26 August 1968).
14. D. Gabor, private communication to GWS, May-June 1968.
15. D. Ansley, Conductron Corp. private communication to GWS, 27 June 1968.
16. L.D. Siebert, *Appl. Phys. Letters* 11, 326-328 (1967); *Proc. IEEE* 56, 1242-1243 (1968).
17a. G.W. Stroke, *Phys. Letters* 28A, 252-253 (18 November 1968).
17b. G.W. Stroke, J.H. Underwood and R.B. Hoover, *Astrophys. J.* (submitted 12 October 1968).
18. R.H. Dicke, *Astrophys. J.* 153, L101-L106 (August 1968).
19. J.H. Underwood, private communication to GWS (July 1968).
20. R.G. Zech, Conductron Corp., Ann Arbor, Michigan, private communication to GWS (November 1968).
21. G.W. Stroke, *Optica Acta* (July 1969) (in press).

REPRINTED PAPERS

A NEW MICROSCOPIC PRINCIPLE

By Dr. D. GABOR

Research Laboratory, British Thomson-Houston Co., Ltd.,
Rugby

IT is known that the spherical aberration of electron lenses sets a limit to the resolving power of electron microscopes at about 5 A. Suggestions for the correction of objectives have been made ; but these are difficult in themselves, and the prospects of improvement are further aggravated by the fact that the resolution limit is proportional to the fourth root of the spherical aberration. Thus an improvement of the resolution by one decimal would require a correction of the objective to four decimals, a practically hopeless task.

The new microscopic principle described below offers a way around this difficulty, as it allows one to dispense altogether with electron objectives. Micrographs are obtained in a two-step process, by electronic analysis, followed by optical synthesis, as in Sir Lawrence Bragg's 'X-ray microscope'. But

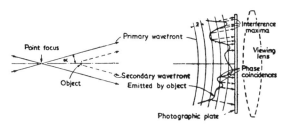

Fig. 1. INTERFERENCE BETWEEN HOMOCENTRIC ILLUMINATING WAVE AND THE SECONDARY WAVE EMITTED BY A SMALL OBJECT

while the 'X-ray microscope' is applicable only in very special cases, where the phases are known beforehand, the new principle provides a complete record of amplitudes *and* phases in one diagram, and is applicable to a very general class of objects.

Fig. 1 is a broad explanation of the principle. The object is illuminated by an electron beam brought to a fine focus, from which it diverges at a semi-angle α. Sufficient coherence is assured if the nominal or Gaussian diameter of the focus is less than the resolution limit, $\lambda/2$ sin α. The physical diameter, determined by diffraction and spherical aberration of the illuminating system, can be much larger. The object is a small distance behind (or in front of) the point focus, followed by a photographic plate at a large multiple of this distance. Thus the arrangement is similar to an electron shadow microscope ; but it is used in a range in which the shadow microscope is useless, as it produces images very dissimilar to the original. The object is preferably smaller than the area which is illuminated in the object plane, and it must be mounted on a support which transmits an appreciable part of the primary wave. The photographic record is produced by the interference of the primary wave with the coherent part of the secondary

Fig. 2. (a) ORIGINAL MICROGRAPH, 1·4 MM. DIAMETER. (b) MICROGRAPH, DIRECTLY PHOTOGRAPHED THROUGH THE SAME OPTICAL SYSTEM WHICH IS USED FOR THE RECONSTRUCTION (d). AP. 0·04. (c) INTERFERENCE DIAGRAM, OBTAINED BY PROJECTING THE MICROGRAPH ON A PHOTOGRAPHIC PLATE WITH A BEAM DIVERGING FROM A POINT FOCUS. THE LETTERS HAVE BECOME ILLEGIBLE BY DIFFRACTION. (d) RECONSTRUCTION OF THE ORIGINAL BY OPTICAL SYNTHESIS FROM THE DIAGRAM AT THE LEFT. TO BE COMPARED WITH (b). THE LETTERS HAVE AGAIN BECOME LEGIBLE

wave emitted by the object. It can be shown that, at least in the outer parts of the diagram, interference maxima will arise very nearly where the phases of the primary and of the secondary wave have coincided, as illustrated in Fig. 1.

If this photograph is developed by reversal, or printed, the loci of maximum transmission will indicate the regions in which the primary wave had the same phase as the modified wave, and the variations of the transmission in these loci will be approximately proportional to the intensity of the modified wave. Thus, if one illuminates the photographic record with an optical imitation of the electronic wave, only that part of the primary wave will be strongly transmitted which imitates the modified wave both in phases and in amplitudes. It can be shown that the 'masking' of the regions outside the loci of maximum transmission has only a small distorting effect. One must expect that looking through such a properly processed diagram one will see behind it the original object, as if it were in place.

The principle was tested in an optical model, in which the interference diagram was produced by monochromatic light instead of by electrons. The print was replaced in the apparatus, backed by a viewing lens which admitted about sin $\alpha = 0.04$, and the image formed was observed and ultimately photographed through a microscope. It can be seen in Fig. 2 that the reconstruction, though imperfect, achieves the separation of some letters which could just be separated in direct observation of the object through the same optical system. The resolution is markedly imperfect only in the centre, where the circular frame creates a disturbance. Other imperfections of the reconstruction are chiefly due to defects in the microscope objectives used for the production of the point focus, and for observation.

It is a striking property of these diagrams that they constitute records of three-dimensional as well as of plane objects. One plane after another of extended objects can be observed in the microscope, just as if the object were really in position.

Racking the microscope through and beyond the point focus, one finds a second image of the original object, in central-symmetrical position with respect to the point focus. The explanation is, briefly, that the photographic diagram cannot distinguish positive and negative phase shifts with respect to the primary wave, and this second image corresponds to the same phase shifts as the original, but with reversed sign.

If the principle is applied to electron microscopy, the dimensions in the optical synthetizer ought to be scaled up in the ratio of light waves to electron waves, that is, about 100,000 times. One must provide an illuminating system which is an exact optical imitation of the electronic condenser lens, including its spherical aberration. To avoid scaling-up the diagram, one has to introduce a further lens, with a focal length equal to the distance of the object from the photographic plate in the electronic device, in such a position that the plate appears at infinity when viewed from the optical space of the point focus. Work on the new instrument, which may be called the 'electron interference microscope', will now be taken in hand.

I wish to thank Mr. I. Williams for assistance in the experiments, and Mr. L. J. Davies, director of research of the British Thomson-Houston Company, for permission to publish this note.

Microscopy by reconstructed wave-fronts[†]

By D. Gabor, Dr.-Ing.

Research Laboratory, British Thomson-Houston Company Ltd., Rugby*

(*Communicated by Sir Lawrence Bragg, F.R.S.—Received* 23 *August* 1948—
Revised 28 *December* 1948—*Read* 17 *February* 1949)

[Plates 15 to 17]

The subject of this paper is a new two-step method of optical imagery. In a first step the object is illuminated with a coherent monochromatic wave, and the diffractio n pattern resulting from the interference of the coherent secondary wave iss uing from the object with the strong, coherent background is recorded on a photographic plate. If the photographic plate, suitably processed, is replaced in the original position and illuminated with the coherent background alone, an image of the object will appear behind it, in the original position. It is shown that this process reconstructs the coherent secondary wave, together with an equally strong 'twin wave' which has the same amplitude, but opposite phase shifts relative to the background.

The illuminating wave itself can be used for producing the coherent background. The simplest case is illumination by a point source. In this case the two twin waves are shown to correspond to two 'twin objects', one of which is the original, while the other is its mirror image with respect to the illuminating centre. A physical aperture can be used as a point source, or the image of an aperture produced by a condenser system. If this system has aberrations, such as astigmatism or spherical aberration, the twin image will be no longer sharp but will appear blurred, as if viewed through a system with twice the aberrations of the condenser. In either case the correct image of the object can be effectively isolated from its twin, and separately observed. Three-dimensional objects can be reconstructed, as well as two-dimensional.

The wave used in the reconstruction need not be the original, it can be, for example, a light-optical imitation of the electron wave with which the diffraction diagram was taken. Thus it becomes possible to extend the idea of Sir Lawrence Bragg's 'X-ray microscope' to arbitrary objects, and use the new method for improvements in electron microscopy. The apparatus will consist of two parts, an electronic device in which a diffraction pattern is taken with electrons diverging from a fine focus, and an optical synthetizer, which imitates the essential data of the electronic device on a much enlarged scale.

The theory of the analysis-synthesis cycle is developed, with a discussion of the impurities arising in the reconstruction, and their avoidance. The limitations of the new method are due chiefly to the small intensities which are available in coherent beams, but it appears perfectly feasible to achieve a resolution limit of 1 Å, ultimately perhaps even better.

INTRODUCTION

The period of steady progress in the resolving power of electron microscopes which was started in 1931 by Knoll & Ruska came virtually to an end in 1946, when Hillier & Ramberg (1947) eliminated the astigmatism of their objective, and achieved a resolving power only insignificantly different from the theoretical limit. The barrier which stopped progress is of a technical nature, but formidable enough to prevent any really essential improvements along the direct line.

The theoretical limit of conventional electron microscopes is about 5Å. It is determined by a compromise between diffraction and spherical aberration in electron objectives, and at the best compromise it is proportional to the fourth root of the aberration constant. Though several suggestions for correction have been put

* Now at Imperial College, London, Electrical Engineering Department.
†Reprinted from *Proc. Roy. Soc.* **A197** (1949) by permission.

forward, they involve such technical difficulties that an improvement by a factor of 2 is the best that can be expected, even optimistically. One can never hope to achieve a resolving power ten times better than the present, which would require a correction of the spherical aberration to about 1 part in 10,000. Such precision can be realized with the technique of the optical workshop, but hardly ever with the means at the disposal of electron optics.

The new method is an attempt to get around the obstacle, instead of across it, by a two-step process, in which the analysis is carried out with electrons, the synthesis by light. The general idea of such a process was first suggested to the author by Sir Lawrence Bragg's 'X-ray microscope' (Bragg 1942; cf. also Boersch 1938). But Bragg's method, in which a lattice is reconstructed by diffraction from an X-ray diffraction pattern, can be applied only to a rather exceptional class of periodic structures. It is customary to explain this by saying that diffraction diagrams contain information on the intensities only, but not on the phases. The formulation is somewhat unlucky, as it suggests at once that since phases are unobservables, this state of affairs must be accepted. In fact, not only that part of the phase which is unobservable drops out of conventional diffraction patterns, but also the part which corresponds to geometrical and optical properties of the object, and which in principle could be determined by comparison with a standard reference wave. It was this consideration which led me finally to the new method.

In order to make the two-step method generally applicable, it had to be combined with a principle apparently not hitherto recognized. If a diffraction diagram of an object is taken with coherent illumination, and a coherent background is added to the diffracted wave, the photograph will contain the full information on the modifications which the illuminating wave has suffered in traversing the object, apart from an ambiguity of sign, which will be discussed later. Moreover, the object can be reconstructed from this diagram without calculation. One has only to remove the object, and to illuminate the photograph by the coherent background alone. The wave emerging from the photograph will contain as a component *a reconstruction of the original wave*, which appears to issue from the object. Conditions can be found in which the remainder can be sufficiently separated from the useful component to allow a true, or very nearly true, reconstruction of the original object.

This principle has been confirmed by numerous experiments. Some of the results are shown in figures 10 to 12 and explained in the last section of this paper.

In light optics a coherent background can be produced in many ways, but electron optics does not possess effective beam-splitting devices; thus the only expedient way is using the illuminating beam itself as the coherent background. This leads us to illumination by a coherent, divergent electron wave, illustrated in figure 1. It will be useful to explain this arrangement first, anticipating the principle of reconstruction which will be proved later.

The apparatus consists of two parts, the electronic analyzer and the optical synthetizer. The analyzer is similar to an electron shadow microscope (Boersch 1939), but with the important difference that it operates with coherent illumination, and under conditions in which the shadow microscope is useless, as the interference diagram has little likeness to the original. An electron gun, combined with a suitable aperture and electron lens system, produces a coherent illuminating beam, as nearly homocentric as possible. Exactly homocentric illumination is of course impossible, because of the unavoidable spherical aberration of electron lenses, but for simplicity we can talk of the narrow waist of the beam as of a 'point focus'. A small object is arranged some small distance before or behind the point focus, and a photographic plate at a comparatively large distance L. The divergence angle of the beam, γ_m, must be sufficient for the required resolution limit d_A, which is by Abbe's relation

$$d_A = \tfrac{1}{2}\lambda \sin \gamma_m.$$

The factor $\tfrac{1}{2}$ will be used in this paper to simplify the discussions, except in numerical calculations, where it will be replaced by the more accurate value 0·6.

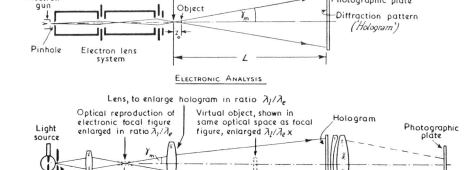

ELECTRONIC ANALYSIS

OPTICAL SYNTHESIS

FIGURE 1. Principle of electron microscopy by reconstructed wave-fronts.

As the photograph of a diffraction pattern taken in divergent, coherent illumination will be often used in this paper, it will be useful to introduce a special name for it, to distinguish it from the diffraction pattern itself, which will be considered as a complex function. The name 'hologram' is not unjustified, as the photograph contains the total information required for reconstructing the object, which can be two-dimensional or three-dimensional.

The hologram must be either printed, or developed by reversal, and the positive is transferred to the optical synthetizer, which is a light-optical imitation of the electronic device. All essential dimensions, which determine the shape of the wave, are scaled up in the ratio of light wave-length λ_l to electron wave-length λ_e. As electrons of about 50 keV energy, with a de Broglie wave-length of about 0·05 Å, are the most useful in electron microscopy, this ratio will be of the order 100,000. It may be noted that the focal length of the electron lenses is not an essential dimension, and need not be scaled up.

To avoid scaling up the photographic plate a further lens is provided, which enlarges it in the ratio λ_l/λ_e in the optical space of the enlarged focal figure. This means that the image of the hologram is moved practically to infinity, i.e. it must be in the focal plane of the collimator lens. In the illustration it has been assumed for simplicity that the angles are the same in the analyzer and in the synthetizer, but it will be shown later that the condition $f = L$ is not essential. Nor is it necessary to use a separate condenser lens system. The condenser and the collimator, which have been shown separate in figure 1 to simplify the explanations, form one optical unit, whose function it is to produce an imitation of the original wave-front in the plane of the hologram. The spherical aberration, and the practically unavoidable ellipticity of the electron lenses must be reproduced with great accuracy, with a tolerance of about one fringe for the marginal rays.

Thus in the new method it is no longer necessary to correct the spherical aberration of electron lenses. The aperture can be opened up far beyond the limit of tolerance in ordinary electron microscopy. It is only necessary to *imitate* the aberrations to the same accuracy as they would have to be corrected to achieve a certain resolution. Thus the difficulty is shifted from electron optics to light optics, where refracting surfaces can be figured to any shape, without the limitations imposed in electron optics by the laws of the electromagnetic field. On the electron-optical side we require only a certain moderate constancy, sufficient to avoid readjustment of the optical system at too frequent intervals.

The technical difficulties of the scheme will not be dealt with in this paper. It may be only mentioned that they involve mechanical and electrical stability, operation with objects much smaller than those hitherto dealt with in electron microscopy, and the problem of obtaining the high current densities required under the additional condition of coherence. For the rest the paper will deal mainly with the general wave-theoretical foundations of the new method.

THE PRINCIPLE OF WAVE-FRONT RECONSTRUCTION

Consider a coherent monochromatic wave with a complex amplitude U striking a photographic plate. We write $U = Ae^{i\psi}$, where A and ψ are real. U may be decomposed into a 'background wave' $U_0 = A_0 e^{i\psi_0}$, and a remainder $U_1 = A_1 e^{i\psi_1}$ which is due to the disturbance created by the object and may be called the secondary wave. Thus the complex amplitude at the photographic plate is

$$U = U_0 + U_1 = A_0 e^{i\psi_0} + A_1 e^{i\psi_1} = e^{i\psi_0}(A_0 + A_1 e^{i(\psi_1 - \psi_0)}) \tag{1}$$

and its absolute value $\quad A = [A_0^2 + A_1^2 + 2A_0 A_1 \cos(\psi_1 - \psi_0)]^{\frac{1}{2}}$.

The density of photographic plates, plotted against the logarithm of exposure, is an S-shaped curve, with an approximately straight branch between the two knees. In this region the transmission of intensity is a power $-\Gamma$ of the exposure. The word 'transmission' and the symbol t will be reserved in this paper for the amplitude transmission, which is in general complex; hence the intensity transmission is tt^*, where the asterisk denotes the complex conjugate. For pure absorption, without phase change, t is a real number, the square root of the intensity transmission. Thus we write for the negative process

$$t_n = (K_n A)^{-\Gamma_n},$$

where K_n is proportional to the time of exposure. In the printing of the negative the exposure is proportional to t_n, hence the transmission of the positive print becomes

$$t_p = [K_p(K_n A)^{-\Gamma_n}]^{-\Gamma_p} = K A^{\Gamma}. \tag{2}$$

where $\Gamma = \Gamma_n \Gamma_p$ is the 'overall gamma' of the negative-positive process. The same type of law applies if reversal development is used.

If now in the reconstruction process we illuminate the positive hologram with the background U_0 alone, a 'substituted wave' U_s will be transmitted, which is, apart from a constant factor

$$U_s = U_0 t_p = A_0 e^{i\psi_0} [A_0^2 + A_1^2 + 2A_0 A_1 \cos(\psi_1 - \psi_0)]^{\frac{1}{2}\Gamma}. \tag{3}$$

The simplest, and as will be seen also the most advantageous choice, is $\Gamma = 2$, which gives

$$U_s = U_0 A^2 = A_0 e^{i\psi_0} [A_0^2 + A_1^2 + 2A_0 A_1 \cos(\psi_1 - \psi_0)]$$

$$= A_0^2 e^{i\psi_0} \left[A_0 + \frac{A_1^2}{A_0} + A_1 e^{i(\psi_1 - \psi_0)} + A_1 e^{-i(\psi_1 - \psi_0)} \right]. \tag{4}$$

Comparing this with (1) one sees that if $A_0 = $ const., i.e. if the background is uniform, the substituted wave contains a component proportional to the original wave U (the first and third terms). This is not in itself a proof of the principle of reconstruction, as any wave can be split into a given wave and a rest. It remains to be shown that the remainder, i.e. the spurious part of U_s, does not constitute a serious disturbance.

This remainder consists of two terms. One of these has the same phase as the background, with an amplitude $(A_1/A_0)^2$ times the amplitude of the background. This term can be made very small if the background is relatively strong, which does not mean that the contrast in the hologram must be poor. Assume for instance $(A_1/A_0)^2 = 0 \cdot 01$, i.e. a secondary intensity which is only 1 % of the primary. This gives $A_1/A_0 = 0 \cdot 1$, and the intensity ratio between the maxima and minima of the interference fringes is $(1 \cdot 1/0 \cdot 9)^2 = 1 \cdot 5$. With $\Gamma = 2$ the intensity transmissions will be in the ratio $1 \cdot 5^2 = 2 \cdot 25$, a very strong contrast. The contrast will fall below the observable limit of about 4 % only for $(A_1/A_0)^2 \leqslant 0 \cdot 0001$, i.e. if the flux scattered by the object into the area of the diagram is less than $0 \cdot 01$ % of the illuminating flux. This remarkable effect of the coherent background has been systematically utilized by Zernike (1948) for the amplified display of weak interference fringes.

The second term of the remainder has the same amplitude $A_1 A_0^2$ as the reconstruction of the original secondary wave, but it has a phase shift of opposite sign relative to the background. It may be called for brevity the 'conjugate-complex' wave. The two twin waves carry the same energy.

The conjugate wave produces a serious disturbance only in rather exceptional arrangements; in most cases the twin waves can be effectively separated. To make this plausible one may think of Fresnel-zone plates. These can be, in fact, considered as holograms of a point object, produced by a point source at infinity. Zone plates act simultaneously as positive and negative lenses, producing two focal points, one at each side of the plate, at equal distance, which can be separately observed. As will be shown later, homocentric illumination produces always such twin images, only, with the source at finite distance, these will be in mirror-symmetrical position with respect to the point source, not to the hologram. In beams which are only approximately homocentric the second image is no longer sharp, but effective separation can be always achieved if the object is sufficiently small, and if certain positions are avoided.

While the twin wave cannot be avoided, the spurious term which is proportional to $(A_1/A_0)^2$ and the distortion due to an uneven background can both be eliminated, or at least effectively suppressed by a modification of the photographic process. In the case of small objects, at least over a large part of the hologram, the photographic density difference between two neighbouring interference maxima is insignificant. This makes it possible to wash out the interference fringes by taking a slightly defocused print of the positive hologram, and processing this print with $\Gamma = 1$. If this print, which has a transmission inversely proportional to $A_0^2 + A_1^2$, is placed in register with the positive, and illuminated by the background wave U_0, the substituted wave becomes

$$U'_s = A_0 \, e^{i\psi_0}[A_0^2 + A_1^2 + 2A_0A_1 \cos{(\psi_1 - \psi_0)}]/(A_0^2 + A_1^2) = e^{i\psi_0}\left[A_0 + \frac{2A_1 \cos{(\psi_1 - \psi_0)}}{1 + (A_1/A_0)^2}\right]$$

$$= e^{i\psi_0}\left[A_0 + 2A_1 \cos{(\psi_1 - \psi_0)} - 2\frac{A_1^3}{A_0^2}\cos{(\psi_1 - \psi_0)} + ...\right], \quad (5)$$

in which the spurious term is of the order $(A_1/A_0)^3$ as compared with the background, and the distortion due to a non-uniform background is eliminated. If one only wants to eliminate the background by itself, one can also use a negative photograph taken in the illuminating beam without the object, processed with $\Gamma = 2$.

To discuss briefly also the case $\Gamma \neq 2$, we put for simplicity $A_0 = 1$ and $A_1/A_0 = a$, and obtain from (3) by binomial expansion

$$U_s = e^{i\psi_0}[1 + \tfrac{1}{2}\Gamma a^2 + \Gamma a \cos{(\psi_1 - \psi_0)} + \tfrac{1}{2}\Gamma(\Gamma - 2)\,a^2 \cos^2{(\psi_1 - \psi_0)} + ...]. \quad (6)$$

In the reconstructed wave the contrast is enhanced in the ratio $\tfrac{1}{2}\Gamma$. But, in addition, one obtains twin waves with phase shifts $2(\psi_1 - \psi_0)$, etc., but with smaller amplitudes. This makes it evident that $\Gamma = 2$ is the best choice, except if the original contrast is so weak that it must be enhanced, even at the cost of faithfulness in the reproduction.

ILLUMINATION BY A HOMOCENTRIC WAVE

In order to study the reconstruction cycle in more detail, it will be advantageous to start with the simple case of homocentric illumination, which can be approximately realized by a sufficiently small pinhole as light source. It will be convenient to restrict the discussion for a start to two-dimensional objects, occupying a part of some closed surface Σ which encloses the point source O. The object at a point P of Σ may be characterized by an amplitude transmission coefficient $t(P)$, which is the

460   D. Gabor

ratio of the complex amplitudes at the two sides of Σ, in proximity of the point P.
t is in general a complex datum, real only for purely absorbing objects. It is, of course,
understood that the concept of a transmission coefficient, real or complex, is not
applicable to an object which is two-dimensional in the mathematical sense. Of
a physical object to which this concept is applicable we must assume that it is at least
several wave-lengths in thickness. Moreover, we must assume that laterally, in the
surface Σ, the function $t(P)$ does not vary appreciably within a wave-length. These
are the conditions for the applicability of the Fresnel-Kirchhoff theory of diffraction.
In electron optics, operating with fast electrons of about $0.05\,\text{Å}$ wave-length, this
condition is always satisfied, as there exists no material object (except nuclei) whose
physical properties change significantly in less than about ten times this wave-length.

    With these qualifications we can apply the Fresnel-Kirchhoff diffraction formula
(cf., for example, Baker & Copson 1939, p. 73). The notations are explained in
figure 2. If the monochromatic source at O is of unit strength, the amplitude in the
illuminating wave is

$$U_0 = \frac{1}{r_0}\,\mathrm{e}^{ikr_0},$$

where r_0 is the distance measured from O, and $k = 2\pi/\lambda$. The presence of an object in
the surface Σ modifies the amplitude at a point Q outside to

$$U(Q) = \frac{1}{2\lambda}\int_{\Sigma} t(P)\,\mathrm{e}^{ik(r_0+r_1)-\frac{1}{2}\pi i}(\cos\theta_0 - \cos\theta_1)\frac{dS}{r_0 r_1}. \tag{7}$$

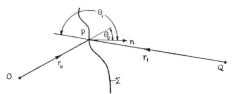

FIGURE 2. Fresnel-Kirchhoff diffraction formula.

    We will now apply this formula to calculate the 'physical shadow' of a plane object
at infinity. The 'physical shadow' includes the diffraction effects, and is to be
distinguished from the 'geometrical shadow' into which it merges at vanishing
wave-length.

    As the beams to be used in practice will have semi-cone angles of 0.05 or less, we
can put $\cos\theta_0 = -\cos\theta = 1$, and consider the factor $1/r_0 r_1$ as constant. We also drop
the constant factor $(1/2\lambda)\,\mathrm{e}^{-\frac{1}{2}\pi i}$, and use equation (7) in the simplified form

$$U(Q) = \int_{\Sigma} t(P)\,\mathrm{e}^{ik(r_0+r_1)}\,dS. \tag{7.1}$$

    Using the notations explained in figure 3, the distance r_0 of a point P in the object
plane $z = z_0$ is

$$r_0 = (x^2 + y^2 + z_0^2)^{\frac{1}{2}} = z_0 + \tfrac{1}{2}(x^2 + y^2)/z_0 - \tfrac{1}{8}(x^2 + y^2)^2/z_0^3 + \ldots.$$

In this section we will use only the first two terms of the expansion.

The observation point Q may be at a distance L in the Z-direction, very large compared with z_0, practically at infinity, so that we can write

$$r_1 = L \sec \gamma - (x \cos \alpha + y \cos \beta).$$

The first terms in the expression for r_0 and r_1 give constant phase factors, independent of x, y, which may be dropped. The remaining essential part may be termed 'the amplitude in the direction ξ, η', and is

$$U(\alpha, \beta) = \iint t(x, y) \exp \{ik[(x^2 + y^2)/2z_0 - (x \cos \alpha + y \cos \beta)]\} \, dx \, dy. \tag{8}$$

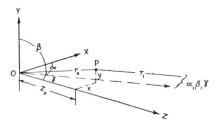

FIGURE 3. Explanation of symbols.

Unless the limits are indicated, integrations in this paper will be always understood to be carried out between infinite limits. As the phase under the integral is valid only for small angles, equation (8) is physically valid only if $t(x, y)$ vanishes rapidly outside a small central area.

It will now be convenient to introduce 'Fourier variables' ξ, η instead of the direction cosines by

$$\xi = \frac{1}{\lambda} \cos \alpha, \quad \eta = \frac{1}{\lambda} \cos \beta. \tag{9}$$

Their connexion with the co-ordinates X, Y in a plane at the large distance $z = L$ is given by

$$X = L \frac{\cos \alpha}{\cos \gamma} = \frac{\lambda L \xi}{[1 - \lambda^2 (\xi^2 + \eta^2)]^{\frac{1}{2}}}, \quad Y = L \frac{\cos \beta}{\cos \gamma} = \frac{\lambda L \eta}{[1 - \lambda^2 (\xi^2 + \eta^2)]^{\frac{1}{2}}}.$$

If the illuminating cone is narrow enough, ξ and η can be taken to represent the co-ordinates in the physical shadow. The geometrical shadow of a point x, y has the Fourier co-ordinates $\xi = x/\lambda z_0, \eta = y/\lambda z_0$. The quantity

$$\mu = \lambda z_0 \tag{10}$$

is the only parameter of the diffraction problem. Its square root can be considered as the characteristic length. Details coarser than $\mu^{\frac{1}{2}}$ will be shown to have shadows more or less similar to themselves, finer details lose all likeness by diffraction.

Using the notations (9) and (10) equation (8) can be written, with the abbreviation $x^2 + y^2 = r^2$,

$$U(\xi, \eta) = \iint [t(x, y) \, \mathrm{e}^{\pi i r^2/\mu}] \, \mathrm{e}^{-2\pi i (x \xi + y \eta)} dx \, dy. \tag{11}$$

462   D. Gabor

Thus the amplitude in the ξ, η direction is the Fourier transform of the function

$$t(x,y)\,e^{i\pi r^2/\mu}$$

in the standard notation of Campbell & Foster (1931). We can at once write down the reciprocal formula

$$t(x,y) = e^{-\pi i r^2/\mu}\iint U(\xi,\eta)\,e^{+2\pi i(x\xi+y\eta)}d\xi\,d\eta. \tag{12}$$

It will be useful to study these reciprocal transformations mathematically, while provisionally disregarding the conditions which must be imposed on the function $t(x,y)$ to give them physical validity. First we put them into a more symmetrical form, by imagining the amplitude $U(\xi,\eta)$ as produced by the passage of the illuminating wave U_0 through a 'shadow object' in the plane ξ, η with a transmission $\tau(\xi,\eta)$. (It may be noted that τ is in general complex; thus the shadow object cannot be replaced by a photographic plate.) That is to say, we put

$$U(\xi,\eta) = U_0(\xi,\eta)\,\tau(\xi,\eta). \tag{13}$$

The background U_0 can be obtained directly from (11) by putting $t = 1$

$$U_0(\xi,\eta) = i\mu\,e^{-\pi i\mu\rho^2}$$

with the abbreviation $\xi^2 + \eta^2 = \rho^2$. We now obtain the symmetrical transformation formulas

$$\tau(\xi,\eta) = \frac{1}{i\mu}\,e^{\pi i\mu\rho^2}\iint t(x,y)\,e^{-\pi r^2/i\mu}\,e^{-2\pi i(x\xi+y\eta)}dx\,dy, \tag{14}$$

$$t(x,y) = i\mu\,e^{\pi r^2/i\mu}\iint \tau(\xi,\eta)\,e^{-\pi i\mu\rho^2}\,e^{+2\pi i(x\xi+y\eta)}d\xi\,d\eta. \tag{15}$$

These may be called the 'shadow transformations', and $t(x,y)$, $\tau(\xi,\eta)$ a pair of 'shadow transforms'. They are, of course, intimately related to Fourier transforms, though simpler in some respects.

The transformations (14) and (15) can be derived from one another by the rule: Interchange t and τ, x and ξ, y and η, i.e. interchange Latin and Greek symbols, and replace i by $-i$, μ by $1/\mu$. Two transformations in succession restore the original. Physically this means that if instead of a photograph we could produce a 'shadow object' with the absorbing and refractive properties of $\tau(\xi,\eta)$, and illuminated this with the background, we should exactly restore the object $t(x,y)$ in its original position. As a photograph cannot imitate the imaginary part of τ, a certain residual wave arises, to which we will return in the next section. But it will be useful to consider first a few examples of shadow transforms.

As in the case of Fourier integrals, the transforms of exponentials of quadratic forms are particularly simple and instructive. It is convenient to write these in the form

$$t(x,y) = \exp[-\pi(A_1 x^2 + 2B_1 x + A_2 y^2 + 2B_2 y)].$$

This is the product of an x- and a y-factor, and as the transform is again the product of a ξ- and an η-factor, it is sufficient to give the transform of

$$t(x) = e^{-\pi(Ax^2+2Bx)}, \tag{16·1}$$

which is $\qquad \tau(\xi) = (1 + i\mu A)^{-\frac{1}{2}} \exp\left[-\dfrac{\pi\mu(\mu A\xi^2 + 2B\xi - iB^2)}{1 + i\mu A} \right].$ (16·2)

Thus the shadow transform of an exponential of a quadratic form is a function of the same type, as in the case of Fourier transforms, but the relation between the parameters is of a different build. For example, if $A = B = 0$, which makes t a constant, τ will be the same constant, while the Fourier transform of a constant is a delta function, which vanishes everywhere except at the argument zero. Moreover, the shadow transform of a harmonic function $(A = 0)$

$$t(x) = e^{2\pi ix/p} \qquad\qquad (16\cdot3)$$

is again a harmonic function $\qquad \tau(\xi) = e^{-i\pi\mu/p^2} e^{2\pi i\mu\xi/p}.$ (16·4)

The period in the shadow is p/μ, which is *the geometrical shadow of the period p*. The only difference is in the phase factor $e^{-i\pi\mu/p^2}$. If the period p is long compared with the characteristic length $\mu^{\frac{1}{2}}$, the phase factor tends to unity, which means that if the object contains no details finer than $\mu^{\frac{1}{2}}$ the physical shadow tends towards the geometrical shadow $\qquad\qquad \tau(\xi, \eta) \to t(\mu\xi, \mu\eta).$

Equations (16·3) and (16·4) contain a simple rule for constructing the shadow transform of an object, by expanding $t(x, y)$ into a Fourier integral with periods p_x, p_y. In the transform the Fourier coefficients will differ from the original only in a phase factor

$$\exp\left[-i\pi\mu\left(\frac{1}{p_x^2} + \frac{1}{p_y^2} \right) \right].$$

As a practical method this may be used with the cautioning remark that infinite trains of periodic functions are not very suitable for the description of small objects, and that the applicability of equations (14) and (15) to the physical process is strictly speaking limited to objects which transmit appreciably only in a region $x/z_0, y/z_0 \ll 1$.

RECONSTRUCTION WITH HOMOCENTRIC ILLUMINATION

Stigmatic illumination is a particularly simple and instructive illustration of the principle of reconstruction which was broadly explained in the first section. It may be recalled that if the hologram is replaced in the original position and illuminated by the background alone, one obtains in addition to the illuminating or primary wave two other waves, one of which is proportional to the original secondary wave emitted by the object, and the other differs from this only by having phase shifts of opposite sign relative to the background. The other small spurious terms may be disregarded for the moment.

It will now be convenient to subtract the background, i.e. the primary wave, both in the object plane, and in the plane of the photographic plate, and to consider instead of t and τ the functions $\qquad t_1 = t - 1 \quad$ and $\quad \tau_1 = \tau - 1.$ (17)

As $t = 1$ corresponds to $\tau = 1$, the functions $t_1(x, y)$ and $\tau_1(\xi, \eta)$ are connected by the relations (14) and (15), the same which connect t and τ. We will talk of t_1 as 'the object proper' and of τ_1 as its shadow.

464   D. Gabor

By equation (6), substituting a photographic plate for the physical shadow means replacing τ_1 by
$$\tfrac{1}{2}\Gamma(\tau_1 + \tau_1^*).$$

Substituting this into the inverse shadow transformation (15), we obtain two terms t_1. The first of these differs from the original object proper only in the factor $\tfrac{1}{2}\Gamma$. But in the second term, derived from $\tfrac{1}{2}\Gamma\tau^*$, the sign of i has been reversed, and this results in a spurious figure in the object plane, superimposed on the correct reconstruction of the object.

We can give a simple interpretation to the wave corresponding to τ^* if we observe that in the equation (14) applied to the object proper t_1

$$\tau_1(\xi, \eta) = \frac{1}{i\mu}\,e^{i\pi\mu\rho^2}\iint t_1(x, y)\,e^{-\pi r^2/i\mu}\,e^{-2\pi i(x\xi + y\eta)}dx\,dy, \qquad (14\cdot1)$$

reversing the sign of i is equivalent to reversing the sign of x, y and of $\mu = \lambda z_0$ and replacing $t_1(x, y)$ by a function

$$t_1(x, y) = t\,(-x, -y). \qquad (18)$$

The transformation has now a parameter $-\mu$ instead of μ, i.e. it corresponds to an object in the plane $-z_0$ instead of in $+z_0$. By equation (18) this object arises from the original by mirroring it on the Z-axis, and changing phase delays into phase advances. Summing up, *the 'twin' wave τ^* corresponds to an apparent 'twin object', in central symmetrical position with respect to the point focus 0, and with opposite phase-shifting properties.*

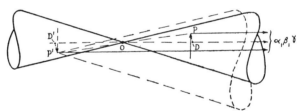

FIGURE 4. The twin images arising in the reconstruction.

Figure 4 is an illustration of the twin objects, from which one can verify this conclusion. The Fresnel-Kirchhoff formula can be interpreted as the sum of elementary spherical waves, originating from the object points P, with amplitudes proportional to $t(P)$. At infinity, in a direction α, β these are plane waves, and their phase difference relative to the background is given by the difference between the ray $\overline{OP}$, and its projection on the direct ray, $\overline{OD}$, apart from the phase shift which arises in the object. Figure 4 makes it clear that the same phase difference, but with opposite sign, would be produced by an object point P', in central symmetrical position to P, if the sign of the phase shift at P' is also reversed.

The interpretation of the residual wave in the reconstruction as a wave emitted by a twin object makes it at once clear that conditions can be found which allow a fairly effective isolation of the reconstructed object, by making use of the limited focal depth of the viewing system. Separation becomes possible if the distance between the twin objects, $2z_0$, exceeds the focal depth D_i, which can be defined as the resolution limit d, divided by the total cone angle utilized in the image formation, $2\gamma_m$. Using Abbe's value d_A for d, the criterion of separation is

$$2z_0 > \frac{d}{2\gamma_m} = \frac{\lambda}{4\gamma_m^2}. \tag{19}$$

If the point focus is not produced by a physical aperture, but by the image of an aperture, formed by an optical system, this is equivalent to the condition that the object must be outside the diffraction region, in which the wave cannot be considered as homocentric.

Outside the focal diffraction region separation is possible, but not complete separation. The twin images will always interfere with one another to some extent, and the interference cannot be regulated at will. This follows from the structure of the transformation equations, which contain only one characteristic length $(\lambda z_0)^{\frac{1}{2}}$, and there is no second length with which to form a dimensionless separation factor. Thus the spurious part of the reconstructed image depends only on the object itself, and on the parameter μ. This disturbance will now be investigated in some detail.

THE SPURIOUS PART OF THE RECONSTRUCTION IN HOMOCENTRIC ILLUMINATION

The simplicity of the transforms (16·1) and (16·2) suggests building up arbitrary plane objects from 'probability spots'. In the limit these tend to two-dimensional delta functions, which can represent any function $t_1(x, y)$, but it is not necessary, nor is it physically justifiable, to pass to this limit. Optical imagery does not operate with points, but with elementary regions of the size of the resolution limit. Inside such a small area the values of $t_r(x, y)$, which describe the reconstituted object, are not independent of one another.

First we carry out the reconstruction cycle for a single probability spot. Assume the transmission in the object plane in the form

$$t(x, y) = 1 - A \exp\left\{-\frac{\pi}{a^2}[(x - x_0)^2 + (y - y_0)^2]\right\} = 1 - A e^{-\pi(r'/a)^2}, \tag{20}$$

where the abbreviation r' has been used for the distance measured from the centre x_0, y_0 of the spot. $(1 - A)$ is the amplitude transmitted at the centre of the spot, at unit background. If the object is a pure absorber, A is real, positive, and less than unity. If the object has pure-phase contrast $|1 - A| = 1$, and $|A|$ is in the limits 0 to 2.

Equations (16·1) and (16·2) give for the physical shadow of (20)

$$\tau(\xi, \eta) = 1 + \frac{i\epsilon A}{1 - i\epsilon} \exp\left(\frac{\pi i \mu \rho'^2}{1 - i\epsilon}\right) \tag{21}$$

with the abbreviations $\qquad \epsilon = a^2/\mu = a^2/\lambda z_0$

and $\qquad \rho'^2 = (\xi - x_0/\mu)^2 + (\eta - y_0/\mu)^2 = (\xi - \xi_0)^2 + (\eta - \eta_0)^2,$

where ξ_0, η_0 is the geometrical shadow of x_0, y_0. The diffraction figure (21) centres around this point. Its character is determined by the dimensionless parameter ϵ. If ϵ is large τ approaches the geometrical shadow $t(\mu\xi, \mu\eta)$. The more important case is $\epsilon \ll 1$, which allows simplifying (21) to

$$\tau(\xi, \eta) = 1 + i\epsilon A\, e^{-\pi(a\rho')^2 + \pi i\mu\rho'^2}. \tag{21.1}$$

The smaller the original spot, the larger its physical shadow.

The photograph substitutes for the complex physical shadow (21.1) the real transmission function

$$\tau_s(\xi, \eta) = |\tau|^\Gamma \doteq 1 + \tfrac{1}{2}\Gamma i\epsilon A\, e^{-\pi(a\rho')^2} e^{\pi i\mu\rho'^2} - \tfrac{1}{2}\Gamma i\epsilon A^*\, e^{-\pi i\mu\rho'^2}. \tag{22}$$

This approximation is valid if $\epsilon^2 \ll 1$.

The inverse transformation (15), applied to the first two terms of (22), restores the original object (20), but the contrast is $\tfrac{1}{2}\Gamma$ times the original. The same transformation applied to the last term of (22) gives the spurious or error term

$$t_e(x, y) = -\tfrac{1}{4}\Gamma i\epsilon A^*\, e^{-\pi(ar'/2\mu)^2}\, e^{-\pi i(r'^2/2\mu)}. \tag{23}$$

This is the amplitude (at unit background) which the twin image produces in the original object plane. The spurious image centres on x_0, y_0, but it has a character quite different from the original. The amplitude t_e falls off only slowly with the distance r' from the centre, the slower the smaller the original spot radius a, while the phase changes rapidly, according to the last factor in (23), in a manner independent of the spot size. Thus the spurious image will manifest itself in a system of fine and weak interference fringes, superimposed on the true reconstruction.

The exact value of the reconstructed transmission function t_r in the case $\Gamma = 2$ may be also given for reference:

$$t_r(x, y) = 1 - A \exp\left[-\pi\left(\frac{r'}{a}\right)^2 \right] - \frac{i\epsilon A^*}{2 + i\epsilon}\exp\left[-\frac{\pi(\epsilon + 2i)}{\mu(4 + \epsilon^2)} r'^2 \right]$$
$$+ \frac{\epsilon^2 A A^*}{1 + \epsilon^2 - 2i\epsilon}\exp\left[-\frac{2\pi\epsilon(1 + \epsilon^2 + 2i\epsilon)}{\mu[(1 + \epsilon^2)^2 + 4\epsilon^2]} r'^2 \right]. \tag{22.1}$$

The first two terms stand for the exact reconstruction, the last two for the spurious amplitude. They differ from (23) only in terms of the order ϵ^2 or higher.

The reconstruction cycle in the case $\Gamma = 2$ is illustrated in figure 5 for a probability spot with a black centre. One must be careful not to go beyond $\Gamma = 2$ if there are sharp contrasts in the object. As shown in figure 6, a lighter centre will appear inside a black ring, and black lines will appear doubled.

Up to this point we have assumed unlimited apertures; consequently there was no lower limit to the spot size a which could be correctly reproduced. The effects of limited resolution can be very simply discussed by assuming a mask used in the taking of the hologram, with an amplitude transmission

$$e^{-\pi(c\rho)^2}.$$

Such graded masks are preferable to sharp apertures, not only from the point of view of mathematical simplicity, but also because they reduce the 'false detail' resulting from the cut-off to a minimum. Their use is well known in structure analysis (Bunn 1945, p. 350).

As the mask is used twice, in the taking of the photograph and in the reconstruction, its total effect is
$$e^{-\pi(\Gamma+1)(c\rho)^2}.$$

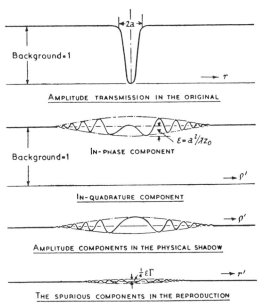

FIGURE 5. Reproduction cycle of a 'probability spot'.

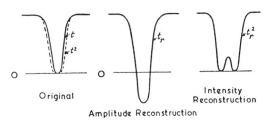

FIGURE 6. Distortion by exaggerated contrast.

We now assume $\Gamma = 2$, so as to obtain correct contrast reconstruction, and we put $3c^2 = b^2$. We have now to reconstruct the object, the probability spot (20), from the physical shadow

$$\tau_s = e^{-\pi(b\rho)^2}\{1 + i\epsilon A\, e^{-\pi(a^2-i\mu)\rho'^2} - i\epsilon A * e^{-\pi(a^2+i\mu)\rho'^2}\}, \tag{24}$$

which differs from (22) only in the masking factor. Introducing the small dimensionless parameter
$$\sigma = b^2/\mu = b^2/\lambda z_0,$$

31-2

one obtains by the transformation (15), neglecting powers of ϵ and σ higher than the first,

$$t_r(x,y) = (1+i\sigma)\exp\left[-\frac{\pi\sigma}{\mu}(1+i\sigma)r^2\right] - \frac{A}{1+(b/a)^2}\exp\left\{-\frac{\pi}{a^2+b^2}[r'^2+\epsilon\sigma(x_0^2+y_0^2)]\right\}$$

$$-\tfrac{1}{2}i\epsilon A^*\exp\{-(\tfrac{1}{2}\pi ir'^2/\mu)-(\pi/4\mu)[\epsilon[(x-x_0)^2+(y-y_0)^2]+\sigma[(x+x_0)^2+(y+y_0)^2]]\}.$$

$$(25)$$

The first is the background term. Apart from a very small diffraction effect, of the order σ^2, it is the geometrical shadow of the mask, projected on the object plane. The second is the 'correct' reproduction term. The chief difference is that the spread of the reproduced spot is now $(a^2+b^2)^{\frac{1}{2}}$ instead of a. Thus b has the meaning of the *resolution limit*, apart from a numerical factor which will be determined later. The factor $[1+(b/a)^2]^{-1}$ before the amplitude expresses the fact that the amplitude decreases in the same ratio as the area of the spot has increased. This loss of contrast for very small objects appears stronger than in the case of ordinary microscopy, where the amplitude falls off with the square root of the area, but the result is the same, as with a strong coherent background the intensity contrast is a linear function of the amplitude.

The error term, in the second row, has a structure different from (23); it no longer centres exactly on the original spot, as it contains a factor which centres on the mirror image of the spot, $-x_0$, $-y_0$. With the abbreviation $r_0^2 = x_0^2+y_0^2$ we can write the error term in a different and very useful form

$$t_e = -\tfrac{1}{2}i\epsilon A^*\exp\left[-\frac{\pi}{4\mu}(\epsilon-\sigma)r'^2\right]\exp\left[-(\tfrac{1}{2}\pi ir'^2/\mu)-(\tfrac{1}{2}\pi\sigma/\mu)(r^2+r_0^2)\right]. \quad (25\cdot1)$$

This is particularly useful in the case $\epsilon = \sigma$, i.e. $a=b$, as in this case the fringe system t_e has an amplitude independent of r'. The amplitude, though not the phase, centres on $x, y = 0$.

This result can serve as a basis for the theory of the spurious part in the reproduction of arbitrary objects, with the aim of obtaining a criterion for objects suitable for two-step microscopy.

A microscope, like any other optical system, can transmit a finite number of data only. Describing an object by a continuous transmission function is an objectionable idealization, as such a function contains an infinity of non-reproducible detail. We come nearer to an adequate description if we divide up the object by a network of lines into cells of the size of the resolution limit, associate a complex datum with each cell, and investigate the transmission of these data through the optical system.

Equation (25·1) suggests that particularly simple results will be obtained if we represent the object by a two-dimensional lattice of probability spots with a spread $a = b$. As illustrated in figure 7, we arrange these spots in a hexagonal lattice, with a distance d between adjacent centres. d is the resolution limit, which we define in a way slightly different from the usual, by postulating that three (instead of two) equal probability spots with spreads $a = b$ can just be resolved if their centres are at distances d from one another, i.e. the minimum in the centre just vanishes. By equation (25) the amplitude in the correct reproduction term follows a law $e^{-\frac{1}{2}\pi(r'/b)^2}$ if $a = b$. In the middle between three centres $r' = d/\sqrt{3}$; thus the condition for d is

$$\exp\left[-\frac{\pi}{2}\left(\frac{d}{\sqrt{3b}}\right)^2 \right] = \frac{1}{3},$$

which gives $d = 1\cdot 45b.$

This is in good agreement with the usual definition of the resolution limit,

$$d = 0\cdot 6\lambda/\sin\gamma_m,$$

if we define γ_m as the angle at which the background amplitude has dropped to $1/\sqrt{3}$, i.e. the background intensity to $1/3$ of its maximum value. Denoting the corresponding radius in the object plane by $R = z\sin\gamma_m$, we have

$$\exp\left[-\frac{2\pi\sigma}{\mu}R^2 \right] = \exp\left[-2\pi\left(\frac{b}{\lambda}\sin\gamma_m\right)^2 \right] = \frac{1}{3},$$

which gives $b = 0\cdot 42\lambda/\sin\gamma_m, \quad d = 0\cdot 61\lambda/\sin\gamma_m.$

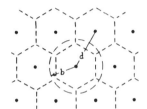

FIGURE 7. Dividing the object into independent elements.

Call N the number of independent elements inside the illuminated field, i.e. number of cells contained in the disk of radius R. Each cell occupies an area

$$\tfrac{1}{4}\sqrt{3}d^2 = 0\cdot 433d^2 = 0\cdot 91b^2;$$

thus the number N is

$$N = \frac{\pi}{0\cdot 91}\frac{R^2}{b} = 3\cdot 45(0\cdot 42)^2\left(\frac{\lambda z_0}{b^2}\right)^2 = 0\cdot 61\frac{1}{\sigma^2}. \tag{26}$$

N can be easily made a very large number, of the order 10^6 to 10^8. This suggests a *statistical evaluation* of the spurious part of the reproduction, by assuming random distribution of the amplitude over the independent elements of the object. It is, of course, understood that this might lead to gross errors in special cases, but it is certainly an acceptable assumption if a great number and variety of objects are considered.

Number the elements from 1 to N. The spurious amplitude in the reconstruction at a point x, y results from the superposition of the error terms of the form (25·1), one for each cell with centres x_n, y_n. The distance of x, y measured from x_n, y_n may be called r'_n. With the simplification resulting from $\sigma = \epsilon$ the resulting error amplitude is

$$t_e(x,y) = \tfrac{1}{2}i\sigma\, e^{-\pi\sigma r^2/2\mu}\sum_{}^{N} A_n^*\, e^{-\frac{1}{2}\pi\sigma r_n^2/\mu}\, e^{-\frac{1}{2}\pi i r_n'^2/\mu}, \tag{27}$$

where we have written r_n for the distance of x_n, y_n from $0, 0$.

470   D. Gabor

The relative distance r'_n between x, y and x_n, y_n occurs here only in the phase factor $e^{-\frac{1}{2}\pi i r'_n/\mu}$. The two probability decay factors fall off slowly. The first of these, before the sum, is the square root of the background attenuation, i.e. it falls off at half the rate of the background amplitude. The second factor centres on $0, 0$ and falls off at the same rate as the first factor. Thus it is admissible to put both equal to unity as an approximation, and simplify equation (27) to

$$ t_e(x,y) = \tfrac{1}{2}i\sigma \sum_{n}^{N} A_n^* e^{-\frac{1}{2}\pi i r'^2_n/\sigma}, \tag{27.1} $$

that is to say, in order to obtain the error amplitude at x, y we have to superimpose at this point a large number N of *undamped* waves, with wave-lengths $2\mu/r'$, emanating from all image points x_n, y_n. This wave-length is always longer than the resolution limit d. Its smallest value is at $r' = 2R$ and is μ/R, while the resolution limit is $0\cdot61\mu/R$.

Introduce now the hypothesis that there is no correlation between the phases of these waves. With this assumption the mean square of the component of t_e in phase with the background, which may be called $t^2_{\text{eff.}}$, is one-half of the sum of the absolute squares of the terms at the right:

$$ t^2_{\text{eff.}} = \tfrac{1}{8}\sigma^2 \sum_{n}^{N} A_n A_n^* = \tfrac{1}{8}\sigma^2 N A_n A_n^* = \tfrac{1}{8}\sigma^2 N A^2_{\text{eff.}}. \tag{28} $$

Here we have introduced the notation $A^2_{\text{eff.}}$ for the mean square secondary amplitudes, $A_n A_n^*$, averaged over the whole field. It is understood that the average level of transmission of the object has to be considered as part of the background, and $A_{\text{eff.}}$ is a measure of the departure from uniformity. Combining (28) with equation (26), $N\sigma^2 = 0\cdot61$, we obtain

$$ t_{\text{eff.}} = 0\cdot28 A_{\text{eff.}}. \tag{29} $$

Equation (29) enables us to formulate a criterion for suitable objects. A background can be considered as practically even if the intensity contrast does not exceed about 5 %, i.e. if the amplitude contrast is less than 2·5 %. This means that for suitable objects we must have

$$ A_{\text{eff.}} \leqslant 0\cdot1 \tag{30} $$

averaged over the whole field. As an example consider a black-and-white object in which the black part, where $A = 1$, covers a fraction κ of the illuminated field, while for the rest $A = 0$. In this case $A_{\text{eff.}} = \sqrt{\kappa}$, and we obtain the simple rule that not more than about 1 % of the illuminated field should be covered with black dots or lines. If, for instance, the object is a disk, half black and half white, its diameter should not exceed one-seventh of the field diameter.

As a second example consider an object with pure phase contrast, but with random distribution of phase delays. We must qualify this by the condition which precedes every application of the Fresnel-Kirchhoff theory; the phase must not vary appreciably between points spaced at less than a wave-length. In other words, the object must appear even and transparent if it is sharply focused. In ordinary microscopy a crinkled sheet of celluloid, or even reticulated gelatine, will satisfy this condition, but not an opal glass with colloidal dispersion. With this qualification in mind we can apply equation (29) and it can be shown that the value of $A_{\text{eff.}}$ is again unity. In the case of pure phase contrast the complex transmission vector $t = 1 - A$ moves on the unit circle, all orientations of t are equally probable. Hence $\bar{t} = 0$, which makes $\bar{A} = 1$, and

$$A^2_{\text{eff.}} = \overline{|A - \bar{A}|^2} = \overline{|A - 1|^2} = \overline{|t|^2} = 1.$$

This means that an object of this type, if it covers the whole field, produces $t_{\text{eff.}} = 0.28$, a very serious disturbance. This result is of interest, because it shows that an irregular transparent support for the object, even if it would be invisible in ordinary microscopy, will make all but the most contrasty or regular features of the object invisible. As it is rather doubtful whether an 'optically flat' or at least acceptably regular supporting membrane can be found in electron microscopy, it appears preferable to use supporting membranes only in a small fraction of the field, or to dispense with them altogether.

IMPROVING THE SEPARATION BY MASKING AND OTHER MEANS

These results lead to the conclusion that high-grade purity in the reproduction cannot easily be achieved even with very small objects, as the spurious intensity is proportional to the square root of the object area. But in the case of small objects special techniques become available, which allow a very effective elimination of the spurious amplitudes. The first of these is the masking of the geometrical shadow in the hologram. The second technique is the masking of the background in the reconstruction process.

The spurious amplitude is objectionable only in the area occupied by the true image. Thus we need eliminate only those rays issuing from the twin object which pass through the object area. As may be seen from figure 4, if the object is small these rays will have substantially the same direction as the primary rays which illuminate the object. This means that we can substantially reduce the spurious amplitude if we mask out the geometrical shadow in the hologram.

This masking process, however, will introduce two new disturbances. First, the mask itself will produce a system of interference fringes. This effect can be reduced to a very low level if a 'probability mask' is used. Secondly, the mask will eliminate some of the data required for a complete reconstruction. Evidently the coarser detail will suffer most, as this is contained in or near to the geometrical shadow area in the hologram, while the finer detail is spread over a larger area outside. But if the object is of the order of the characteristic length $\mu^{\frac{1}{2}}$ or smaller, the suppressed detail becomes insignificant. Thus masking of the shadow is a very effective method for improving the reproduction of very small objects.

In the second method the background, i.e. the primary wave, is suppressed *after* it has traversed the hologram. This can be done by producing a real image of the point source by means of the reconstructing lens in figure 1, and arranging a small black mask at this point, preferably a probability mask. This arrangement is similar to the well-known 'schlieren' method. The result is, that instead of an amplitude in the object plane

$$1 - t_c - t_e,$$

where c stands for 'correct' and e for 'error', we now obtain

$$-t_c - t_e,$$

neglecting the diffraction effects at the mask. Hence an absorbing object will now appear bright on a dark background, as in 'dark-field illumination'. While in the ordinary or 'bright field' method the intensity is approximately

$$1 - 2t_c - 2t_e,$$

the 'dark field' intensity is $\quad t_c^2 + 2t_e t_c + t_e^2.$

One can consider t_e^2 as the spurious background, while $2t_e t_c$ is the interference product of the two images. The spurious background is now the square of its previous value, proportional to the coverage fraction instead of to its square root, and becomes negligible for objects which cover only a few percent of the illuminated field. There remains, however, the interference product $2t_e t_c$. This contributes nothing to the background, as it is zero everywhere outside the object, where $t_c = 0$. In the object area it represents merely a small modulation of the correct density values. In the case of black-and-white objects this effect is negligible, as the outlines remain unchanged. How far it can distort graded objects is a matter for further investigation.

A combination of the two methods, i.e. masking the geometrical shadow *and* the primary wave, appears to be particularly promising in the case of small objects.

A third, somewhat laborious method for improving the separation is taking a series of reconstructions, with different values of μ. While the true image always remains the same, the spurious image varies, and can thus be discriminated. A fourth method will be discussed later, in connexion with non-homocentric illumination.

ILLUMINATING WAVES WITH ASTIGMATISM AND SPHERICAL ABERRATION

Following a method first introduced by Debye, we build up a general coherent illuminating wave of plane wavelets, normal to the direction α, β, γ, with an amplitude $A d\Omega$ in the infinitesimal solid angle $d\Omega$:

$$A(\alpha, \beta) \exp\{ik[x \cos \alpha + y \cos \beta + z \cos \gamma - p(\alpha, \beta)]\} \, d\Omega. \tag{31}$$

The amplitude A is assumed as real, the phase factor e^{-ikp} expressing the advance of the phase compared with the direct ray through the origin O. Assuming that O coincides with the 'mean paraxial focus' of the beam, let the phase function p be

$$p(\alpha, \beta) = \tfrac{1}{2} A_s (\cos^2 \alpha - \cos^2 \beta) + \tfrac{1}{4}(C_x \cos^4 \alpha + 2C_{xy} \cos^2 \alpha \cos^2 \beta + C_y \cos^4 \beta). \tag{32}$$

The first term is the phase advance due to astigmatism, the second is 'elliptical' spherical aberration. It has been assumed for simplicity that the elliptical errors of second and fourth order have the same principal axes x, y.

With the polar angles γ, θ, connected with α, β by

$$\cos \alpha = \sin \gamma \cos \theta, \quad \cos \beta = \sin \gamma \sin \theta, \tag{33}$$

p can be put into the form

$$p(\gamma, \theta) = \tfrac{1}{2} A_s \sin^2 \gamma \cos 2\theta$$
$$+ \tfrac{1}{4} \sin^4 \gamma [\tfrac{3}{8}(C_x + C_y) + \tfrac{1}{4}C_{xy} + \tfrac{1}{2}(C_x - C_y)\cos 2\theta - \tfrac{1}{4}[C_{xy} - \tfrac{1}{2}(C_x + C_y)]\cos 4\theta]. \tag{34}$$

The fourth-order term now appears as the sum of spherical aberration and two astigmatism terms, one elliptical, the other with fourfold periodicity. If the lens is round

$$C_x = C_y = C_{xy} = C_s,$$ (35)

and the fourth-order astigmatic terms vanish. C_s is the constant of spherical aberration. Its meaning is illustrated in figure 8, which shows the ray structure of a beam. The geometric-optical approximation is well justified in the most important practical applications of the present theory, as it is proposed to use beams with apertures about ten times larger than in ordinary electron microscopy, where the diffraction disk is of the same order as the geometrical aberrations. As the minimum cross-section of the beam increases with the third power of the aperture, and the diffraction effect is inversely proportional to the first power, it will represent a small correction only, of the order of 10^{-4} of the geometrical dimensions.

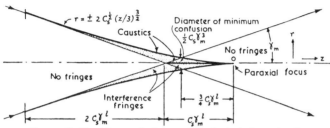

FIGURE 8. Focal figure of a beam with spherical aberration C_s.

If the aperture angle is γ_m, all rays cross the axis in the axial caustic, a line of length $C_s\gamma_m^2$ behind the paraxial focus O. The diameter of the beam in the Gaussian plane $z = 0$ is $2C_s\gamma_m^3$, but at the cross-section of minimum confusion, at $z = -\frac{3}{4}C_s\gamma_m^2$, it is four times less. The minimum cross-section is the intersection of the envelope or outer caustic, a rotational surface with an equation $r = \pm 2C_s^{\frac{1}{2}}(z/3)^{\frac{3}{2}}$, with the cone of maximum divergence, $r = \pm (z + C_s\gamma_m^2)\gamma_m$. This cone and the outer caustic divide up the beam into four regions of different character, of which two are dotted, to indicate that they contain interference fringes. The first of these is inside the envelope but outside the cone. The rays cross in every point of this region. The second is the region surrounding the axial caustic, limited by the envelope and by the cone of maximum divergence, which has three rays crossing in every point. The interference fringes in both regions are so sharp and contrasty as to make objects placed into them almost invisible; thus the whole dotted volume must be ruled out as a possible location for objects. In the remaining two regions, at the right and left, there is only one ray to every point, and they represent smoothly graded backgrounds, suitable for microscopic objects. In the region at the left the illumination density is largest near the edge; in the second, at the right, the density has a maximum on the axis.

If, in addition, as is always the case in electron optics, the beam is also astigmatic, figure 8 can still serve as an illustration, but only for the principal sections of the beam, and these must be imagined as displaced longitudinally by $\pm A_s$. Thus O will be now in the middle between the two focal lines, at right angles to one another and to the beam axis, and a distance $2A_s$ apart.

Returning to the wave-optical representation, summing the wavelets (31) gives for the complex amplitude at the point x, y, z

$$U_0(x, y, z) = \iint A(\gamma, \theta) \exp\{ik[(x\cos\theta + y\sin\theta)\sin\gamma + z\cos\gamma$$
$$- \tfrac{1}{2}A_s\sin^2\gamma\cos 2\theta - \tfrac{1}{4}C_s\sin^4\gamma]\}\sin\gamma\,d\gamma\,d\theta. \quad (36)$$

We have used here the simplifications arising from equation (35), and these will be assumed also in the following formulae to simplify the discussion, but the results will be of such a nature as to permit their extension without difficulty also to the more general case expressed by equations (32) and (34).

Introduce under the integral sign in (36) the Fourier variables $\xi = (\cos\alpha)/\lambda$, $\eta = (\cos\beta)/\lambda$ and $\rho = (\xi^2 + \eta^2)^{\frac{1}{2}}$. The exact transformation equations are

$$\sin\gamma = \lambda\rho, \quad \cos\gamma = (1 - \lambda^2\rho^2)^{\frac{1}{2}}, \quad d\Omega = \sin\gamma\,d\gamma\,d\theta = \frac{\lambda^2\,d\xi\,d\eta}{(1 - \lambda^2\rho^2)^{\frac{1}{2}}}.$$

We again assume a sufficiently narrow beam to justify neglecting $\sin^2\gamma = \lambda^2\rho^2$ in the denominator of the last expression. In the phase factor, however, we must take into consideration terms up to the fourth order in ρ, and write

$$\cos\gamma = (1 - \lambda^2\rho^2)^{\frac{1}{2}} = 1 - \tfrac{1}{2}\lambda^2\rho^2 - \tfrac{1}{8}\lambda^4\rho^4.$$

In this approximation

$$U_0(x, y, z) = \lambda^2 e^{ikz}\iint A(\xi, \eta)\exp\{2\pi i[x\xi + y\eta - \tfrac{1}{2}z\lambda\rho^2 - \tfrac{1}{2}A_s\lambda(\xi^2 - \eta^2)$$
$$- \tfrac{1}{8}(z + 2C_s)\lambda^3\rho^4]\}\,d\xi\,d\eta. \quad (36\cdot1)$$

This integral, like the exact expression (36), can be easily evaluated at large distances R from the origin, in a direction α, β. One obtains

$$U_0(R, \alpha, \beta) = -i\frac{\lambda}{R}A(\gamma, \theta)e^{ik(R-p)}, \quad (36\cdot2)$$

where p is given by equations (32) or (34). The factor $-i$, which expresses an advance of the wave-front by $\tfrac{1}{4}\lambda$ as compared with the components (31) arises in the transition from plane to spherical waves, and is familiar in diffraction theory. Equation (36·2) supplies the background to the physical shadow of an object, which we are now going to calculate.

The object, in a plane $z = z_0$, may be characterized, as before, by the complex transmission function $t(x, y)$. Using the fundamental premissa of the Fresnel-Kirchhoff diffraction theory, we assume that the amplitude immediately before the object is that of the undisturbed illuminating wave, $U_0(x, y, z_0)$, and $t(x, y)$ times this immediately behind it. We must now give the angular variables in the illuminating wave by suffixes '0' ('original') to distinguish them from the variables of the outgoing wave, without suffixes.

The problem is building up the outgoing wave from the diffraction products of the plane wavelets which compose the original wave. The Fresnel-Kirchhoff formula in the simplified form (7·1) can be again applied, but with the modification that the wave $r_0^{-1}\,e^{ikr_0}$ must now be replaced by the sum of the wavelets (34). There is no change in the meaning of r_1, the distance of the observation point $Q(R,\alpha,\beta)$ from the object point P. Thus the Fresnel-Kirchhoff formula now assumes the form

$$U(Q) = U(R,\alpha,\beta)$$

$$= \frac{1}{i\lambda R}\iiiint t(x,y)\,A(\alpha_0,\beta_0)\exp\{ik[r_0(\alpha_0,\beta_0)+r_1(\alpha,\beta)]\}dx\,dy\,d\alpha_0\,d\beta_0, \quad (37)$$

where
$$r_0(\alpha_0,\beta_0) = x\cos\alpha_0 + y\cos\beta_0 + z_0\cos\gamma_0 - p(\alpha_0,\gamma_0),$$
$$r_1(\alpha,\beta) = R - x\cos\alpha - y\cos\beta - z_0\cos\gamma.$$

Expressing the angles by the Fourier variables ξ,η and ξ_0,η_0 we obtain, with the same approximations as in (36·1),

$$U(R,\xi,\eta) = \frac{\lambda}{iR}\,e^{ikR}\iint t(x,y)\,dx\,dy$$

$$\iint A(\xi_0,\eta_0)\exp\{2\pi i\,[x(\xi_0-\xi)+y(\eta_0-\eta)-\tfrac12 z_0\lambda(\rho_0^2-\rho^2)-\tfrac18 z_0\lambda^3(\rho_0^4-\rho^4)$$
$$-\tfrac12 A_s\lambda(\xi_0^2-\eta_0^2)-\tfrac14 C_s\lambda^3\}]\,d\xi_0\,d\eta_0. \quad (38)$$

The symmetry of this expression is disturbed by the last two terms, but it is at once restored if we go over to the 'physical shadow', by dividing the amplitude $U(R,\alpha,\beta)$ into the background $U_0(R,\alpha,\beta)$ as given by equation (36·2):

$$\tau(\xi,\eta) = \iiiint t(x,y)\,\frac{A(\xi_0,\eta_0)}{A(\xi,\eta)}\exp[2\pi i\{x(\xi_0-\xi)+y(\eta_0-\eta)-\tfrac12 z_0\lambda(\rho_0^2-\rho^2)$$
$$-\tfrac12 A_s\lambda[(\xi_0^2-\xi^2)-(\eta_0^2-\eta^2)]-\tfrac18\lambda^3(z_0+2C_s)(\rho_0^4-\rho^4)\}]\,dx\,dy\,d\xi_0\,d\eta_0. \quad (39)$$

This is the formula for the physical shadow at infinity of an object at $z=z_0$, illuminated by a beam with fourth-order aberrations, but which can be evidently extended to aberrations of any order. It is the equivalent of the transformation formula (14) for homocentric illumination, but it cannot be put into the form of an integral over the object plane, as the integration over the angular variables cannot be carried out in terms of the transcendentals recognized in analysis. On the other hand, it can be immediately reduced to a double integral over the angular variables by means of the Fourier transform $T(\xi,\eta)$ of $t(x,y)$ which is

$$T(\xi,\eta) = \iint t(x,y)\,e^{-2\pi i(x\xi+y\eta)}\,dx\,dy,$$

which converts equation (42) into

$$\tau(\xi,\eta) = \iint T(\xi-\xi_0,\eta-\eta_0)\,\frac{A(\xi_0,\eta_0)}{A(\xi,\eta)}\exp[\pi i\{z_0\lambda(\rho^2-\rho_0^2)$$
$$+A_s\lambda[(\xi^2-\xi_0^2)-(\eta^2-\eta_0^2)]+\tfrac14\lambda^3(z_0+2C_s)(\rho^4-\rho_0^4)\}]\,d\xi_0\,d\eta_0. \quad (40)$$

This transformation may be illustrated by a few simple examples. If $t=1$, i.e. if there is no object, T is a delta function

$$T(\xi-\xi_0,\eta-\eta_0) = \delta(\xi-\xi_0,\eta-\eta_0),$$

which means that the integral (40) is the value of the integrand for $\xi_0=\xi$, $\eta_0=\eta$, which is unity, as before.

476   D. Gabor

If $t(x, y)$ is a harmonic function of x, y with periods $1/a, 1/b$

$$t(x, y) = e^{2\pi i(ax + by)}. \tag{41.1}$$

T is again a delta function, but shifted to the point a, b

$$T(\xi - \xi_0, \eta - \eta_0) = \delta(\xi - \xi_0 - a, \eta - \eta_0 - b),$$

and we have again to take the value of the integrand, but this time at $\xi_0 = \xi - a$, $\eta_0 = \eta - b$. The physical shadow is

$$\tau(\xi, \eta) = \frac{A(\xi - a, \eta - b)}{A(\xi, \eta)} \exp\{\pi i a(2\xi - a)[\lambda(z_0 + A_s) + \tfrac{1}{4}\lambda^3(z_0 + 2C_s)(2\xi^2 - 2\xi a + a^2)]\}$$

$$\exp\{\pi i b(2\eta - b)[\lambda(z_0 - A_s) + \tfrac{1}{4}\lambda^3(z_0 + 2C_s)(2\eta^2 - 2\eta b + b^2)]\}. \tag{41.2}$$

We have met the first factors under the exponential in the shadow transformation with homocentric illumination. But the period in the shadow is no longer a constant, that is to say, the shadow of a sinusoidal grid is not of the same type as the original. If, for example, $b = 0$, i.e. the grid is parallel to y, the spacing between two maxima is

$$1/(\lambda z_0 a)[1 + A_s/z_0 + \tfrac{1}{4}\lambda^2(1 + 2C_s/z_0)(2\xi^2 - 2\xi a + a^2)].$$

The first factor is the geometrical shadow of the period $1/a$, the second is the correction arising from astigmatism and spherical aberration, and also from the fourth-order term which expresses the departure of a spherical wave-front from a paraboloid. In all practical applications z_0 will be of the order $C_s\gamma_m^2$, and z_0 can be neglected against $2C_s$. Thus the astigmatism and spherical aberration of a beam can be determined from two holograms of a sinusoidal grid, taken in two positions, at right angles to one another. But the method is not very sensitive. Near the edge of the field where $\xi \gg a, \eta \gg b$, the spacing of two neighbouring maxima will be a fraction

$$1\Big/\Big(1 + \frac{C_s}{z_0}\gamma_m^2\Big)$$

of the geometrical spacing. But as z_0 will be of the order $C_s\gamma_m^2$ if good photographs are to be obtained, this fraction will be of the order unity. This shows that a sinusoidal grid, even if it were available, would not be a very suitable test object. Spherical aberration can be much better determined from the physical shadow of a thin wire, but the discussion of this case cannot be carried out in elementary terms, and may be omitted.

RECONSTRUCTION IN THE PRESENCE OF SPHERICAL ABERRATION AND ASTIGMATISM

Assume that a photograph has been taken of the physical shadow of an object, according to equations (39) or (40). We have seen that, if the background is relatively strong, this is equivalent to substituting for τ_1 its real part, $\tfrac{1}{2}(\tau_1 + \tau_1^*)$, where, as before, τ_1 relates to the 'object proper' without the background. In order to find the spurious term in the reconstructed object, we must apply to τ_1^* the transformation inverse to (39). But this is rather complicated, while an interpretation in terms of 'twin images' is easy, and leads to much simpler and clearer results.

An expression for τ_1^*, the complex conjugate of the physical shadow τ_1 is obtained from (39) by reversing the sign of i. Assume now, as before, in the plane $z = z_0$ a twin object with a transmission function

$$t_1'(x, y) = t_1^*(-x, -y).$$

Renaming the integration variables $-x, -y$ instead of x, y, one obtains for τ_1^* the expression

$$\tau_1^*(\xi, \eta) = \iiiint t_1'(x, y) \frac{A(\xi_0, \eta_0)}{A(\xi, \eta)} \exp\{2\pi i[x(\xi_0 - \xi) + y(\eta_0 - \eta) + \tfrac{1}{2}z_0\lambda(\rho_0^2 - \rho^2)$$
$$+ \tfrac{1}{2}A_s\lambda[(\xi_0^2 - \xi^2) - (\eta_0^2 - \eta^2)] + \tfrac{1}{8}\lambda^3 (z_0 + 2C_s)(\rho_0^4 - \rho^4)]\}\,dx\,dy\,d\xi_0\,d\eta_0. \quad (42)$$

This is the physical shadow of an object t_1' in the plane $-z_0$, according to equation (39), but with the important difference that the sign of A_s and C_s has been also reversed. The physical significance of this becomes clearer if instead of τ_1^* we consider the complementary wave U_1' which arises in the reconstruction, and which is obtained from (42) by multiplying it with the background (36·2). The result can be written

$$U_1'(R, \xi, \eta) = \frac{\lambda}{iR} e^{ikR} \iint t_1'(x, y)\,dx\,dy \iint A(\xi_0, \eta_0) \exp\{2\pi i\,[x(\xi_0 - \xi) + y(\eta_0 - \eta)$$
$$+ \tfrac{1}{2}z_0\lambda(\rho_0^2 - \rho^2) + \tfrac{1}{8}z_0\lambda^3(\rho_0^4 - \rho^4) + \tfrac{1}{2}A_s\lambda(\xi_0^2 - \eta_0^2) + \tfrac{1}{4}C_s\lambda^3\rho_0^4]\}$$
$$\times \exp\{-2\pi i[A_s\lambda(\xi^2 - \eta^2) + \tfrac{1}{2}C_s\lambda\rho^4]\}\,d\xi_0\,d\eta_0. \quad (43)$$

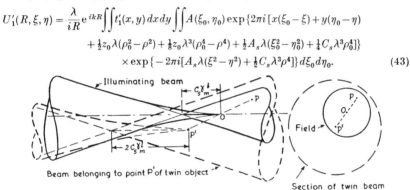

FIGURE 9. The twin object in a beam with spherical aberration.

Comparing this with equation (41), it can be seen that the first two lines represent the emission of an object t_1' in the plane $-z_0$, but *illuminated by a wave in which the signs of the astigmatism A_s and of the spherical aberration C_s are reversed*. This assures complete symmetry in the illumination of the object and its twin. But the emitted wave is modified by the phase factor in the last line. This means that the wavelet issuing from any element $t_1'(x, y)\,dx\,dy$ of the twin object has astigmatism $2A_s$ and spherical aberration $2C_s$. Thus *in the presence of astigmatism or spherical aberration the twin object which appears in the reconstruction will be no longer sharp, but will appear as if viewed through a system with twice the aberrations of the condenser system.* One could, of course, view the twin object instead of the original by means of a viewing system with aberrations of the opposite sign, but not both simultaneously.

This result is illustrated in figure 9, which allows also an elementary verification. The illuminating beam envelope is shown in continuous lines, the beam appearing to issue from a point P' of the twin object in interrupted lines. The axial caustic of this beam is always twice the caustic of the illuminating beam. This can be immediately understood if one imagines the axial caustic as the locus of the centres of homocentric beams, each emitting rays only in a certain cone. For each of these partial beams there exists a sharp twin point to P, on the line joining P with its centre. Equation (43) proves that this geometric-optical reasoning is in fact justified.

Figure 9 shows also that the beam associated with any point of the twin object intersects the object plane in an area four times larger than the field. From this we can infer at once that if the illumination were even, the spurious amplitude in the object plane would bear the same relation to the correct amplitude as in the case of homocentric illumination, i.e. equation (29) would apply again. In fact the illumination is very uneven in a beam with spherical aberration in cross-sections not very far from the caustic, and on this is based a fourth method of improving the separation, in addition to the others which have been discussed in a previous section. Masking is not very efficient in the presence of spherical aberration, as the geometrical shadow of a point object is a radial line, the projection of the axial caustic. This becomes small only if the object is in the axis, but in electron optics it is not possible to fix small objects by means of a transparent support in the middle of the field.

This fourth method for improving the separation is to place the object in a position where it receives less than the average of illumination density. To explain this briefly, define as 'coefficient of illumination', J, the ratio of the mean intensity over a small object area to the mean over the *whole* illuminated field. If the object has the average intensity transmission tt^* and covers a fraction κ of the field, the fraction of the total flux issuing from the object is $tt^*\kappa J$. Exactly the same flux emerges also from the twin object. But of this only a fraction $\frac{1}{4}\kappa J$ will pass through the object. The factor J is here the same as defined from the direct illumination of the object, because, as may be seen in figure 9, the small twin objects interfere with one another in the direction in which they are directly illuminated. Passing from the intensities to the root mean square amplitudes one obtains a separation factor proportional to $\sqrt{(\kappa J)}$, i.e. $\sqrt{J}$ times what we have previously obtained for uniform, homocentric illumination. Thus by placing small objects in *relatively dark* parts of the field, where $J < 1$, one improves the separation, by reducing the spurious background in the object area. Correspondingly more light is sent by the twin object to other regions of the field, but the spurious amplitude is of course harmless if it falls well outside the reconstructed object.

It may be noted that relatively weak illumination does not affect the contrast in the reconstructed object, so long as it is not submerged by ghosts, scattered light, and impurities arising from uneven development of the photograph.

COHERENCE CRITERIA

Up to this point we have assumed an absolutely coherent monochromatic illuminating wave, originating from a point source, but distorted by passing through a lens system. Absolute coherence means interference fringes of any order, but it means of course zero intensity. In practice we must strike a compromise between these two conflicting claims. The best compromise is obtained if the degree of coherence is just sufficient to produce an interference pattern from which the object can be reconstructed with the required resolution limit.

A necessary criterion of coherence can be immediately formulated, without any regard to details of the hologram. Imagine that an absolutely coherent illuminating beam is moved during the exposure parallel to itself, so that a representative point of it, e.g. the mean paraxial focus, fills a circular disk with diameter d_c. But this is equivalent to moving the object within a disk of the same diameter, as only the relative position of beam and object matters for the physical shadow at infinity, and from such a 'wobbled' hologram we could at best reconstruct an image with a resolution limit d_c. Thus we obtain the necessary condition that the Gaussian or nominal diameter of the illuminating disk, d_c, must not exceed the Abbe limit d_A

$$d_c \leqslant d_A = \tfrac{1}{2}\lambda/\sin\gamma_m. \tag{44}$$

But we can show that this necessary condition is also sufficient, because it will produce holograms practically indistinguishable from one taken with an absolutely coherent beam, within a plate radius corresponding to the maximum angle γ_m. Express in equation (47) the wave-length by de Broglie's relation as

$$\lambda = h/p,$$

where p is the momentum associated with the wave. This relation is valid for photons as well as for electrons or any other particles. Interpreting $p\sin\gamma_m$ as the maximum transversal momentum p_t of the particles in the beam, we write (47) in the form

$$d_c 2p_t \leqslant h. \tag{44·1}$$

Confront this with Heisenberg's inequality

$$d_c' 2p_t' \geqslant h, \tag{45}$$

where d_c' is the maximum transversal uncertainty of position of particles in the beam in the Gaussian focal plane, and $2p_t'$ is the maximum uncertainty in the transversal momentum. Consider first the case that the beam is limited by a physical aperture in the plane considered, i.e. $d_c = d_c'$. Heisenberg's principle states that if the particles composing the beam are specified to the limit (45), they are indistinguishable, that is to say, they produce effects, such as interference fringes, which cannot be distinguished from one another by observation within the cone-angle corresponding to that value of p_c' which changes the inequality into an equality. Comparing (44·1) and (45) we see that if $d_c = d_c'$, we must have $p_t < p_t'$, thus the interference fringes inside the cone γ_m are *a fortiori* the same for all beam particles.

But if d_c is not a physical aperture, but the Gaussian image of one, formed by an optical system, the criterion still holds, because $d_c\sin\gamma_m$ is an invariant in Gaussian optics. If the criterion (44) were not sufficient, it would be possible to break through Heisenberg's principle by placing a suitable lens system in front of the physical aperture to produce observable differences in the fringe system, which would make the particles to some extent distinguishable.

These very general considerations are of course uncertain to a factor of the order unity. In order to obtain a more quantitative idea of the changes which are produced in the hologram by departure from absolute coherence, consider the simple case of illumination through a physical aperture of diameter d, and investigate its effect on the fringe system produced by a point object on the axis, at a distance z_0 from the aperture. Each point of the illuminating aperture produces a fringe system concentric with the axis which connects this point with the point object. These fringe systems are incoherent with one another, hence their intensities must be summed. At the edge of the hologram the angular spacing of two fringes is $\lambda/z_0 \sin \gamma_m$. Two fringe systems will just wipe out one another if they are displaced by half this amount. This will be the case if the spacing of the two point sources is $\frac{1}{2}\lambda/\sin \gamma_m$, which is just the Abbe limit d_A.

With Zernike (1948), we define the 'degree of coherence' D_c, as the range of intensity difference between maxima and minima in the fringe system at the marginal angle γ_m, divided by the corresponding quantity if the same light flux issues from a point source at the centre of the aperture. Assuming that the intensity variation in the fringe system is sinusoidal, one obtains

$$D_c = \iint \cos (\pi x/d_A) \, dx \, dy \Big/ \iint dx \, dy, \qquad (46)$$

where the integration has to be carried out over the area of the illuminating aperture of diameter d. The integrand $\cos (\pi x/d_A)$ expresses the fact that two points spaced in the X-direction by d_A just oppose one another. The integration gives

$$D_c = J_0\left(\tfrac{1}{2}\pi \frac{d}{d_A}\right) + J_2\left(\tfrac{1}{2}\pi \frac{d}{d_A}\right), \qquad (47)$$

where J_0 and J_2 are the Bessel functions of zero and second order. Some values are

d/d_A	0	0·5	0·75	1·0	1·25	1·5	1·75	2·0
D_c	1	0·925	0·837	0·723	0·590	0·448	0·312	0·181

This justifies the expectation that the fringes system at the edge of the hologram will be rapidly effaced if the diameter of the light source appreciably exceeds the Abbe limit.

The coherence condition (47) represents a severe limitation of the available intensities, and it is the chief reason why the applications of the method of reconstructed wave-fronts will be probably restricted to light, with wave-lengths not very far from the visible, and to electrons. X-rays, protons and other particles will have to be excluded, as no sufficiently intense sources are available. Even in the case of electrons rather long exposures will be necessary, unless the present-day technique is improved.

THE OPTICAL RECONSTRUCTION

So far we have assumed in the formulae, for simplicity, that the reconstruction is carried out with the same wave-length as used in the production of the diffraction pattern. Let us now distinguish the first wave-length by λ', the second by λ'', and use the primes ' and '' also for distinguishing the data A_s, C_s in the analyzer and in the synthetizer. The same formal distinction will be used also for z_0' and z_0'', but here a word of explanation is required. z_0' is a datum of the analysis; it is the actual distance of the object from the mean paraxial focus of the illuminating beam. But there is no physical object in the synthetizer, and z_0'' means merely the plane on which the viewing system must be focused in order to obtain a true, or at least the truest possible image of the original object.

The result of the analysis, the physical shadow, now to be called τ', is described by equation (42). We write down this equation again, but replace the Fourier variables ξ, η by the angles α, β, γ. For reasons of symmetry it will be convenient to attach the prime ' not only to the data of the analyzer, but also to the co-ordinates $x, y, \alpha, \beta, \gamma$ and $\alpha_0, \beta_0, \gamma_0$ used in the analysis. We write

$$\tau'(\alpha', \beta') = \iiiint \frac{A'(\alpha_0', \beta_0')}{A'(\alpha', \beta')} t'(x', y')$$

$$\times \exp\{2\pi i [Q(\alpha_0', \beta_0') - Q(\alpha', \beta')]\} \frac{d(\cos \alpha_0')\, d(\cos \beta_0')}{\cos \gamma_0'} dx'\, dy', \quad (48)$$

where the phase function Q is

$$Q(\alpha, \beta') = \frac{1}{\lambda'} [x' \cos \alpha' + y' \cos \beta' - \tfrac{1}{2} z_0' \sin^2 \gamma_0' - \tfrac{1}{2} A_s'(\cos^2 \alpha' - \cos^2 \beta') - \tfrac{1}{8}(z_0' + 2C_s') \sin^4 \gamma'].$$

$$(49)$$

The same equation applies to the synthesis, i.e. to the reconstruction of an object t'', with all primes ' changed into ". The fact that the hologram obtained in the analysis is used in the reconstruction is expressed by

$$\tau''(\alpha'', \beta'', \gamma'') = \tau'(\alpha', \beta', \gamma'), \tag{50}$$

where the angles α', β', γ' and $\alpha'', \beta'', \gamma''$ belong to corresponding points of the hologram. The relation between them is given by the geometries of the analyzer and of the synthetizer.

Consider first the simple case, illustrated in figure 1, in which the focal length f of the collimator lens in the synthetizer, which moves the hologram optically to infinity, is the same as the throw L in the analyzer. In this case the angles α', β' and α'', β'' are the same, and their primes can be disregarded. It can be seen by inspection of equation (48) that it is transformed into the corresponding equation for $\tau'' = \tau'$ if we put

$$x' = \frac{\lambda'}{\lambda''} x'', \quad y' = \frac{\lambda'}{\lambda''} y'', \quad A_s' = \frac{\lambda'}{\lambda''} A_s'', \quad C_s' = \frac{\lambda'}{\lambda''} C_s'', \quad z_0' = \frac{\lambda'}{\lambda''} z_0'', \tag{51}$$

and

$$\tau''(x'', y'') = \tau'\left(\frac{\lambda'}{\lambda''} x'', \frac{\lambda'}{\lambda''} y''\right). \tag{52}$$

The transformation of the integration variables is purely formal. The next two equations postulate the scaling up of the aberrations A_s', C_s' in the synthetizer, and the last of the conditions (51) states that one must focus on the plane z_0'' in order to see the object t'' given by equation (52).

Consider now the more general case

$$f = kL, \tag{53}$$

i.e. we use a collimator lens of focal length k times the throw in the analyzer, always assuming of course that the hologram is in the focal plane of the lens. (This covers also the case in which the hologram used in the synthesis is an m times enlarged replica of the original; in this case the parameter k which figures in the following equations has the value f/mL.) The angles $\alpha' \ldots$ and $\alpha'' \ldots$ are now connected by the relations

$$\frac{\cos \alpha'}{\cos \gamma'} = k \frac{\cos \alpha''}{\cos \gamma''}, \quad \frac{\cos \beta'}{\cos \gamma'} = k \frac{\cos \beta''}{\cos \gamma''}. \tag{54}$$

The solution of these equations can be written in the form

$$\cos \alpha' = k \cos \alpha'' [1 - \tfrac{1}{2}(k^2 - 1) \sin^2 \gamma'' - \tfrac{3}{8}(k^2 - 1)^2 \sin^4 \gamma'' - \ldots]. \tag{54·1}$$

Only the first two terms of the expansion will be required. Introduce these into equation (48), where for simplicity we put $\lambda' = \lambda''$, to separate the change of geometry from the change of wave-length. The essential properties of the transformation can be deduced from the phase function Q, equation (49), which now assumes the form

$$\lambda Q = k \left(x' \cos \alpha'' + y' \cos \beta'' \right) - \tfrac{1}{2}k^2 z_0' \sin^2 \gamma'' - \tfrac{1}{2}k^2 A_s'(\cos^2 \alpha'' - \cos^2 \beta'')$$

$$- \tfrac{1}{8}k^4 \left(\frac{z_0'}{k^2} + 2C_s' \right) \sin^4 \gamma'' - \tfrac{1}{2}k(k^2 - 1)(x' \cos \alpha'' + y' \cos \beta'') \sin^2 \gamma''$$

$$+ \tfrac{3}{8}z_0' k^2 (k^2 - 1) \sin^4 \gamma'' + \tfrac{1}{2}A_s' k^2 (k^2 - 1) \sin^2 \gamma'' (\cos^2 \alpha'' - \cos^2 \beta''). \tag{55}$$

The terms in the first row and the first term in the second correspond to an exact reproduction, the others represent errors which arise only if $k^2 \neq 1$. Considering the first four terms only, equation (48) transforms into an identical equation for τ'' instead of τ' by putting

$$kx' = x'', \quad ky' = y'', \quad k^2 A_s' = A_s'', \quad k^4 C_s' = C_s'', \quad k^2 z_0' = z_0'', \tag{56}$$

and

$$t''(x'', y'') = t'\left(\frac{x''}{k}, \frac{y''}{k} \right). \tag{57}$$

This means that in order to see an image which is a k times enlarged replica of the original we must scale up the astigmatism k^2 times, the spherical aberration k^4 times, and focus the viewing system on a plane $z_0'' = k^2 z_0'$.

But this image will appear with certain aberrations, which are indicated by the new terms in (55). The second term in the second row represents a *coma*. The first term in the last row is an addition to the spherical aberration, which can be incorporated in C_s''. The last term shows that the astigmatism A_s' of second order in the analyzer has produced astigmatism of the fourth order in the analyzer, i.e. a spherical aberration of the elliptical type.

All these error terms can be kept very small unless $k^2 \ll 1$. It can be shown that the best positions of the object are near $z_0' = -C_s' \sin^2 \gamma_m'$, hence x', y' will be of the order $C_s' \sin^3 \gamma_m'$, even if the object is in a marginal position. Hence the coma term in (55) will be of the order

$$k(k^2 - 1) C_s' \sin^3 \gamma_m' \sin^3 \gamma_m'' \simeq \frac{k^2 - 1}{k^2} C_s' \sin^6 \gamma_m',$$

i.e. unless $k^2 \ll 1$ this will be a very small term, except in extreme cases when the spherical aberration $C_s' \sin^4 \gamma_m'$ is of the order of several hundred fringes. In such cases the coma might amount to a few fringes, and coma compensation in the viewing system may become necessary.

The last term in (55) is of the order

$$\frac{k^2 - 1}{k^2} A'_s \sin^4 \gamma'_m,$$

which is again very small unless $k^2 \ll 1$. A'_s in good electron lenses is 10^{-4} or even less of C'_s; thus even if the spherical aberration is of the order of a thousand fringes, this term will represent a fraction of a fringe only.

Thus it is admissible to make the length of the optical synthetizer appreciably different from the throw in the electronic analyzer. It may be particularly advantageous to make $k < 1$, that is to say, not to make use of the full magnification λ''/λ' which is about 100,000, but only of a part of it. The rest can be supplied by the viewing system. This has the advantage that one can work with smaller lenses, though with proportionately larger numerical aperture. Assuming, for instance, $C'_s = 1$ cm. and $\sin \gamma'_m = 0.05$, the minimum diameter of the electron beam is 0.625μ, and if one makes $k = 1$ one requires an optical system capable of handling a light beam with 6·25 cm. minimum diameter. It will be advantageous to reduce this to one-half, or even to one-quarter, as optical systems with numerical apertures of 0·1 to 0·2 present no difficulties if the lenses need not be large.

To sum up, if in the optical synthetizer the data of the electronic condenser system are scaled up according to

$$A''_s = k^2 \frac{\lambda''}{\lambda'} A'_s, \quad C''_s = k^4 \frac{\lambda''}{\lambda'} C'_s, \tag{58}$$

the transversal dimensions of the object will appear scaled up in a ratio $k\lambda''/\lambda'$ and the longitudinal dimensions in the ratio $k^2\lambda''/\lambda'$. Thus *the geometrical or k-part of the transformation is of the type as produced by optical instruments, with a longitudinal magnification equal to the square of the transversal, while the λ-part is a uniform scaling-up, not realizable by ordinary optical imagery.*

The accuracy with which the conditions (58) have to be fulfilled can be best stated in terms of fringes. The maximum admissible deviation of a wave-front from the spherical shape without loss of resolving power has been estimated by Glaser (1943) as 0·4 of a wave-length, by Bruck (1947) as one wave-length. The second can be considered as the more reliable estimate. Thus the condition (58) for C''_s must be observed to an accuracy of one fringe. Assuming again $C'_s = 1$ cm. and a resolution limit of 1Å, one requires by Abbe's rule an aperture $\sin \gamma'_m = 0.025$, and with the more accurate numerical factor 0·6, $\sin \gamma'_m = 0.030$. This gives 200 or 400 fringes at the edge of the field, according to which numerical factor one adopts. Thus the spherical aberration in the optical model must imitate C'_s to about one fringe in 200 or in 400.

The astigmatism tolerance at the edge of the field is about a quarter fringe. In carefully manufactured electron objectives A_s is of the order of a few microns, and it can be reduced by the compensation methods introduced by Hillier & Ramberg (1947) by at least one order of magnitude. This is necessary for realizing the full resolving power of present-day electron microscopes. In terms of fringes, the astigmatism in carefully manufactured but not compensated electron lenses amounts to a few fringes at apertures of 0·003, and if this is opened up ten times, to realize

32-2

a ten times improved resolving power, the distortion will be of the order of a few hundred fringes. Thus A'_s must be also imitated in the optical synthetizer to an accuracy of one part in a few hundred.

One could think of imitating the data of the electron-optical system by first carefully measuring A'_s and C'_s and computing an optical system with these data. But this is hardly a practicable method. Apart from the difficulties of measuring to the required accuracy, by the time the computation is finished and the optical replica is made the data of the electron-optical system are likely to have changed by far more than the error tolerance. It will be much preferable to make the astigmatism and the spherical aberration of the synthetizer variable, and adjust them until certain known parts of the object, such as the support, or certain standard test objects appear with maximum sharpness. The spherical aberration can be made variable by shifting a fourth-order plate, the astigmatism by crossed cylindrical lenses or by tilting lenses. Expert opticians will be doubtlessly able to work out a schedule to carry out the three adjustments of focus, astigmatism and spherical aberration in a systematic way. Thus only a moderate degree of constancy is required of the electron-optical system, sufficient at least for a series of reconstructions, without too frequent readjustments.

EXPERIMENTAL TESTS

Experiments were started almost as soon as the idea of reconstruction first emerged. They confirmed the soundness of the basic principle, but pointed to the necessity of elaborating and modifying the original, somewhat primitive views on the mechanism of reconstruction, which have been described elsewhere (Gabor 1948). The experiments were later continued in order to test the conclusions from the quantitative theory described in this paper.

In these tests analysis and synthesis were both carried out with visible light, though not always with the same wave-length. The arrangement for taking holograms was substantially as shown in the upper part of figure 1, but with optical instead of with electron lenses. A condenser threw an image of a high-pressure mercury arc (of the 'compact' type, with tungsten electrodes) through a colour filter on an aperture of about 0·2 mm. diameter. The lines used were 4358Å (violet), and 5461Å (green), isolated by Wratten light filters nos. 47 and 61. In the earlier tests a microscope objective was used to produce an image of this aperture, about 40 times reduced, i.e. with a nominal diameter of about 5μ, which formed the 'point source'. The objects were mostly microphotographs, sandwiched with immersion oil between two polished glass plates. In the earlier experiments the distance between the point source and the object was about 50 mm., the distance from the object to the photographic plate 550 mm., thus the geometrical magnification was about 12.

The photographic plate was held in position against three locating pins. Originally it was planned to develop the holograms by reversal, to make sure of exactly identical positions in the analysis and in the synthesis. In the negative-positive process the printing was carried out on the same locating pins. These precautions proved unnecessary in those experiments in which not only the Gaussian but also the physical diameter of the source was of the order of the resolution limit, which proves that in these cases the theory of homocentric illuminating beams is a satisfactory approximation. But they were required later, in experiments with very strong spherical aberration in the illuminating beam. Reversal development, however, was found unnecessary, and the far more flexible negative-positive photographic process was used throughout. The negative hologram was usually processed with $\Gamma = 1.2$ to 1.6, and the positive with $\Gamma = 0.7$ to 1.6, so that a wide range of overall gammas could be tested. When it was confirmed that an overall gamma of 2 gave the best results, this was realized as closely as possible.

In the reconstruction the positive hologram was sandwiched with immersion oil between polished glass plates, which had to be carefully selected. It was backed by a viewing lens, which was an achromatic doublet, cemented and bloomed, with a focal length of 175 mm. and a linear aperture of 47 mm. The spherical aberration was 3 fringes at infinite conjugates. The diameter which satisfies the quarter-wave tolerance can be estimated at 27 mm., and the numerical aperture figures given below are based on this 'effective diameter'. The reconstructed image was viewed in a microscope, and photographed on plates introduced into the eyepiece.

Figure 10, plate 15, is a record of one of these earlier experiments. The figure at the left is a direct photograph of the original, which was a microphotograph of the names of the three founders of the wave theory of light. It was taken through the viewing system, with the same optics as used for the reconstruction. The top figure is the central part of the hologram, and the one at the right, is the reconstruction. All three were taken with the violet mercury line 4358Å. The effective numerical aperture was 0.025, thus the resolution limit $0.6 \times 0.436/0.025 = 10\,\mu$. This is $\frac{1}{150}$ of the diameter of the reproduced part of the microphotographs, and corresponds about to the gap between the 'Y' and the 'G' in 'HUYGENS'.

Though in its best parts the reconstruction almost attains the resolution of the direct photograph, the picture is very 'noisy'. This is due only to a smaller part to the essential disturbance created by the twin image, to a greater part it is due to specks of dust, and inhomogeneities in the two microscope objectives. It may be noted that these very troublesome effects, unwelcome concomitants of the great phase-discriminating power of the methods using a coherent background, cannot be expected to appear in an electronic analyzer. However imperfect an electron lens may be from the point of view of theoretical optics, it can contain neither dust nor 'schlieren', as the electromagnetic field smoothes itself out automatically, and in this respect any electron lens is superior to all but the best optical lenses.

In order to avoid these inessential disturbances, in some later experiments the optical surfaces were reduced to a minimum. In the experiments of which figures 11 and 12, plates 16 and 17, are records, the source was a pinhole of $3\,\mu$ diameter, pierced into tinfoil with a very fine needle. Thus no glass surfaces other than those of the microphotographs were involved in the taking of the hologram. In the reconstruction the optics was also reduced to a minimum by cutting out the second microscope. The spacing between the object and the viewing lens was reduced to 180 mm., the distance between the lens and the plate increased to 700 mm., so that a fourfold enlargement of the object was produced by the viewing lens, sufficient for

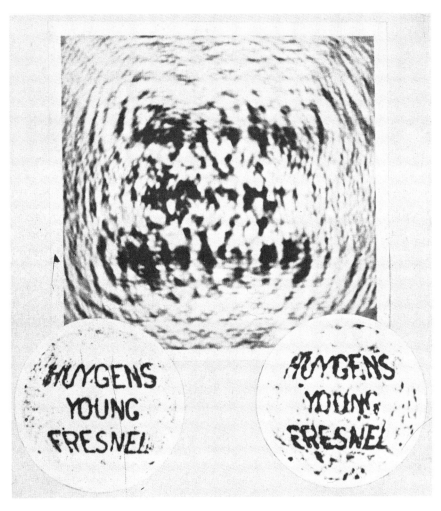

original   hologram   reconstruction

FIGURE 10. Optical reconstruction cycle. The original was a microphotograph of 1·5 mm. diam. Illuminated with $\lambda = 4358$ Å through pinhole 0·2 mm. diam., reduced by a microscope objective to 5μ nominal diameter, at 50 mm. from object. Geometrical magnification 12. Effective aperture of lens used in reconstruction 0·025. Noisy background chiefly due to imperfections of illuminating objective.

hologram

original   reconstruction

FIGURE 11. Reconstruction cycle with pinhole illumination. The letters in the original were inscribed in a rectangle 0.65×0.5 mm. Illumination with $\lambda = 4358$ Å through pinhole of 5μ diam. at 18 mm. from object. Geometrical magnification 10. Effective aperture used in reconstruction 0.075.

AN INTRODUCTION TO COHERENT OPTICS AND HOLOGRAPHY   299

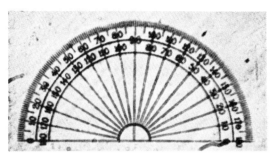

original

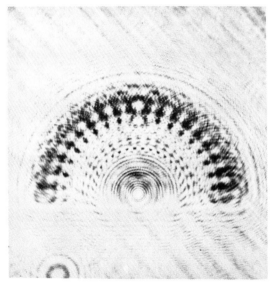

hologram

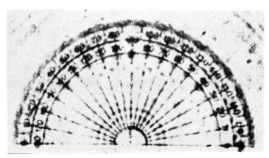

reconstruction

FIGURE 12. Reconstruction with pinhole illumination and wave-length change. The original was a micro-protractor of 1 mm. diam. Same conditions as in figure 11, but the wave-length used in the reconstruction was $\lambda = 5461$ Å.

direct photography on not unduly slow plates. Further enlargement was obtained in some cases by taking the hologram with the violet line, but reconstructing it with the green line..

The effective numerical aperture in this experimental series was 0·075, and the theoretical resolution limit 3·5 μ. This is about $\frac{1}{350}$ of the diameter of the part of the microphotograph which is reproduced in figure 11 and which contains ten great names in the theory of light. The resolution is just about sufficient to resolve the hole in an 'A'. The theoretical resolution of the reconstruction is less, because the pinhole source of 3 μ, used both in the analysis and in the synthesis, is of the same order. It can be estimated at about 5·5 μ, by the thumb rule of orthogonal composition of errors. This resolution has been in fact very nearly achieved in the case of figure 11 and also in figure 12. It can be also seen that the background is very much more even than in figure 10. The residual disturbance is mostly essential, and due to the twin object. In these experiments the twin object could be separately focused, and as regards sharpness could not be distinguished from the 'true' image.

Experiments for testing the theory in the case of illuminating beams with large spherical aberration are in progress, but they have already confirmed its main results.

CONCLUSION

The new principle can be applied in all cases where coherent monochromatic radiation of sufficient intensity is available to produce a divergent diffraction pattern, with a relatively strong coherent background. While the application to electron microscopy promises the direct resolution of structures which are outside the range of ordinary electron microscopes, probably the most interesting feature of the new method for light-optical applications is the possibility of recording in one photograph the data of three-dimensional objects. In the reconstruction one plane after the other can be focused, as if the object were in position, though the disturbing effect of the parts of the object outside the sharply focused plane is stronger in coherent light than in incoherent illumination. But it is very likely that in light optics, where beam splitters are available, methods can be found for providing the coherent background which will allow better separation of object planes, and more effective elimination of the effects of the 'twin wave' than the simple arrangements which have been investigated.

I thank Mr L. J. Davies, Director of Research of the British Thomson-Houston Company, for permission to publish this paper and Mr J. Williams for assistance in the experimental work.

REFERENCES

Baker, B. B. & Copson, E. T. 1939   *The mathematical theory of Huygens' principle*. Oxford: Clarendon Press.
Boersch, H.   1938   *Z. techn. Phys.* **19**, 337.
Boersch, H.   1939   *Z. techn. Phys.* **20**, 346.
Bragg, W. L.   1942   *Nature*, **149**, 470.
Bruck, H.   1947   *C.R. Acad. Sci., Paris*, **224**, 1553.
Bunn, W.   1945   *Chemical crystallography*. Oxford University Press.
Campbell, G. A. & Foster, R. M.   1931   *Fourier integrals for practical applications*. Bell Telephone System Monograph, B. 584. New York.
Gabor, D.   1948   *Nature*, **161**, 777.
Glaser, W.   1943   *Z. Phys.* **121**, 647.
Hillier, J. & Ramberg, E. G.   1947   *J. Appl. Phys.* **18**, 48.
Zernike, F.   1948   *Proc. Phys. Soc.* **61**, 147.

THE PROCEEDINGS OF
THE PHYSICAL SOCIETY

Section B

VOL. 64, PART 6	1 June 1951	NO. 378 B

Microscopy by Reconstructed Wave Fronts: II

By D. GABOR
Department of Electrical Engineering, Imperial College, London

MS. received 11th September 1950, and in amended form 12th December 1950

ABSTRACT. The theory of diffraction microscopy is completed and extended in different directions. In this two-step method of image formation the object is reconstructed by optical means from a diffraction diagram, taken in coherent illumination with light or with electrons. The ' projection method ', originally described, and the ' transmission method ', recently proposed by Haine and Dyson, are two variants which can be treated by one theory. The process of image formation, the coherence requirements, and the conditions for a good reconstruction are discussed in detail. It is shown that the reconstructed image of extended objects suffers from some spurious detail, but this can be largely suppressed in the ' dark-field ' method of reconstruction, in which the illuminating wave is cut out after it has passed through the diffraction diagram.

§ 1. INTRODUCTION

THE principle of diffraction microscopy was described in a previous paper (Gabor 1949, to be referred to as I) with a detailed discussion of one of the schemes for putting it into practice. In this 'projection method' the microscopic object is illuminated by a beam of light or of electrons which issue from a small aperture, either directly or through a lens system without any further lenses between the object and the photographic plate, where it forms a diffraction pattern called a 'hologram'. It was shown that if this photograph, suitably processed, is illuminated by a replica of the original wave, the original object is reconstructed in space, together with a ghost or 'conjugate object', which is in general diffuse and distorted. It has been shown theoretically and experimentally that with objects of a suitable type with sufficiently large clear spaces between the dark parts, the effect of the spurious part of the reconstructed image need not be serious.

Electron microscopy was contemplated from the start as the chief, though not the only application of the method of reconstructed wave fronts. In the course of their experiments, which will be described in a separate paper, Haine and Dyson (1950) became aware of certain considerable practical inconveniences, and proposed a modified optical scheme which they suggested to the author. This is called the 'transmission method', because it operates with an electron microscope of the transmission type, only slightly modified, in the taking of the hologram. While in the projection method the whole electron-optical system is between the small illuminating aperture and the object, in the transmission scheme it is between the object and the photographic plate. This scheme has various advantages, the most obvious being that the object can be inspected and adjusted before taking the hologram by the well-tried methods of ordinary electron microscopy, and must be only defocused to a certain extent for the final exposure. Another equally important advantage is the wide field. In the projection method the illuminating beam had already its final, wide divergence angle, and a correspondingly small cross section in the region where the object could be situated. In the transmission scheme the divergence of the illuminating beam is small, and the field correspondingly large. Moreover, as will be shown later on, these advantages are not obtained at the price of other disadvantages. It was feared originally—and this was the reason why this suggestion was not followed up earlier—that this scheme would require a more complicated optical reconstructing device. But, as will be seen, the apparatus is by no means more complicated than the one required for the projection scheme, moreover it offers the additional advantage that a real image of the illuminating aperture is accessible in it. This makes it possible to cut out the unmodified part of the illuminating wave, the 'background', after it has passed through the hologram, or to shift its phase by a Zernike phase plate, and thus to inspect the reconstructed image by the 'dark-field' or by the phase-contrast method.

The practical modifications which have become necessary naturally suggested reviewing the theory on a somewhat broader basis, and rounding it off by a discussion of operating conditions not dealt with in I.

§ 2. THE PRINCIPLE OF WAVE-FRONT RECONSTRUCTION

Consider a monochromatic, coherent wave with amplitude $U = A \exp(i\psi)$ striking a photographic plate. Assume that the plate is developed by reversal, or printed with an overall 'gamma' Γ†. On the 'straight' branch of the Hurter and Driffield curve its amplitude transmission t will be $t = KA^{\Gamma} = K(UU^*)^{\Gamma/2}$, where K is a constant. Imagine now that U is split into a background or illuminating wave U_0 and a disturbance or secondary wave U_1 due to the object

$$U = U_0 + U_1 = A_0 \exp i\psi_0 + A_1 \exp i\psi_1 = \exp i\psi_0 [A_0 + A_1 \exp \{i(\psi_1 - \psi_0)\}].$$

On the other hand, if we illuminate the photograph hologram with the background wave alone we obtain immediately behind the emulsion a wave $U_s = t U_0$. The simplest, and also the most advantageous, case is $\Gamma = 2$. Assuming this value

$$U_s = t U_0 = KA^2 A_0 \exp i\psi_0 = KA_0^2 \exp i\psi_0$$
$$\times [A_0 + A_1^2/A_0 + A_1 \exp \{i(\psi_1 - \psi_0)\} + A_1 \exp \{-i(\psi_1 - \psi_0)\}]. \quad \ldots\ldots(1)$$

For simplicity assume that the background is uniform in intensity, i.e. $A_0 = $ constant, and compare the last two equations. The factors before the brackets are identical, apart from a constant factor. Inside the bracket, we see that $A_0[1 + (A_1/A_0)^2]$ has been substituted for the background amplitude A_0. If we assume—and this is a very essential condition—that the background is strong relative to the amplitudes scattered by the object, the change is insignificant. The next terms $A_1 \exp i\psi_1$ are identical in the two expressions.

---

† The 'gamma' is, by definition, the slope of the Hurter and Driffield curve, i.e. of the plot of the 'density' (the logarithm of the reciprocal intensity transmission) against the logarithm of the 'exposure' (the product of intensity and time).

This means that the original secondary wave has been reconstructed, with the correct phase, and, neglecting the small term A_1^2/A_0, with the correct relative amplitude. But the last term is new. It represents a wave with the same amplitude as the secondary wave U_1, but with a phase of opposite sign relative to the background.

This is perhaps made clearer by the diagrammatic explanation in Figure 1. If the secondary wave is relatively weak, the resulting wave amplitude is very nearly $\overline{OP}$, which is equal to the background amplitude, plus the in-phase component of the secondary. In the photographic process with $\Gamma = 2$ this amplitude, squared, appears as transmission, as shown in the lower figure. The background amplitude is assumed unity, in order to bring the two diagrams to the same scale. The transmitted amplitude is now very nearly the background amplitude, plus *twice* the in-phase component of the secondary wave. The result can be interpreted as shown, i.e. that the reconstructed wave differs from

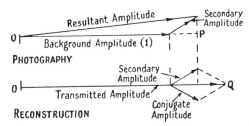

Figure 1.   Vector diagram, explaining the principle of reconstruction in the case $\Gamma = 2$.

the original only in a 'conjugate amplitude', with equal in-phase and opposite in-quadrature component to the secondary. This might appear an arbitrary interpretation since the in-quadrature component has been suppressed, but it will be shown that it corresponds very closely to the way in which the eye will interpret the secondary amplitude: as a wave issuing from a 'conjugate' or 'ghost' object.

In order to show this we can proceed in a somewhat more general way than in I : we still assume that the illuminating wave issues from a point source, but we drop the assumption, made in I, that the photographic plate is at infinity. The passage to illuminating waves other than spherical can be easily made at a later stage.

Consider a point source S, a point object O_1, and a point P of the photographic plate. P receives a direct ray SP, and an indirect ray SO_1P, with a phase difference between them which is proportional to the difference of the lengths of the broken and of the straight line, plus some phase change which the ray may have suffered in the diffraction at the object. Their interference produces some resulting amplitude at P. But exactly the same amplitude would have been produced with a phase difference of opposite sign between the two rays. We can construct two such rays with opposite phase difference by first of all reversing the direction of propagation, i.e. imagining a spherical wave converging on S instead of diverging from it. Let QO_2 be such a converging wave front. If this is so determined that $\overline{SO_1} + \overline{O_1P} - \overline{SP} = \overline{O_2P} - \overline{PQ}$ the rays from O_1 and from O_2 will have opposite phase difference relative to the direct ray. (If O_1 is not a point,

2 F–2

but a small object which produces a certain phase delay, one has to postulate that O_2 produces an equal phase advance.) A simple calculation gives that for small angles α the relation between r_1 and r_2 is

$$1/r_1 + 1/r_2 = 2/R, \qquad \ldots\ldots(2)$$

that is to say the conjugate points O_1 and O_2 are optical conjugates, relative to the spherical wave front through P, considered as a mirror. It may be noted that this is true only in first approximation; for larger values of α, O_2 will depend on α and not only on the radius $R = \overline{SP}$; for constant R it is not a point but an aberration figure, which is not the same as the aberration figure of the spherical mirror. But there are two special cases of particular interest in which the conjugate of a point is a sharp point, without reference to the point P. The first is approached if the object point O_1 is very near to the source 'S relative to its distance from the plate. This is the *projection case* which was discussed in I; the conjugates are symmetrical with respect to the point source. The second case is realized if the object is very near to the plate relative to its distance from the source. It will be shown that this is the case in the *transmission method*: in this case the conjugates are mirror images with respect to the plate.

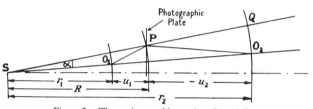

Figure 2. The conjugate object points O_1 and O_2.

One might ask whether some advantage may not be gained by an intermediate arrangement, for instance by placing the object midway between the source and the plate, so that the conjugate is removed to infinity. But nothing is gained by this, because, as may be seen from equation (2), the two conjugate amplitudes at P are always equal. There is no way of getting rid of the conjugate object; it is an unavoidable consequence of the 'phase ignorance' of the photographic plate (cf. Bragg 1950).

Equation (2) can be written, with the notations as in Figure 2,

$$1/u_1 + 1/u_2 = 2/R.$$

This is an optically invariant relation, that is to say if in the reconstruction one images the optical space of the source and the hologram in such a way that the spherical wave from the source to the plate has some other curvature, the two conjugates remain mirror images relative to the transformed spherical wave front.

In equation (1) it was assumed that the wave used in the reconstruction has the same phase differences between different points of the photographic plate as the one used in the taking. We can now drop this assumption, and show that, at least in this approximation, the substitution of one illuminating wave for another has the same effect as an optical transformation. Assume, as illustrated in Figure 3, that the plate was originally illuminated by S' from a distance R', while in the reconstruction this is changed to R''. Imagine now that we place before the hologram a thin negative lens, with a power $-1/F = 1/R' - 1/R''$ and a positive lens with equal and opposite sign behind it. Thus the total effect is *nil*, but as the negative lens has restored the original curvature of the wave front, it is seen immediately that in the optical space between the two lenses the conjugates have their original positions, O_1' and O_2'. Hence the substitution of the wave front has the same effect as the positive lens behind the hologram, and we obtain for instance for the first point the transformation equation

$$1/u_1' - 1/u_1'' = 1/F = 1/R' - 1/R''.$$

In particular, if the new illuminating wave is plane, the hologram acts like a lens with power $1/R'$ in transforming the original position of the object. This, with some other interesting optical properties of holograms has been recently pointed out by Rogers (1950).

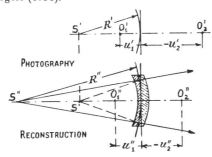

Figure 3.   Substituting a different illuminating wave in the reconstruction has the same effect as an optical transformation.

If the illuminating wave is not spherical the conjugate object is no longer sharp, but marred by *twice* the number of aberrations corresponding to the deviations of the wave front from a sphere. This is illustrated in Figure 4 (Plate*). For the theory we must refer to I.

§ 3. THEORY OF IMAGE FORMATION IN THE TRANSMISSION METHOD

Figure 5 illustrates the optics used in the 'projection' and 'transmission' methods. In both cases three electron lenses are required to produce the necessary high demagnification of the pinhole in the first scheme, and the high magnification of the object in the second. It can be seen immediately that the transmission method is a sort of virtual projection method, if the last optical space is considered, which contains the photographic plate, in the H-plane. But it will be shown that it is simpler to consider instead the first optical space, which contains the pinhole and the object. This leads in a much more straightforward way to a simple understanding of the reconstructing process, and it will also show the essential identity of the two methods, which we have emphasized in advance by choosing the same symbol z_0 for the source–object distance in the first scheme, and for the defocusing distance in the second.

The notations are explained in Figure 6. H_0 is the gaussian conjugate of the H-plane, i.e. of the photographic plate; it is a small distance z_0 off the object plane, and will be reckoned positive as illustrated, i.e. with objective 'overfocused'.

* For Plates see end of issue.

454   D. Gabor

The calculation consists of two essential steps. In the first we calculate the amplitudes in the H_0-plane, which we call the 'virtual hologram'. In the second we take into account the aberrations of the optical system and calculate the actual hologram. The first may be called $U^0(x_0, y_0)$, the second $U(X, Y)$.

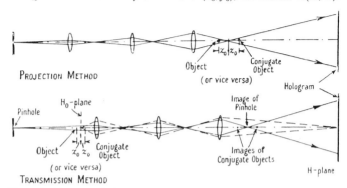

Figure 5. Electron-optical system in the two methods of producing the hologram.

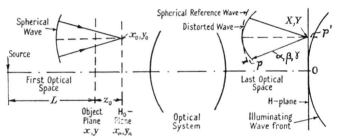

Figure 6. Notations for the theory of the transmission method. The z-axis is the optic axis. α, β, γ are in turn the angles of the normals of the plane waves with the x, y, z axes. In the geometrical approximation they become rays.

To calculate the first we use Kirchhoff's principle in the simplified form valid for small off-axis angles

$$U^0(x_0, y_0) = \frac{1}{z_0} \iint U_0^0(x, y) t(x, y) \exp(ikr_1) \, dx \, dy. \qquad \ldots \ldots (3)$$

Here $U_0^0(x, y)$ is the illuminating amplitude in the object plane, $t(x, y)$ is the amplitude transmission coefficient of the object, in general a complex quantity, $k = 2\pi/\lambda$ is the propagation constant, and r_1 is the distance of the point of reference x_0, y_0, z_0 in the H_0-plane from a point $x, y, 0$ in the object plane. We expand

$$r_1 = z_0 + \tfrac{1}{2}[(x - x_0)^2 + (y - y_0)^2]/z_0 - \tfrac{1}{8}[(x - x_0)^2 + (y - y_0)^2]^2/z_0^3 + \ldots.$$

It can be verified *a posteriori* that the defocusing distances z_0 suitable for practical use are large enough to justify the application of Kirchhoff's formula in the simplified form (3), moreover the third term in the expansion of r_1 can be dropped. We are *a fortiori* entitled to drop the corresponding term in the phase of the illuminating wave U_0^0, which is, apart from a constant factor,

$$U_0^0 = \exp \tfrac{1}{2} ik(x^2 + y^2)/L.$$

We thus obtain for the amplitude in the virtual hologram

$$U^0(x_0, y_0) = \exp\left(\frac{ik}{2z_0}r_0{}^2\right)\int\int t(x,y)\exp\left[\tfrac{1}{2}ik\left(\frac{1}{z_0}+\frac{1}{L}\right)r^2 - i\frac{k}{z_0}(xx_0+yy_0)\right]dx\,dy,$$

$$\ldots\ldots(4)$$

where we have used the abbreviations $x^2 + y^2 = r^2$ and $x_0{}^2 + y_0{}^2 = r_0{}^2$.

This virtual hologram is imaged on the photographic plate through the optical system, with certain ray limitations, and with certain geometrical aberrations. These can be described, as is well known, by specifying the phase distortion of a wave which is spherical in the first optical space and converges on a point x_0, y_0 of the H_0-plane.

To simplify the formulae we assume the magnification to be unity. This is no restriction, as ultimately it will be convenient to refer all optical data back to the object space. A plane wavelet, which is a component of the distorted wave in the last optical space (corresponding to the spherical wave in the first space which converges on x_0, y_0), will arrive at a point X, Y with a phase advance

$$(x_0 - X)\cos\alpha + (y_0 - Y)\cos\beta + p(\alpha, \beta, x_0, y_0) + p'(X, Y).$$

Here we have split the phase distortion into two components, illustrated in Figure 6; p is the distance, measured in radial direction, between the distorted wave and a spherical wave of reference which coincides with it on the radius drawn parallel to the axis. This is a function of the angular variables α, β and of the point x_0, y_0. The second term p' is the advance of the wave as a whole. This is a function of X, Y only, and represents the distortion of a wave which was plane and normal to the axis in the first space. But as it depends on X, Y only, it adds only a phase factor $\exp(ikp')$ to the amplitude, which has no effect on the photographic density, and hence on the whole process. Therefore we can drop p' from the start in all the following formulae.

We neglect image distortion, field curvature and third-order astigmatism, but take into account first-order astigmatism, spherical aberration and coma. It is well known that astigmatism on the axis, arising from ellipticity or misalignment of the electrodes, is a very important error in electron lenses, and in view of this it would not be justified to assume the 'spherical' aberration to be rotationally symmetrical. We only assume for simplicity that its principal axes coincide with the astigmatic axes, and write

$$p = \tfrac{1}{2}A_s(\cos^2\alpha - \cos^2\beta) + \tfrac{1}{4}(C_x\cos^4\alpha + 2C_{xy}\cos^2\alpha\cos^2\beta + C_y\cos^4\beta)$$

$$+ B(x_0\cos\alpha + y_0\cos\beta)\sin^2\gamma.$$

A_s is the axial separation of the two astigmatic foci, C_x, C_y and C_{xy}, which have the dimension of a length, are the constants of the aperture error; in the case of rotational symmetry $C_x = C_y = C_{xy} = C_s$ is the constant of the spherical aberration. B is the coma coefficient.

The effect of the aperture in the optical system may be represented by a transmission factor $\exp\{-(\gamma/\gamma_m)^2\}$. Such a 'gaussian' cut-off is analytically very convenient, as has been shown in I. It will be also convenient, as in I, to replace the angular variables α, β, γ by 'Fourier variables' $\xi = (\cos\alpha)/\lambda$, $\eta = (\cos\beta)/\lambda$, $\rho = (\xi^2 + \eta^2)^{1/2} = (\sin\gamma)/\lambda$.

With these conventions and notations, applying again Kirchhoff's formula in the simplified form, we obtain for the amplitude in the H-plane

$$U(X, Y) = \int\int U^0(x_0, y_0) \, dx_0 \, dy_0$$

$$\times \int\int \exp\left\{-(\rho/\rho_m)^2 - 2\pi i[(x_0 - X)\xi + (y_0 - Y)\eta + p]\right\} d\xi \, d\eta.$$

It is easy to check by Fourier's formula that, if there is no distortion, i.e. $p = 0$, and if there is no ray-limitation, this transformation restores U^0, that is, in this case $U(X, Y) = U^0(X, Y)$.

We now substitute U^0 from equation (4) and obtain the amplitude in the hologram in the form of a sixfold integral

$$U(X, Y) = \int\int\int\int\int t(x, y) \exp\left\{\frac{\pi i}{\lambda}\left[\left(\frac{1}{z_0} + \frac{1}{L}\right)(x^2 + y^2)\right.\right.$$

$$+ \frac{1}{z_0}[x_0^2 + y_0^2 - 2(xx_0 + yy_0)]\right]\right\} \exp\left\{-2\pi i[(x_0 - X)\xi + (y_0 - Y)\eta\right.$$

$$+ \tfrac{1}{2}A_s\lambda(\xi^2 - \eta^2) + \tfrac{1}{4}\lambda^3(C_x\xi^4 + 2C_{xy}\xi^2\eta^2 + C_y\eta^4)$$

$$+ B\lambda^2(x_0\xi + y_0\eta)\rho^2]\right\} \exp\left[-(\rho/\rho_m)^2\right] dx \, dy \, dx_0 \, dy_0 \, d\xi \, d\eta. \quad \ldots\ldots(5)$$

But this can be immediately reduced to a fourfold integral, because x_0 and y_0 occur in the exponentials only in the first and second order, and this part of the integral can be readily calculated by the formula

$$\int\int \exp\frac{i\pi}{\lambda z_0}(x_0^2 + y_0^2) \exp\left\{-2\pi i\left[x_0\left(\frac{x}{\lambda z_0} + \xi(1 + B\lambda^2\rho^2)\right)\right.\right.$$

$$+ y_0\left(\frac{y}{\lambda z_0} + \eta(1 + B\lambda^2\rho^2)\right)\right]\right\} dx_0 \, dy_0$$

$$= i\lambda z_0 \exp\left[-\pi i\lambda z_0\left\{\left(\frac{x}{\lambda z_0} + \xi(1 + B\lambda^2\rho^2)\right)^2 + \left(\frac{y}{\lambda z_0} + \eta(1 + B\lambda^2\rho^2)\right)^2\right\}\right].$$

$$\ldots\ldots(6)$$

Hence

$$U(X, Y) = \int\int\int\int t(x, y) \exp\left[\pi i\left(\frac{r^2}{\lambda L} - \lambda z_0\rho^2\right)\right] \exp\left[-\left(\frac{\rho}{\rho_m}\right)^2\right]$$

$$\times \exp\left\{-2\pi i[(x - X)\xi + (y - Y)\eta]\right\} \exp\left\{-2\pi i[\tfrac{1}{2}A_s\lambda(\xi^2 - \eta^2)\right.$$

$$+ \lambda^3[\tfrac{1}{4}(C_x\xi^4 + 2C_{xy}\xi^2\eta^2 + C_y\eta^4) - Bz_0\rho^4] + B\lambda^2(x\xi + y\eta)\rho^2]\right\} dx \, dy \, d\xi \, d\eta.$$

$$\ldots\ldots(7)$$

This is the simplest form in which the exact * solution can be put. It gives the amplitude in an image, focused or defocused, in any microscope, ordinary or electronic, with a gaussian aperture, up to errors of the third order, for stigmatic, coherent illumination of the object from a distance L.

The interpretation is simple if we take the factors one by one. Comparing with (5) the coma now appears transferred to the object plane, i.e. x, y appear in it instead of x_0, y_0. The spherical aberration has suffered a change, which in

---

* It is exact, strictly speaking, only in so far as Kirchhoff's formula can be considered as exact. But this approximation is valid in electron optics with an accuracy almost unknown in light optics, as the wavelength of fast electrons is small compared with the dimensions of any material object except atomic nuclei.

the case of rotational symmetry is equivalent to a decrease of C_s by $4Bz_0$. This will be seen later to be absolutely negligible. Thus the phase distortion due to aberrations, in the last factor of equation (7) is appreciably the same as if the system were focused on the object. The only appreciable effect of the defocusing appears in the factor $\exp(-\pi i \lambda z_0 \rho^2)$ which may be called the 'defocusing factor'.

Instead of operating with the amplitude U, it is advantageous, as in I, to operate with a 'shadow object' which, placed in the hologram plane and illuminated with the background U_0, would produce the amplitude U immediately behind it. This 'shadow object' has the *complex* transmission coefficient $\tau(X, Y) = U(X, Y)/U_0(X, Y)$, which differs from the hologram in the imaginary part only. The background amplitude U_0 can be calculated by the method of stationary phase, and is

$$U_0(X, Y) = \frac{L}{L+z_0} \exp\left\{2\pi i \left[\frac{X^2 - Y^2}{2\lambda(L+z_0)} - \tfrac{1}{2}A_s \frac{X^2 - Y^2}{\lambda(L+z_0)^2}\right.\right.$$
$$\left.\left. - \frac{\tfrac{1}{4}(C_x X^4 + 2C_{xy}X^2 Y^2 + C_y Y^4) - B(L+z_0)R^4}{\lambda(L+z_0)^4}\right]\right\}, \qquad \ldots\ldots (8)$$

where we have written $X^2 + Y^2 = R^2$. The terms of the exponential after the first are found to be absolutely negligible, and even the first is small. In practice one can consider the illumination in the transmission method as plane, but we will retain the first term, for some later conclusions.

A considerable further simplification of equation (7) is possible only in the case when the defocusing is large compared with the astigmatic separation A_s and with the length of the caustic in the spherical aberration figure, i.e. if $z_0 \gg A_s$ and $z_0 \gg C_s \gamma_m^2$. In this case a somewhat lengthy calculation gives the formula for the transmission of the 'shadow object'

$$\tau(X, Y) = \frac{1}{i\lambda z_0} \exp\left(\frac{\pi i}{\lambda z_0}R^2\right) \iint t(x, y) \exp\left(\frac{\pi i}{\lambda z}r^2 - \frac{W^2}{(z_0 \gamma_m)^2}\right)$$
$$\times \exp\left[-\frac{2\pi i}{\lambda z_0}(Xx + Yy)\right] \exp\left\{-2\pi i \left[\tfrac{1}{2}A_s \frac{(X-x)^2 - (Y-y)^2}{\lambda z_0^2}\right.\right.$$
$$\left.\left. + \tfrac{1}{4}C_s \frac{W^4}{\lambda z_0^4} + B \frac{[X(X-x) + Y(Y-y)]W^2}{\lambda z_0^3}\right]\right\} dx\, dy, \qquad \ldots\ldots (9)$$

where we have introduced the abbreviation $W^2 = (X-x)^2 + (Y-y)^2$. The passage from (7) to (9) can be expressed by the simple rule: Replace every pencil of wave normals converging in X, Y by a single *ray* from x_0, y_0 to X, Y. It represents therefore the geometrical approximation to the solution. A lengthy investigation, which may be omitted, shows that this approximation is in fact valid almost up to the point when z_0 approaches the tip of the caustic. A further approximation, in which the 'ray' is drawn not simply through x_0, y_0, but is determined by the condition of stationary phase, is valid practically without restriction in diffraction microscopy where the geometrical errors in the caustic region strongly outweigh the diffraction spread of the wave. This however requires the solution of a cubic equation for every ray, and the explicit solution is too complicated to be discussed with advantage.

It may be noted that if the geometrical errors are neglected, equation (9) becomes identical with the equation (14) of I which gives the 'shadow transform'

D. Gabor

of an object, illuminated by a point source from a distance z_0, with the plate at infinity, apart from the ray-limiting factor $\exp\left[-W^2/(z_0\gamma_m)^2\right]$. But it is of course just the ray limitation which constitutes the chief difference between the projection method and the transmission method. In the projection method a point of the photographic plate receives radiation from *all* the object, if there is sufficiently fine detail present, while in the transmission method it is, broadly speaking, only a disc of radius $z_0\gamma_m$ in the object which contributes to the amplitude at one point of the hologram. Hence, in spite of the formal similarity, which we have emphasized by using the same symbol z_0 in its two different meanings in both papers, there is a very essential difference in the actual holograms.

For a general discussion we can as well start from the exact formula (7) as from the approximation (9). For simplicity put $L = \infty$, i.e. assume parallel illumination. In this case the background U_0, given by equation (8), is reduced to unity, and $U(X, Y)$ gives the transmission of the shadow object as well as the amplitude at the hologram. The transmission of the photograph, assuming correct processing, as explained in § 1, is proportional to the real part of τ, hence, in this case, of U. But it can be seen on inspection of equation (7), with $L = \infty$, that taking the real part is equivalent to *two* objects, with transmissions $t(x, y)$ and $t^*(x, y)$ at $\pm z_0$ from the plane H_0, and reversing the sign of the aberrations, A_s, C_r, etc. for the second. This is illustrated in Figure 7, which shows the

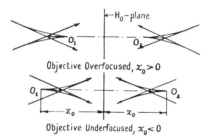

Figure 7. The conjugate objects in the transmission method.

position of the spherical aberration caustics of a point object and its conjugate. After the reconstruction one of the objects will be sharp—it does not matter which of the two—and the other will be affected by twice the aberrations. This is a point to which we will return later.

It may be instructive to look at the relation between object and hologram from a different point of view. Note that in equation (7), which gives the amplitude in the hologram, the coordinates x, y of the object plane occur only in an integral

$$\iint t(x,y)\exp\frac{\pi i r^2}{\lambda L}\exp\left[-2\pi i(x\xi + y\eta)(1 + B\lambda^2\rho^2)\right]dx\,dy. \qquad \ldots\ldots(10)$$

For simplicity put again $L = \infty$, parallel illumination, and replace ξ, η by the new variables

$$\xi' = \xi(1 + B\lambda^2\rho^2); \qquad \eta' = \eta(1 + B\lambda^2\rho^2). \qquad \ldots\ldots(11)$$

The expression (10) now appears in the form

$$\iint t(x,y)\exp\left[-2\pi i(x\xi' + y\eta')\right]dx\,dy,$$

which is the Fourier transform of $t(x, y)$ in the new variables ξ', η', in the standard notation of Campbell and Foster (1931). Writing Ft for the Fourier transform, we have thus for (10) the expression $Ft[\xi(1 + B\lambda^2\rho^2), \eta(1 + B\lambda^2\rho^2)]$. One can say that the coma has produced a certain distortion of the transform in Fourier space. Substituting this into (7), where for simplicity we write the aperture error in the rotationally symmetrical form, we obtain

$$U(X, Y) = \iint Ft[\xi(1 + B\lambda^2\rho^2), \eta(1 + B\lambda^2\rho^2)] \exp\{-2\pi i[\tfrac{1}{2}\lambda z_0\rho^2$$
$$+ \tfrac{1}{2}A_s\lambda(\xi^2 - \eta^2) + \tfrac{1}{4}C_s\lambda^3\rho^4]\} \exp[-(\rho/\rho_m)^2] \exp[2\pi i(X\xi + Y\eta)]\, d\xi\, d\eta.$$

This again is a Fourier integral in the standard form (a 'left-transform' if (10) is called a 'right-transform'). Inverting it, we obtain the Fourier transform of U, to be written FU,

$$FU(\xi, \eta) = Ft[\xi(1 + B\lambda^2\rho^2), \eta(1 + B\lambda^2\rho^2)] \exp[-(\rho/\rho_m)^2] \exp[-2\pi i(\tfrac{1}{2}\lambda z_0\rho^2 + P)]$$
$$\dots\dots(12)$$

where P is the phase advance due to astigmatism and spherical aberration only; the coma is contained in the factor Ft. The result means that the Fourier transform of the amplitude at the hologram is obtained by simple multiplication of the (coma-distorted) Fourier transform of the object. In other words, a simply periodic component of the object transmission of density $\exp 2\pi i(x\xi' + y\eta')$ will be transformed in the H-plane into another simply periodic component $\exp[2\pi i(X\xi + Y\eta)] \exp[-(\rho/\rho_m)^2 - 2\pi i(\tfrac{1}{2}\lambda z_0\rho^2 + P)]$, where the connection between ξ, η and ξ', η' is given by (11). Provided that there are no aberrations other than those here considered, this is true in every plane, whether the image is focused or not. Defocusing merely shifts the phases of the different components by the factor $\exp(-\pi i\lambda z_0\rho^2) \simeq \exp[-\pi i\lambda z_0\rho'^2(1 - 2B\lambda^2\rho'^2)]$, where in the second expression we have made use of the fact that in all practical cases $B\lambda^2\rho^2 = B\gamma^2 \ll 1$.

In spite of its formal simplicity this Fourier interpretation of an optical transformation is in general only of limited usefulness—though Duffieux (1950) has given many interesting examples to the contrary—because the intensity has to be calculated at every point by summing the amplitudes of the Fourier components and taking their absolute square. But this is different in diffraction microscopy if it is properly carried out, that is, if the amplitude of the uniform background is large compared with all other amplitudes. A uniform background $U_0{}^0(x, y) = 1$ is by (12) transformed into $U_0(X, Y) = 1$, also uniform, and the leading term in the intensity is simply the sum of the real parts of the Fourier components. Thus we obtain a new view of diffraction microscopy, as reconstruction from the *real part* of the periodic components of the amplitude in the hologram. In a special case this has been already indicated in I.

§4. THE RECONSTRUCTION

We have already seen in §1 that it is not necessary to illuminate the hologram in the reconstruction with an exact replica of the original wave. This is of great importance in the transmission method. As we have seen in the last section, the hologram can be considered as obtained with the background, such as it was, in the first optical space; only the geometrical and ray-limitation errors of the optical system need be taken into consideration, but not the phase distortion p' of the background. As the illumination in the first optical space is substantially parallel, we can illuminate the hologram in the reconstruction by a plane wave. But we must of course correct the first-order astigmatism, spherical aberration and the coma. It may be noted that because of the mirror symmetry of the conjugates in the first optical space we can use a correcting system with errors of the same or opposite sign to that of the original errors, depending on which conjugate one wants to correct.

So far we have assumed the magnification as unity. We now give it the value M, which can be obtained electronically, in the microscope, or by optical magnification of the hologram. Let us give, as in I, one prime to the parameters used in the taking, e.g. λ' for the de Broglie wavelength, two primes to those in the reconstruction, e.g. λ'' for the wavelength of light. We can now obtain all parameters of the reconstruction by inspection of equation (7), or of the simpler equation (9), by postulating that all phases, measured in periods or in fringes, must be the same in the reconstruction as in the taking. We thus obtain for instance from the first factor before the integral sign in (9), the condition $R^2/\lambda' z_0' = M^2 R^2/\lambda'' z_0''$ or $z_0'' = M^2 z_0' \lambda'/\lambda''$.

It is convenient to introduce the 'excess magnification' m by $m = M\lambda'/\lambda''$. This indicates by how much the actual magnification exceeds the ratio of the light and electron wavelengths. In practical applications this will be probably of the order unity, as λ''/λ' is about 100,000 and this order of magnification is required for high-resolution photographs. With this notation we obtain, by inspection of (7) or (9)*,

$$z_0'' = m^2 z_0' \lambda''/\lambda' \; ; \quad A_s'' = m^2 A_s' \lambda''/\lambda' \; ; \quad C_s'' = m^4 C_s' \lambda''/\lambda' \; ; \quad B'' = m^2 B'. \quad \ldots\ldots (13)$$

The spherical aberration constant of electron microscope objectives C_s' is of the order of a few centimetres, hence with m unity we obtain a value of C_s'' of the order of several kilometres. But it is not of course necessary to carry out the correction in the optical space of the hologram. If one introduces, behind the hologram, a demagnifying optical system, in which m is for instance $1/10$, the spherical aberration constant in this space comes down to reasonable dimensions.

Thus the reconstruction apparatus will have to include a demagnifier, at least if M exceeds 10,000. Moreover it is advantageous to make this demagnification variable in certain limits. As the spherical aberration required for correction varies with the fourth power of the magnification, a variation of m in the ratio $1:2$ is sufficient to cover a range of C_s' from 1 to 16.

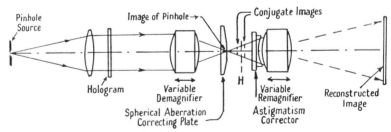

Figure 8. Optical reconstructing device.

Figure 8 is a schematic illustration of a reconstructing device on these lines. The hologram is illuminated by a parallel beam. This is brought to a point focus by means of a variable demagnifying system. The accessibility of this

* These formulae have been obtained also in I, but note that z_0 had of course a different meaning in the projection method.

point focus is an important advantage for several reasons. One is that, as Dyson has pointed out in connection with reconstruction devices for projection holograms, this focus is the proper place for a fourth-order correcting plate; elsewhere the plate would introduce strong coma, which would have to be corrected. The other is that in this scheme it is possible to cover up the point focus, preferably with a dark spot graded according to a probability law, and thus use the dark-field method, which will be discussed at the end of this paper. Coma correction can be effected by shifting the fourth-order plate slightly in axial direction.

In the optical space after the demagnifier the conjugate images appear approximately at equal distances from the image H of the hologram. The astigmatism can be corrected by two cylindrical lenses of opposite sign, each of which can be rotated. These are followed by a remagnifier, also variable, which produces the final image on a photographic plate.

Instead of using the hologram, processed with $\Gamma = 2$ in transmission, one could also process it with $\Gamma = 1$ and back it by a mirror, that is use it in reflection. This allows the first lens to be used both for collimation and for demagnification, but the advantages of this scheme are somewhat doubtful, especially in view of the loss of light introduced by a beam-splitting device.

Summing up, the reconstructing device in the transmission method is by no means more complicated than in the projection scheme, and it has the added advantage of an accessible real image of the light source, which could have been realized in the projection scheme only by the introduction of almost prohibitive complications.

§ 5. COHERENCE CRITERIA

In the theory we have so far assumed a monochromatic point source which emits absolutely coherent radiation. The effects of replacing this by a real, small source have been discussed on general lines, and in the special case of the projection method in I. It will be necessary to apply the theory to the transmission method, and also to discuss the question, most important practically, of intensity and exposure.

One can distinguish three coherence criteria, which may be characterized as 'transverse', 'longitudinal' and 'chromatic'. The first concerns the admissible diameter of the pinhole source, the second the constancy of position of the H_0-plane, and the third the spectral width of the radiation.

Transverse Coherence

We can obtain the criterion for the admissible source diameter from equation (7), which gives for the amplitude in the hologram plane due to an infinitesimal area of the object

$$u(X, Y)dx\,dy = t(x,y)\exp\left(\frac{\pi i}{\lambda L}r^2\right)dx\,dy$$

$$\times \int\int \exp\left(-\pi i\lambda z_0\rho^2\right)\exp\left\{-2\pi i[(x-X)\xi+(y-Y)\eta+p]\right\}d\xi\,d\eta. \quad \ldots\ldots(14)$$

The background amplitude is, on the other hand, by (8)

$$U_0(X, Y) = \exp[2\pi i(R^2/2\lambda L + \ldots)],$$

where we have quoted only the relevant part, the rest being entirely negligible. For satisfactory coherence we require that the phase of u does not change by more than half a fringe, that is by π, if the illuminating spot moves in a circle of diameter d_s. Instead of moving the source we can as well move the object plane and the H_0-plane by equal amounts, that is we add equal amounts to x and X, or to y and Y. It is seen that the expression under the integral sign in (14) depends only on $x - X$ and on $y - Y$, and hence remains unchanged. (The coma makes a negligible difference, as may be seen if equation (9) is used instead of (7).) Thus the phase difference between u and U_0 which arises from the moving of the source is $(\pi/\lambda L)(r^2 - R^2) = (\pi/\lambda L)[(x - X)(x + X) + (y - Y)(y + Y)]$. The worst case is if x, y and X, Y are in line with the axis $0, 0$ and the movement of the spot, of value d_s is also in the same line. In this case we obtain the criterion

$$(2d_s/\lambda L)|r - R| \leqslant 1. \qquad \ldots\ldots(15)$$

Now $|r - R|$ must be interpreted as the radius of the circle in the H_0-plane in which the information about the point-object is contained, down to the detail d_A, where $d_A = \lambda/2\gamma_m$ is the Abbe resolution limit. If the defocusing distance z_0 is large compared with A_s and the length of the spherical aberration caustic $C_s\gamma_m^2$, the information can be considered as contained in a cone of angle γ_m, so that $|r - R| = z_0\gamma_m$. This, substituted into (15), gives the coherence criterion

$$d_s \leqslant \frac{L}{z_0}\frac{\lambda}{2\gamma_m} = \frac{L}{z_0}d_A \quad (z_0 \ll A_s, C_s\gamma_m^2). \qquad \ldots\ldots(16)$$

If however z_0 is small we must take into consideration that the information is contained in a region which is essentially that of the geometrical aberration figure, because under the conditions of diffraction microscopy the geometrical errors can be considered as large compared with the diffraction spread on the beam. Considering for simplicity only spherical aberration, the radius of this region is $C_s\gamma_m^3$ in the gaussian plane, and $\frac{1}{4}C_s\gamma_m^3$ in the plane of minimum confusion. Substituting this into (15) we obtain a new criterion, valid for small $z_0 - s$, which it is not necessary to write down explicitly. It is sufficient to remark that from the point of view of transverse coherence the effective defocusing distance z_0 in equation (16) can never be less than one-quarter of the length of the caustic, i.e. $z_0 \geqslant \frac{1}{4}C_s\gamma_m^2$. This may also be stated in the form that the pinhole of diameter d_s, seen from the object, must never subtend an angle larger than the wavelength divided by the diameter of minimum confusion.

Longitudinal Coherence

One of the most important practical difficulties in diffraction microscopy, or, for that matter, of any method of improving the resolving power of electron microscopes, is the required high constancy of focus. It may be remembered that electron microscopy operates with focal lengths of the same order as optical microscopy, a few millimetres, while achieving resolutions about 100–200 times better. Moreover electron lenses are not as stable as glass lenses, but are subject to fluctuations, and they are not achromatic. Hence magnetic electron microscopy has become possible only by very careful stabilization of the lens currents, to about one part in 20,000. This is not so critical in 'unipotential' electrostatic microscopes, where the focal length remains fixed. Even this was possible only through the relatively very large focal depth of electron objectives, due to the small aperture angles. But any further progress sharpens the requirements regarding stability, as the focal depth decreases with the square of the aperture angle.

The limits for the variation of z_0, that is of the gaussian plane which is focused, can be immediately seen from equation (14). This contains z_0 only in the factor $\exp(-\pi i \lambda z_0 \rho^2)$ under the integral sign. The effect of the variation of z_0 by Δz is greatest when ρ is greatest. If z_0 itself is sufficiently large, the maximum effective value of ρ is $\rho_m = \gamma_m/\lambda$. Postulating that Δz shall cause a phase shift by less than half a fringe we obtain the *sufficient* criterion

$$\Delta z \leqslant \pm \tfrac{1}{2}\lambda/\gamma_m{}^2 = d_A/\gamma_m, \qquad \dots\dots(17)$$

which means simply that the variations must remain inside the focal depth. This criterion is sufficient *and* necessary for large $z_0 - s$. There is no need to investigate the somewhat less stringent necessary conditions for small $z_0 - s$, where the object merges into the caustic, because, as will be seen later, such a setting has considerable practical disadvantages.

The practical consequences of the criterion (17) will be discussed by Haine in a separate publication, with particular reference to the magnetic electron microscope. But it may be noted that even with absolute stabilization of the lenses, or with unipotential electrostatic systems, a limit will be reached, at about 1 to 2 A. resolution, where the energy spread of the electrons will bar further progress, unless achromatic lenses are used. The possibilities of achromatic electron lenses have been discussed by the author in a separate paper (Gabor 1951).

Chromatic Coherence

The criterion for chromatic coherence can also be obtained from equation (14), in which the most important λ-dependent factor is again, if z_0 is not too small, $\exp(-\pi i \lambda z_0 \rho^2)$. Applying the same criterion as before, we obtain for the largest admissible relative change of wavelength $|\Delta\lambda/\lambda| \leqslant \lambda/z_0\gamma_m{}^2 = 2d_A/z_0\gamma_m$. Even with the best resolutions contemplated, and the largest $z_0 - s$ which may be compatible with exposures of reasonable length, this gives values well inside the limits to which present-day electron microscope supplies are stabilized. There is therefore no need to take this criterion into consideration, except if achromatized unipotential electrostatic lenses are used, so that the criterion (17), which otherwise overshadows it, is eliminated.

The Coherent Current

Consider the scheme of illumination, illustrated in Figure 9. The information on a very small object is contained, in the H_0-plane, in a disc of radius $z_0\gamma_m$. In order to operate the method correctly, this area must receive a coherent primary wave. This is the condition formulated in equation (16). (In I it is shown that this condition actually corresponds to $72 \cdot 3\%$ coherence.) It will now be shown that the electron current through this disc, the 'coherent current', is entirely limited by the conditions of emission.

We read from Figure 9, combined with equation (16), the relations $\gamma_s = z_0\gamma_m/L = (d_A/d_s)\gamma_m$ and combining this with Abbe's relation we obtain

$$\gamma_s d_s = \gamma_m d_A = \tfrac{1}{2}\lambda. \qquad \dots\dots(18)$$

This means that in coherent beams the Smith–Lagrange invariant has a definite value, equal to half the wavelength. We now apply this to electrons. Calling the velocity V, and the maximum transversal velocity component V_t we have $\gamma_s = V_t/V$. On the other hand, by de Broglie's relation, $\lambda = h/mV$. Substituting this into (18) we obtain $d_s = h/2mV_t$. This is quite generally true for any cross section, if we put in, at the right-hand side, the maximum tranverse velocity in that cross section. The area is $S_c = \frac{1}{4}\pi d_s^2 = \frac{1}{16}\pi(h/mV_t)^2$, where c stands for coherent.

Well-known methods are available in electron optics for the limitation of V_t, hence it might appear that we can increase, for instance, the coherent area of the cathode beyond any limit. But it will now be shown that if the maximum transverse velocity is reduced, the current approaches a definite limiting value. Assume at the cathode surface a Maxwellian distribution with a charge density

$$C \exp(-mv^2/2kT)\,dv_x dv_y dv_z = 2\pi C \exp\{-m(v_n^2 + v_t^2)/2kT\}v_t dv_t dv_n.$$

If V_t is the maximum transverse velocity admitted into the beam, the current through the area S_c is

$$I_c = \frac{\pi}{16}\left(\frac{h}{mV_t}\right)^2 2\pi C \int_0^\infty \exp(-mv_n^2/2kT)v_n dv_n \int_0^{V_t} \exp(-mv_t^2/2kT)v_t dv_t.$$
$$\dots\dots(19)$$

We express this as the product of an 'effective coherently emitting area' S_{eff} with the total emission density of the cathode, that is, we put

$$I_c = S_{eff} 2\pi C \int_0^\infty \exp(-mv_n^2/2kT)v_n dv_n \int_0^\infty \exp(-mv_t^2/2kT)v_t dv_t. \quad\dots\dots(20)$$

From (19) and (20)

$$S_{eff} = \frac{\pi}{16}\left(\frac{h}{mV_t}\right)^2 \frac{\int_0^{V_t}\exp(-mv_t^2/2kT)v_t dv_t}{\int_0^\infty \exp(-mv_t^2/2kTv_t dv_t)} = \frac{\pi}{16}\left(\frac{h}{mV_t}\right)^2\{1-\exp(-mV_t^2/2kT)\}$$
$$\leqslant \pi h^2/32\,mkT,$$

the limit being reached when the cut-off is at very small velocities, so that $mV_t^2/2kT \ll 1$. This means that the maximum coherently emitting area of a thermionic cathode depends on nothing but its temperature.

This area is very small, and so are the coherent currents I_c which can be obtained from it, as shown by the Table for tungsten cathodes.

Temperature T (° K.)	2400	2500	2600	2700	2800	2900	3000
Emission density (amp/cm²)	0·116	0·298	0·717	1·63	3·54	7·31	14·1
Max. area S_{eff} (10^{-14} cm²)	1·42	1·36	1·31	1·26	1·21	1·17	1·14
Coherent current I_c (10^{-14} amp.)	0·165	0·405	0·94	2·06	4·30	8·55	16·1

It may be noted that the coherent current is quite independent of the resolution limit. But its limitation makes itself very strongly felt at high resolutions, for two reasons. In the H_0-plane this current is spread over an area $\pi(z_0\gamma_m)^2 = (\pi/4)(\lambda z_0/d_A)^2$, hence the current density in the object space decreases with the square of the resolution, if the defocusing distance z_0 is held constant. For reasons explained in the next section it may be even necessary to increase z_0 with increasing resolving power, hence conditions may be even worse. Second, the magnification must be increased proportionally to the resolving power, if d_A is to be kept in some given proportion to the grain size of the photographic plate, which gives a further quadratic factor. Thus the exposure times will have to be increased, other things being equal, with at least the fourth power of the resolution. We shall not discuss this in detail, as this will be done in a separate paper by Mr. M. E. Haine, but need only remark that with tungsten cathodes and the best photographic plates available the required exposure times become of the order of an hour if a resolution of better than about 2 A. is aimed at.

It is therefore very desirable to find emitters with higher emission density than tungsten cathodes. Autoelectronic emission is an obvious suggestion. Benjamin and Jenkins (1940) found that current densities at least 1,000 times greater than the thermionic emission of tungsten could be maintained for long times. This would reduce the exposure times to seconds instead of hours. Considerable development work will, however, be needed to adapt these very sensitive point cathodes to demountable devices and, as their velocity spread is probably of the order of some volts, they will probably be useful only in conjunction with achromatic lenses, which form the subject of a special investigation.

§ 5. CONDITIONS FOR THE TAKING OF HOLOGRAMS

Assuming that the electron source, the supply voltage and the magnification are suitably fixed, the only important remaining parameter is the defocusing distance z_0. There are several considerations which determine this, of which we mention three.

The first consideration is intensity. We have already seen that, for example, the current density at the object is inversely proportional to the square of z_0, though it cannot be increased beyond any limit, because with decreasing z_0 the 'effective z_0' never falls below one-quarter of the length of the caustic. Thus it would be desirable to immerse the object into the aberration figure. Such sharp focusing is difficult and precarious, and it conflicts also with the two other considerations.

It is the essence of the method of diffraction microscopy that as much information as possible must be put into the clear spaces, where the background is strong, that is it must appear in the form of diffraction fringes surrounding the object. (These are always called Fresnel fringes by electron microscopists though light opticians do not seem to approve of the term.) A typical object suitable for diffraction microscopy, with relatively large clear spaces of average width D, is shown in Figure 10. The clear space is best utilized if the fringes

Figure. 9   Illustrating the concept of the coherent current.

Figure 10.   Typical object suitable for diffraction microscopy.

from both sides just cover it, that is if $z_0 \gamma_m \simeq \frac{1}{2} D$, or, using Abbe's relation $z_0 \simeq D d_\lambda / \lambda$. For example, if $D = 1000$ A. and $d_\lambda = 5$ A., z_0 is 10 microns, and for $d_\lambda = 1$ A. z_0 is only 2 microns. It can be already seen that at high resolutions this condition becomes unimportant.

A third condition for z_0 is obtained from considering the reconstruction, in which the conjugate object must disturb the reconstructed object as little as possible. If one image is made sharp, the other appears with aberrations doubled, as shown in Figure 11 for the case of spherical aberration. Two cases must be distinguished, $z_0 > 0$ and $z_0 < 0$. Theory (cf. Picht, 1931, p. 163) and the optical experiments illustrated in Figure 4 show clearly that the second is the more favourable case. In front of the pointed end of the caustic there extends a region of appreciable length in which the intensity has a rather sharp maximum on the axis. In the case $z < 0$, on the other hand, the point O_1 is in the hollow region of the aberration figure of its conjugate O_2, where the intensity has a flat minimum at the axis. The reconstruction in Figure 4 was obtained with the object in this region, experiments with the opposite position failed because of very large density differences in the photograph, which went far beyond the straight region of the Hurter and Driffield curve.

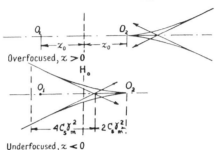

Figure 11. The conjugate objects in the reconstruction, after correcting the errors in one of them.

Within a region which has twice the length of the axial caustic this region is surrounded by a bright annular region, containing a fine system of interference fringes. These vanish at a certain point, and one can see from Figure 11 that the condition for this is

$$-z_0 > 3C_s\gamma_m^2 = \tfrac{3}{4}C_s(\lambda/d_A)^2. \qquad \ldots\ldots(21)$$

As an example let $C_s = 1$ cm. (magnetic microscope), $\lambda = 0.05$ A. (corresponding to 60 kev.) and $d_A = 5$ A. Then $z_0 > 0.75\mu$, much less than the criterion stated above. But with $d_A = 1$ A. the new criterion gives $z_0 > 19\mu$, which is much more. It is seen that the condition (21) becomes stringent only at high resolutions. It appears advisable always to take the larger of the two figures.

It may be asked whether reconstruction is not possible from a sharply focused photograph. *One* sharp image can, of course, be always obtained, but this is of little use if it is overlaid by the diffuse image of the conjugate. Even the reconstruction of high-contrast objects is doubtful, because in the region of the caustic high-contrast objects produce a complicated and sharp system of interference fringes, which can be expected to make identification very difficult.

§6. THE DARK-FIELD METHOD OF RECONSTRUCTION

As already explained in I, a hologram has the advantage over an ordinary photograph, taken in incoherent illumination, that it can be treated as a physical object rather than as a photograph, and can be inspected in bright-field, dark-field or with the phase-contrast method.

It has been also shown that small objects, with dimensions smaller than the characteristic length $(\lambda z_0)^{1/2}$ give reconstructions in which the disturbance by the conjugate or ghost is not serious, and may even be negligible. But in electron microscopy this condition is not always easily satisfied. As a sufficiently transparent support is not available, the object must be stable in itself during the long exposures, which is not always the case with thin filamentary objects as illustrated in Figure 10.

One of the least favourable cases, which occurs frequently, is the reconstruction of the more or less straight and sharp edge of an extended object. For simplicity we will discuss the case of an absorbing half-plane only, bounded by a straight edge. The reconstruction gives one sharp image, but behind this, at a distance $2z_0$, appears the 'ghost plane', with an extended Fresnel-fringe system, which may be so contrasty that it masks even the images of smaller objects which project from the edge, and which, in themselves, would be very suitable objects for diffraction microscopy. It will now be shown that this very unfavourable case can be greatly improved by the dark-field method. As already explained, in this method the direct or illuminating wave is removed after it has passed the hologram, by placing a small, preferably 'apodized' dark spot at a real image of the pinhole source. The 'apodization', i.e. grading, removes the diffraction fringes which might arise at a sharply limited spot.

For simplicity we disregard the geometrical aberrations in the following calculation. The expression (7) for the amplitude in the hologram now simplifies to

$$U(X, Y) = \iiiint t(x,y) \exp\left[-(\rho_m^{-2} + \pi i \lambda z_0)\rho^2 \right]$$
$$\times \exp\left\{ -2\pi i[(x-X)\xi + (y-Y)\eta] \right\} dx\, dy\, d\xi\, d\eta,$$

where we have separated the Fourier factor under the integrand. The transmission in the hologram, at least in those regions which are neither too much underexposed, nor overexposed, is proportional to $U + U^*$, where U^* is the complex conjugate of U.

The amplitude in the reconstruction, in the plane of one of the images, is obtained by the same formula, by adding the 'refocusing factor' $\exp(\pi i \lambda z_0 \rho^2)$ under the integral sign. Applying this to $U + U^*$ we obtain an amplitude

$$U_{\text{rec}} = \iiiint t(x,y) \exp\left[-(\rho^2/\rho_m^2) \right] \exp\left[-2\pi i(x-X)\xi + (y-Y)\eta \right] dx\, dy\, d\xi\, d\eta$$
$$+ \iiiint t(x,y) \exp\left[-(\rho_m^{-2} - 2\pi i \lambda z_0)\rho^2 \right] \exp\left\{ 2\pi i[(x-X)\xi + (y-Y)\eta] \right\} dx\, dy\, d\xi\, d\eta.$$
$$\dots\dots(22)$$

The first line represents the reconstructed object, the second the ghost.

We now assume the object to be a half-plane, so that $t = 0$ from $x = -\infty$ to 0, and $t = 1$ for positive values of x. It will however be simpler to make $t = \pm \frac{1}{2}$ for $x \lessgtr 0$, and add the uniform level $\frac{1}{2}$ afterwards. The integrals in (22) now become well-known Fourier integrals (cf. Campbell and Foster 1931) and we obtain for the first expression in (22)

$$\tfrac{1}{2} + \tfrac{1}{2}\,\mathrm{erf}\,(\pi\rho_m X) = \tfrac{1}{2} + \pi^{-1/2}\int_0^{\pi\rho_m X} \exp - z^2 dz \qquad \dots\dots(23)$$

and for the second, assuming $\lambda z_0 \rho_m^2 \gg 1$,

$$\tfrac{1}{2} + \tfrac{1}{2}\,\mathrm{erf}\left[\frac{\pi\rho_m X}{(1 - 2\pi i \lambda z_0 \rho_m^2)^{1/2}}\right] = \tfrac{1}{2} + \tfrac{1}{2}(1+i)\left[\mathrm{C}\left(\frac{\pi X^2}{2\lambda z_0}\right) - i\mathrm{S}\left(\frac{\pi X^2}{2\lambda z_0}\right) \right]$$
$$\dots\dots(24)$$

where erf is the error function, and C, S are the Fresnel integrals.

2 G–2

This result is however, as was said before, valid only in the range of moderate exposures, where the background can be considered as strong. If the half-plane is completely absorbing, there is no background on the shadow side, and the amplitude is obtained by taking the absolute square of the 'edge wave', which is the difference between (24) and its value without diffraction, that is for $\lambda = 0$. Thus the intensity at the shadow side is $\frac{1}{4}[(C + S - 1)^2 + (C - S)^2]$ while on the bright side the intensity is obtained, without appreciable error, by adding the real part of (24) to the background, and squaring, giving $[\frac{3}{4} + \frac{1}{4}(C + S)]^2$. The intensities thus obtained are plotted in Figure 12. It has been assumed for simplicity that the resolution limit is very small compared with the characteristic length $(\lambda z_0)^{1/2}$, so that the correctly reconstructed part of the image, corresponding to (23) is a step-function.

It can be seen from Figure 12, and this has been also found in numerous

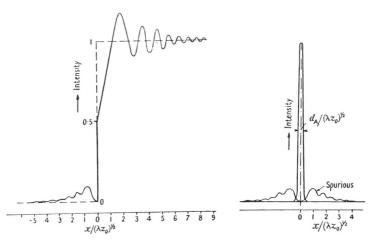

Figure 12. Reconstruction of a straight edge in the bright-field method.

Figure 13. Reconstruction of a straight edge in the dark-field method.

experiments, that the reconstruction is by no means satisfactory. The fringes at the bright side are too prominent. One can suppress them by contrasty photography, but this suppresses also useful detail.

But if the background is suppressed, the picture changes completely. There is no amplification of the fringes, the intensity, as shown in Figure 13, consists of the superposition of the two edge waves, one focused, the other unfocused, both with equal light sums, that is equal areas under the curves. But if the resolution limit is small compared with $(\lambda z_0)^{1/2}$, the intensity in the sharp image can be enormously larger than the spurious image. Thus the correct outline of the object becomes visible, with very high contrast, and under suitable conditions the image may not be noticeably different from a dark-field image in which the ghost is entirely absent. Thus by combining bright-field, dark-field, and perhaps phase-contrast observations, it may be possible to observe much true detail even in otherwise rather unsuitable objects. This is well borne out by provisional observations in optical experiments.

ACKNOWLEDGMENT

A part of this paper was written while the author was a member of the Research Laboratory of The British Thomson-Houston Company Ltd., Rugby, and the photographs in Figure 4 were taken in collaboration with Mr. I. Williams. The author wishes to thank Mr. L. J. Davies, Director of Research, the British Thomson-Houston Company, for permission to publish these photographs.

REFERENCES

BENJAMIN, M., and JENKINS, R. O., 1940, *Proc. Roy. Soc.* A, **176**, 262.

BRAGG, W. L., 1950, *Nature, Lond.*, **166**, 399.

CAMPBELL, G. A., and FOSTER, R. M., 1931, *Fourier Integrals for Practical Applications, Bell Telephone System, Monograph* B 584.

DUFFIEUX, P. M., 1950, *Réunion d'Opticiens* (Paris : Ed. Revue d'Optique), contains a list of the works of this author between 1935 and 1948.

GABOR, D., 1949, *Proc. Roy. Soc.* A, **197**, 454; 1951, *Proc. Phys. Soc.* B, **64**, 244.

HAINE, M. E., and DYSON, J., 1950, *Nature, Lond.*, **166**, 315.

PICHT, J., 1931, *Optische Abbildung* (Braunschweig : Vieweg).

ROGERS, G. L., 1950, *Nature, Lond.*, **166**, 237.

Figure 4. Reconstruction from a hologram taken with the projection method, in the presence of strong spherical aberration. The photographs show three cross sections of the illuminating beam: the original object, the hologram, the reconstructed image and the 'conjugate object'. It is seen that details of the object which are very much finer than the diameter of minimum confusion are reconstructed. (Plate appears on page 324.)

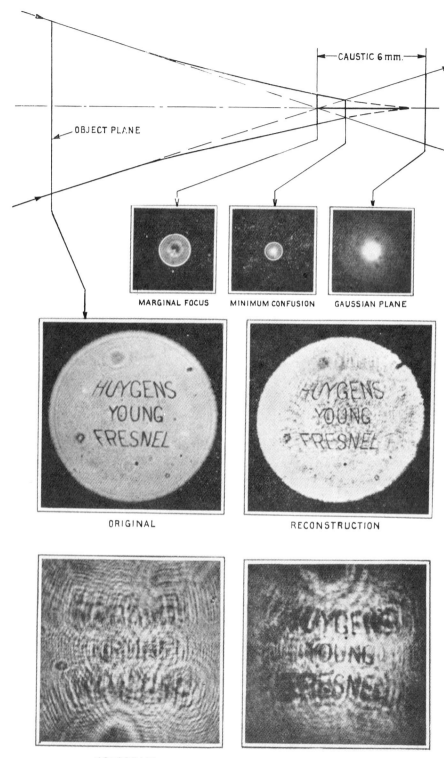

CAUSTIC 6 mm.

OBJECT PLANE

MARGINAL FOCUS   MINIMUM CONFUSION   GAUSSIAN PLANE

ORIGINAL   RECONSTRUCTION

HOLOGRAM   SECOND IMAGE

AUTHOR INDEX

Numbers in parentheses are reference numbers and indicate that an author's work is referred to, although his name is not cited in the text. Numbers in italics show the page on which the complete reference is listed.

Abbe, E., 13, *15*, 73, 74, *96*
Aktar, S. A., *238*
Alexandrov, E. B., 202 (1), *228*
Ansley, D., *259*
Armitage, J., 192 (86), 196 (86), *230*
Armstrong, J. A., 112, *162*, *163*
Arsac, J., *15*, 164, *180*

Baez, A. V., 12 (22, 25), 13 (25), *14*, 114, 115 (36, 37), *162*
Baker, B. B., 272, *302*
Bakhrakh, L. D., 188 (2), 194 (2), *228*
Ballard, G. S., 189 (116), *231*
Ballik, E. A., 48 (3), *69*
Barnett, N. E., 188 (39), 191 (39), *229*
Beddoes, M. P., *238*
Benjamin, M., 319, *323*
Bergstein, L., 226, *228*
Billings, B. H., *238*
Blanc-Lapierre, 13, *15*, 66 (15), *69*
Blum, J., *235*
Bobrinyev, V. I., *236*
Bodo, G., 13 (28), *15*, 144 (57), *163*
Boersch, H., 144, *163*, *228*, 267, *302*
Bolstad, J. O., *233*
Bonch-Bruevich, A. M., 202 (1), *228*
Bond, W. L., 48 (3), *69*
Born, M., 21 (6), *25*, 40, *69*
Borodina, S. B., 187 (3), *228*
Bouix, M., *15*
Bourgeon, M. H., *234*
Bracewell, R., *15*
Bragg, W. L., 109 (33), 114 (33), 144, 148, *162*, *163*, 267, *302*, 306, *323*
Brillouin, L., *15*
Brinton, J. B., Jr., *234*
Briones, R. A., 150 (50), *162*
Brooks, R. E., *96*, 150, *162*, 188 (4, 5), 190 (4, 5), 202 (43), 205 (43), *228*, *229*
Brown, G. M., 205, *228*
Bruck, H., 295, *302*

Brumm, D., 11 (42), *15*, 74 (5), 81 (5), 85 (11), 90 (5), 91 (5), 92 (5), 95 (14), 96 (14), *96*, 105 (86), 106 (30), 108 (89), 116 (29, 30), 117 (30), 119 (29), 121 (86), 128 (30), 131 (30), 132 (30), 133 (30), 138 (44), 144 (30), 148 (30), 149 (29, 30), 158 (30), *162*, *163* 233, *237*
Buerger, M. J., 13, *15*, 144, *162*
Bulabois, J., *234*
Bunn, W., *302*
Burch, J. M., *96*, 149, *162*, 204, *228*, *236*
Burckhardt, C. B., 199, *228*, *234*
Burke, J. F., 247, *259*

Campbell, G. A., 274, *302*, 313, 321, *323*
Carcel, J. T., 188 (9), *228*, *238*
Carpenter, R. L., *233*
Carter, W. H., 219 (10, 11), *228*, *235*
Cathey, W. T., Jr., 108, *163*, *236*
Chabbal, R., *180*
Champagne, E. B., *233*
Chau, H. H. M., *237*
Cheslidze, T. Ya., 205 (132), *232*
Clemmow, P.C., 5 (6), *14*
Clifford, K. I., *233*
Closets, F., *235*
Cochran, G., 151 (71), *163*
Collier, R. J., *96*, 196 (12, 101), 197 (101), 204 (13), *228*, *231*, *235*
Connes, J., 151 (76), *163*
Copson, E. T., 272, *302*
Corcoran, V. J., *238*
Croce, P., 13 (39), *15*, 72, 73, 79, *96*, 102 (21, 22), 105, *162*, 253, *259*
Cutrona, L. J., 12 (19), *14*, 73, 88, 89, *96*, 105, *162*, 192, 193, *228*

Davy, J., *236*
De, M., *233*, *235*
DeBitetto, D. J., *233*, *238*

325

SUBJECT INDEX

A

Abbe double-diffraction principle, 73
Abbe image formation, 72–74, *see also* Double diffraction
Abbe resolution criterion, 74–77, 116, 209, 268
  in holography, 116
  relation to degree of coherence in holography, 292, 317–319
  to spatial resolution in object, 125
Aberrations, *see* Wavefront aberration
Acoustical holography, 191–193
Addition of diffraction patterns, 90–96
  contrast with multiplication, 73, 79–90
  holographic, 90–96
  in image synthesis, 90–96
  by intensity interferometry, 90–96
  successive, in single hologram, 90–96
Addition of wavefronts
  holographic, 90–96, 202–208
  by intensity interferometry (holographic), 90–96, 202–208
Advantages of holography, 181–259, *see also* Applications of holography
  in image coding and decoding, 90–96, 127–137, 226–227, 250–256
  in image deblurring, 226–227, 252–256
  in imaging through turbulent and distorting media, 211–218
  in information storage, 198–201
  in interferometry and nondestructive testing, 202–208
  in microscopy (high-resolution wide field), 185, 220–221
  in optical computing, 254–256, *see also* Holographic computing
  in pattern and character recognition, 79–86, 181–183, 195–197
  in particle size analysis, xii, 188–189
  in printing (contactless, micro-circuit masks), 210–211
  in radar (side-looking microwave radar) imaging, 192–193, 218
  in three-dimensional imaging, 187–195
  in x-ray imaging
    astronomy, 226–227, 252–256
    crystallography, 127–137, 186

Alkali-halide crystals, use in holography, 184, 199–200
Ambiguity function, 89
Angular spectrum of plane waves, 2–8, *see also* Boundary value solutions of diffraction problems
Aperture synthesis (holographic), 226–227, 252–256
Apodization
  a posteriori, analogy with *a posteriori*, correlative reconstruction-compensation for source-effects, in holography, 131–132
  in holography, 131–132, 321
Applications of holography, 181–259 (table, 182), *see also* Holographic: cinematography, Fourier-transform division, imaging through turbulent and distorting media, image deblurring, optical image processing, image restoration, interferometry, machining, memory, microscopy, optics, television
Artificial holograms, *see* Hologram
Associative properties of holograms, 183, *see also* Holographic memories, Character recognition
Atomic beam light source, 54
Autocorrelation function, 174–177
  analysis of light sources, 42
  cosine Fourier-transform of spectrum, 42
  definition, 174
  graphical illustration, 176
  in power spectrum, 47
Automatic reading, by optical correlation filtering, 83

B

Beam splitting
  with double-exposed holograms, 90–96
  in holography, *see also* Holography
  "in-line" (Gabor) holography, absence of problems, xii, 11, 105
  in interferometry, *see also* Interferometry

331